William L. Carter
1980

MATHEMATICALLY SPEAKING

MATHEMATICALLY SPEAKING

Morton Davis

City College of
The City University of New York

HARCOURT BRACE JOVANOVICH, INC.

New York / San Diego / Chicago / San Francisco / Atlanta
London / Sydney / Toronto

To Gloria,
    Jeanne,
    and
    Joshua

Cover art and interior illustrations: Bill Negrón

Technical drawings: Vantage Art, Inc.

Picture Credits appear on page 480.

PREFACE

Mathematically Speaking was written especially for liberal arts students who do not require extensive mathematical training but who are intellectually curious about the subject. For such an audience this book is meant as a stimulating introduction to the fascinating world of mathematical thinking.

Indeed, it is *not* our intention here to provide a watered-down version of any formal mathematics course. Such a book would be too superficial to be of practical use and would fail to capture the student's imagination. Rather, the purpose of this book is to convey some idea of the power of mathematical analysis and to show that mathematics can be esthetically appealing. We do not try to develop technical skills for their own sake; in the calculus chapter, for example, we are primarily concerned with broad ideas, whereas the routine techniques of differentiation and integration, which make up the bulk of the traditional calculus course, are omitted almost entirely.

Concepts are usually introduced by means of a familiar example before being formalized. Then, once we have given a clear definition of a concept, we employ several additional examples to strengthen the student's grasp of the concept and to give some idea of the breadth of its applications. Whenever possible, we provide appealing and unusual settings (such as sports, tournaments, airline scheduling, the stock market, law enforcement, marketing, and banking) for the examples and for the exercises at the ends of sections.

We have included a number of routine exercises at the end of each section to amplify the text and test the student's understanding. In most sections there is also a set of more challenging problems, called "Extras for Experts." And finally, the "Puzzles to Ponder" call for clear thought rather than calculations; many of these are problems that do not have straightforward, simple answers, or classic paradoxes that defy resolution. Since our main goal is to stir the student's imagination and motivate further reading, we have not hesitated to pose problems or paradoxes without resolving them if they are sufficiently interesting. Solutions are given to the odd-numbered exercises and to most "Extras for Experts" and "Puzzles to Ponder" beginning on page 439.

Instructors will have much flexibility in choosing topics from this book. However, some of the chapters necessarily depend on others. Thus set theory should precede probability; probability should precede statistics, game theory, and Markov chains; and analytic geometry should precede calculus. Otherwise, the chapters are essentially independent.

The accompanying Instructor's Manual provides answers to those exercises not answered in the textbook. In addition, there are test questions and supplementary comments and instructional material for each chapter.

I would like to thank the following reviewers, whose comments on the manuscript were most helpful: Douglas Crawford, College of San Mateo; Robert Gold, Ohio State University; Earl Hasz, Metropolitan State College; Steven Kahan, Queens College; Helen Salzberg, Rhode Island College; Erik Schreiner, Western Michigan University; Karen Schroeder, Bentley College; Seymour Shuster, Carleton College; and Joseph Troccolo, Cleveland State University. I also wish to express my gratitude to my editors, Marilyn Davis and Judy Burke. Their invaluable suggestions about style and choice of subject matter, their technical competence, and their patience and attention to detail are surely beyond the mere requirements of duty. Finally, my thanks go to designer Marilyn Marcus, to art editors Barbara Salz and Yvonne Steiner, and to production manager Robert Karpen.

MORTON DAVIS

CONTENTS

1 WHY SPEAK MATHEMATICALLY?

INTRODUCTION 2

2 THE FOUNDATION

SET THEORY 20

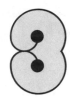

METHOD IN OUR MADNESS

PROBABILITY 46

THE ART OF INFERENCE

STATISTICS 106

GAMES ADULTS PLAY

GAME THEORY   152

TO PICK AND CHOOSE

AN APPLICATION TO VOTING   200

ON CONSTANT CHANGE

MARKOV CHAINS 246

NUTS AND BOLTS

COMPUTERS 272

BUT CAN THEY THINK?

COMPUTER LEARNING AND GAMES   314

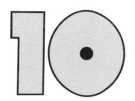

A MARRIAGE OF CONVENIENCE

ANALYTIC GEOMETRY   356

MATHEMATICS OF CHANGE

CALCULUS 398

MATHEMATICALLY
SPEAKING

1

WHY SPEAK MATHEMATICALLY?

INTRODUCTION

A representation of the Ptolemaic ▷
(geocentric) model of the universe
(1245 A.D.).

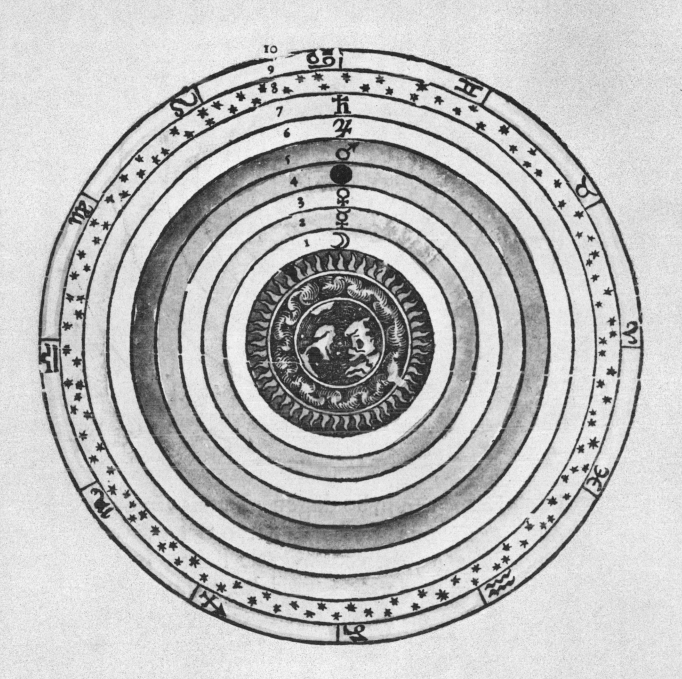

The various branches of mathematics have usually been created in an attempt to solve some problem. One may want to predict the weather, figure out when to bluff in poker, or estimate how often a telephone switchboard will be overloaded. Right now we'll loosely describe how a mathematician approaches such problems, and later on we'll look at some actual applications.

When we talk about problem solving, we will often use puzzles for our examples. These puzzles may be fun, but they're not frivolous; you attack them in the same way you attack deep mathematical problems. Still, there are differences between puzzles and the problems with which mathematicians wrestle. Puzzles usually have "elegant" solutions, while real mathematical problems need not. Real mathematical problems are more abstract than puzzles, and their implications may be so diverse that they aren't even guessed at by the people who solve them. And it takes an apprenticeship to do serious mathematics. Unlike puzzles, which can be simply stated if not easily solved, mathematics requires that you learn a special vocabulary and the rules of the game: the definitions and axioms.

But for all the differences, serious mathematics and puzzles have much in common, and the pleasure of solving puzzles is familiar to the mathematician. Creativity in mathematics and problem-solving ability are different skills, but the two often go together.

If you want to analyze a problem, it's often a good idea to start by building what is known as a *model*. A model is nothing fancy; it's simply the "bare bones" of the problem—what it looks like after you've stripped away the unimportant details. For instance, imagine that on Saturday a mother wants to shop for food, cook dinner, drop off her son at the library, pick him up, drop him off at a baseball game at 4:00 P.M., and be at the beauty parlor at 11:00 A.M., while her husband wants to buy roof tiles, fix the roof,

and go bowling. If the family has one car, how are they to plan the day?

What's important here? A number of things. Certain tasks must be done before others (the mother must buy food before cooking, and her husband must buy tiles before fixing the roof); some jobs must be done at a fixed time; two jobs that require a car can't be done at the same time; each job takes a certain amount of time; and so on.

To solve such a problem you might list all the jobs and the time required to complete them and make note of which jobs require that other jobs be done first, which must be done at a certain time, which can't be worked on together, and so on. Certain other details about the jobs, such as the names of the family members, what they're wearing, and whether the family is to eat hot dogs or steak for dinner, are not related to the problem you're trying to solve, and you would ignore them. This *reduced* version of the original problem is called a *model*.

This process of reducing a problem to its essentials may yield benefits you haven't expected, because in the process of solving one problem you may be solving others as well. When you construct a building, you're involved with a problem that's very similar to the one the family faced. There are many different tasks that are interrelated in many intricate ways: You can't build a wall until the bricks are delivered, you can't wire the fifth floor until at least part of the third floor is finished, and so on. A very simple version of the same problem is faced by someone who is writing a book and must decide the order in which chapters will be read (so that all the material required to understand a particular chapter will be covered in the earlier chapters). In fact, all such problems can be reduced to one basic model, and all of them may be solved at the same time. (There is a computer technique called *critical path scheduling* that does exactly that.)

Of course, you may not know what is important and what is not. It took some time for scientists to figure out that the gravitational force between two heavenly bodies depends only on the distance between them and their mass (the property you measure indirectly when you step on a scale) and not on their size, velocity, color, or shape. But after some thoughtful observation you will try to express the essential relationships of the problem in terms of mathematical equations, and these, once solved, will predict what will happen or state what you should do. If it turns out that the advice you obtained is good or that the predictions are accurate, you've probably constructed a model that reflects the real world pretty well; otherwise, you must go back and try again.

When you finish constructing a useful model, what will it do for you? For one thing, it will often keep you out of trouble by indicating that certain "obvious" truths about the world are false. (The earth isn't flat, and your "solid" armchair is a swirl of rapidly moving electrons.) It will simplify the analysis of a problem because it allows you to concentrate on the essentials. And finally, it can save you from doing the same work twice. Let's look at some specific examples to illustrate each of these points.

A Model Is a More Reliable Guide Than Pure Intuition

Smith and Jones each pick a positive whole number at random (so that all numbers have the same chance of being chosen). What is the probability that Smith's number will be greater than Jones's?

With no more information than this the answer must be $\frac{1}{2}$, since Smith and Jones pick their numbers in the same way. But now suppose after both numbers are chosen you find that Smith picked 5. How does this new information affect your answer?

Since there are only four numbers less than 5 but infinitely many greater, Jones's number is almost certainly going to be greater than Smith's. In fact, *whatever number Smith chooses,* there will only be finitely many numbers that are smaller and infinitely many numbers that are greater. So peeking at Smith's number and ignoring Jones's almost guarantees that Jones's number will be larger. But on the other hand, by looking at Jones's number first you can make it seem that Smith's number almost certainly must be larger!

So it comes to this: At first Smith and Jones have equal chances of picking the larger number. But after they choose, you can make either one of their numbers larger by inspecting the other one first.

Clearly, something is wrong with this reasoning. It turns out that you *cannot* pick a positive whole number at random, but this is not obvious to someone who is just using "common sense." While picking a random number is usually a harmless procedure, in this case it's meaningless. We will see why in a later chapter, when we construct a probabilistic model. With such a model you can avoid such "reasonable" blunders; without one it's hard to see what went wrong.

Once you've deduced the basic laws that govern a situation and reduced them to a mathematical model, that model is generally a more reliable guide than pure intuition. As a simple test, try answering the following four questions using only your "common sense" and see how well you do:

1. A dollar earns 8% interest (compounded annually). How much will the dollar be worth in 180 years?

2. Every day you visit a neighbor who lives 3 miles away. When there's no wind, you walk 3 miles per hour, so the round trip takes 2 hours. When the wind is blowing in the direction of your neighbor, you walk 2 miles per hour in one direction and 4 miles per hour in the other. Since you gain in one direction what you lose in the other, the effect of the wind is canceled and you spend your usual 2 hours on the round trip. True or false?

3. A traveler in the desert approaches two wells, one of which he knows to be poisoned. Two families inhabit the desert; the members of one of the families always tell the truth, while the members of the other family always lie. The traveler asks a person who is known to belong to one of these two families a single question to which the answer is "yes" or "no." Can he possibly discover from the answer to this single question whether the well was poisoned? Can he discover from the answer whether the person told the truth or lied? Can he do both?

4. You have $100 and are betting repeatedly that a fair coin will turn up heads. Each time you bet half the money in your possession and either win or lose that amount. If you lose as many bets as you win, will you always come out winning, come out losing, or break even, or is it possible to tell?

The answers to these questions are as follows. How well did your intuition do?

1. In 180 years your dollar will grow to over a million dollars.

2. The effect of the wind does *not* cancel out. Walking against the wind for 3 miles at 2 miles per hour takes $1\frac{1}{2}$ hours, and walking with the wind at 4 miles per hour takes $\frac{3}{4}$ of an hour—a total of $2\frac{1}{4}$ hours.

3. The traveler, pointing to a well, might ask, "What would you answer if I were to ask, 'Is this the poisoned well?'" If you think it through, you'll find liars and truth tellers answer identically: yes if the well is poisoned and no otherwise. By asking, "What would you answer if I were to ask you, 'Are you a liar?'" he can find out if he's talking to a liar or a truth teller, but there's no way of finding out whether the well is poisoned *and* whether the person is a liar with a single question.

4. You always end up behind! When you win you multiply your capital by $\frac{3}{2}$, and when you lose you multiply your capital by $\frac{1}{2}$. When you win once and lose once you multiply your capital by $\frac{3}{4}$, and when you lose N times and win N times you multiply your capital by $\left(\frac{3}{4}\right)^N$. The order of wins and losses is irrelevant. (Try it out.)

II. *A Mathematical Model Simplifies Your Analysis*

About 245 years ago the great mathematician Leonhard Euler observed several bridges connecting a river island with the surrounding mainland, in the configuration shown in Figure 1. He wondered whether it was possible to take a walk in such a way that each bridge would be crossed exactly once—no more and no less. This problem is now referred to as the Königsberg bridge problem.

For convenience we will reformulate the problem in a slightly different but equivalent form. Look at the diagram in Figure 2; crossing each one of the Königsberg bridges exactly once is

Figure 1

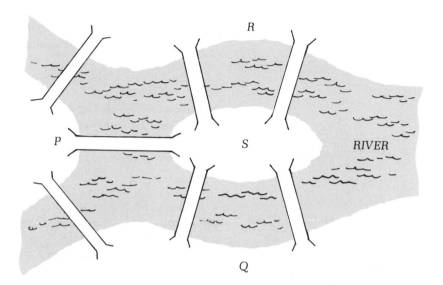

Figure 2

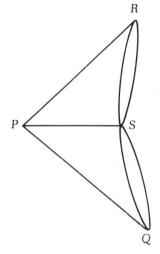

Can you draw this figure, starting anywhere and finishing anywhere, without lifting your pen from the paper or repeating a line? (This is another form of the Königsberg Bridge Problem.)

Figure 3

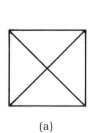
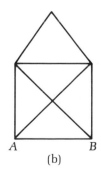

(a)   (b)

equivalent to drawing this figure without lifting your pen from the paper or repeating a line. Before seeing whether this can be done, let us ask the same question about two simpler diagrams, the ones shown in Figure 3. If we answer the question for these diagrams in the right way, we will find that we also have obtained the answer to the Königsberg bridge problem.

Onc way to approach the diagrams in Figure 3 is simply to go ahead and start drawing. For the diagram in part (b)—the square with the hat—you should hit on a solution quickly enough, and that will be that. But for the other diagram, the plain box, you certainly won't find a solution, since there is none to be found. But how could you have known this in advance? And even if you are persuaded by constant failures that you can't draw this particular figure, how would you analyze the problem if the figure was altered slightly?

Euler not only solved this problem but also solved a great many similar ones (including the Königsberg bridges) at the same time. By looking at the problem in the "right" way—separating the essential elements from the inessential ones—he reduced it to its bare bones, and in that form it was easily handled.

What properties of the figures are inessential? Clearly, the lengths of the lines don't matter; if the figures were half or double their size, the basic problem would be the same. If the lines were curved or wiggly instead of straight, it wouldn't matter either. What does matter is the way that lines cross one another. If you were to draw the figures on a rubber sheet that you stretched and pulled without tearing or cutting, the figures would become distorted but would remain essentially the same in the sense that you could draw the distorted figure only if you could also draw the original one. (Incidentally, this type of problem comes from a branch of mathematics called *topology*; for the reasons suggested, topology is nicknamed "rubber sheet geometry.")

Euler realized that the key to this problem lay in examining the vertices (the points where two or more lines cross) and counting the number of lines that emanate from them. He observed that *a vertex which is neither the starting point nor the final point of the drawing must have an even number of lines emanting from it.* The reason for this is not hard to see; if you enter the vertex on one path, you must leave by another. If we call a vertex with an odd number of lines emanating from it an *odd vertex,* it follows there can be only two odd vertices: the starting and final points of the drawing. Therefore:

A figure with more than two odd vertices cannot be drawn in a continuous curve without either duplicating a line or lifting your pen from the paper.

With this analysis it immediately becomes clear that the hatless square (Figure 3a) cannot be drawn, since each of the four corners is an odd vertex. For the square with the hat the Euler analysis not only suggests the figure can be drawn; it tells you how to go about it. There are two odd vertices in the figure (*A* and *B*), and one of them must be the start of the drawing and the other must be the end of it; there is no other possible way to draw the figure.

Observe that Euler's conclusions apply not only to the original problem but to a much larger class of problems. Specifically, if you look at the original diagram of the Königsberg bridge problem and think of the four land masses as the vertices and the bridges as the lines joining them, all four "vertices" turn out to be odd. Therefore there is no solution (see Figure 2 on page 9 again).

III. *A Good Model Is Economical*

A model can be a labor-saving device in more than one way. Proper notation often allows you to solve problems by manipulating symbols instead of thinking them through each time, and this can be a great advantage. Algebraic problems that high school students solve routinely today were considered formidable before the Arabs devised an efficient notation for algebra, and the Romans had trouble with arithmetic because of the clumsiness of their number system. To see why notation is important, let's look at some familiar examples.

A basic law of arithmetic, the commutative law of multiplication, says the order in which you multiply numbers doesn't matter, that is, *ab* is always equal to *ba*. You often apply this algebraic law mechanically; for example, you know that $23 \times 37 = 37 \times 23$, even though it's not *intuitively* obvious that a school with 23 classrooms each containing 37 students has the same population as one with 37 classrooms each containing 23 students. (If you ask a child to multiply 7 by 9 after he's multiplied 9 by 7 he'll often take as long to do the second operation as the first.)

A "magic" trick that mystifies children provides a second example:

1. Take a number from 2 to 10 and square it.
2. Subtract 1.
3. Divide your answer by one less than your original number.
4. Subtract your original number.

The answer is 1.

While our ability to predict this answer may be a mystery to others, the algebraist immediately sees that this series of operations can be translated into an equation,

$$\frac{x^2 - 1}{x - 1} - x = 1$$

and that this equation is true for *any* number x from 2 to 10. Once you learn which operations are legitimate, you needn't keep track

A Babylonian tablet showing an early attempt to calculate the area of a circle. The method used was to circumscribe a square, whose area was known, about the circle.

of the numbers any more; you simply let the notation and manipulations do the work for you.

A final example that shows the power of a good system of notation is found in elementary algebra. You may recall that the expression 3^4 is a shorthand way of writing the product of four 3's: $3^4 = (3)(3)(3)(3)$. From this definition you can check that $(3^3)(3^4) = 3^7$, since the product of three 3's multiplied by the product of four 3's is a product of seven 3's. And, in fact, if M and N are any positive whole numbers, then $(3^M)(3^N) = 3^{M+N}$.

Now if we are given the expression 3^{-1}, what are we to make of it? Our original definition is of no use here, since you can't write a product consisting of minus one factors. But if you *mechanically* apply the formula $(3^M)(3^N) = 3^{M+N}$, which worked for positive whole numbers, here (where it doesn't properly belong), you would conclude that $(3^2)(3^{-1}) = 3^1$, since $2 + (-1) = 1$. Now $3^2 = 9$ and $3^1 = 3$, so if the formula is to hold, it must be true that $9(3^{-1}) = 3$ or $3^{-1} = \frac{3}{9} = \frac{1}{3}$. And it turns out to be very useful in more advanced algebra to define 3^{-1} as $\frac{1}{3}$.

What has happened here may seem odd, but it is not at all unusual. You first derive a valid law that only applies under certain restricted conditions. Then you formally apply the law in cases where it is not valid and derive something new that turns out to be meaningful. Fractions and negative numbers may seem meaningless in some applications (you can't have half a student in a class or win an award minus three times), but they turn out to be useful in others. The laws governing negative numbers and fractions may often be suggested by the notation even before you find an application. It's almost as though you've created a structure that has a mind of its own and that goes on, after being created by you, to create something new by itself.

Models can save labor in other ways too. By changing a few names a model used for one purpose can also be used for an entirely different one. We will see in the chapter on Markov chains (Chapter 7) that the sociologist interested in class mobility (the probability that children of one social class will eventually belong to another), the gambler calculating the chance of winning $150 before going broke, the epidemiologist concerned with a spreading disease, and the physicist studying gas diffusion are all really facing one problem in different disguises.

The following paradox is an example of a single basic problem arising in two entirely different ways: Suppose Smith, Jones, and Green arrange a tournament to decide which of them is the fastest runner. They run a series of three three-way races and agree that if one of them beats another in two of the three races, he will be considered the better of the two. If Smith is better than Jones and Jones is better than Green, does it follow that Smith will be better than Green? Not at all! If the respective times in the three races for Smith, Jones, and Green are (18, 19, 20), (19, 20, 18), and (20, 18, 19), then Smith is faster than Jones, who is faster than Green, who is faster than Smith.

Essentially the same situation arises in a different form in Chapter 6, on voting. Smith, Jones, or Green is to be elected president. A series of three two-man elections is to be held to see who is most popular. It can turn out that Smith is more popular than Jones, who is more popular than Green, who is more popular than Smith. The common thread in both cases is what mathematicians call *intransitivity*—of speed in one case and of popularity in the other.

After each section of this book you will find a set of Exercises. These give you a chance to test your understanding of the concepts discussed in the section and to explore ways of applying them. In addition to these routine problems, we will sometimes have some harder ones, called "Extras for Experts." And at times we will include deeper problems that we term "Puzzles to Ponder." These include problems of a more philosophical nature, paradoxes, and so on. Don't expect to come up with a simple or final answer—just try to enjoy the puzzle.

exercises

1 Suppose you want to build a model to help you analyze each of the situations described below. Which of the factors mentioned do you think would be included in the model and which would be discarded as irrelevant?

The Problem

Possible Factors

(a) You want to predict what the price of the stock of a company listed on the New York Stock Exchange will be at various times during the following year.

(i) The price of the stock throughout the previous year, (ii) the company's earnings last year, (iii) the trend of the company's earnings during the last four years, (iv) the economic state of the country as a whole, (v) the weather during the following year, (vi) the length of women's skirts, (vii) the number of shares that were bought and sold by officers of the company, (viii) the season of the year, (ix) the size of the competition, (x) the size of the stock dividend, (xi) the appearance of sunspots.

(b) You want to determine how many tollbooths should be used on a toll bridge at various times during the day so that the lines don't get too long, on the one hand, and the toll collectors aren't standing idle too often, on the other.

(i) The name of the river spanned by the bridge, (ii) the cost of the toll, (iii) the existence of ticket-toll books, (iv) the time of day, (v) the number of exact change lanes, (vi) the manufacturers of the cars that cross the bridge, (vii) the number of passengers per car, (viii) the color of the cars, (ix) the number of local sports arenas, (x) congestion on neighboring roads, (xi) the frequency of accidents, (xii) the density of traffic during rush hour, (xiii) the times most sports events finish.

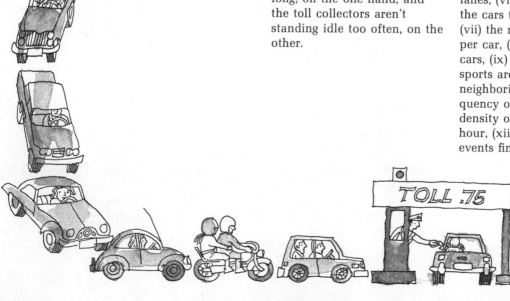

(c) You want to know how various animal species and types of vegetation are mutually dependent and how the extinction of one species affects others. You also want to know the vulnerability of each species to manmade hazards.

(i) The amount of rainfall, (ii) what each species eats, (iii) the number of offspring for each species, (iv) the likelihood that an offspring will survive for each species, (v) the ability of the members of each species to substitute one kind of food for another, (vi) the attitude of the fashion world toward animal skins, (vii) the use of pesticides, (viii) the relationships between predator and prey, (ix) the color of each animal and the color of the surrounding countryside, (x) the state of the nation's economy, (xi) the conditions that affect a species' predators, prey, and other species competing for the same food.

2 A *directed graph* is a collection of points, called *vertices*, and arrows that connect some of these vertices in a particular direction. In the diagram in Figure 4 there are five vertices (A, B, C, D, and E), and there are connections going from A to E, A to D, A to C, B to D, and D to B. For instance, the vertices might represent five people and the arrows might indicate which of the people owe money to others in the group.

Because many different problems may be expressed in terms of such graphs, considerable theoretical work has been done to analyze them. In each of the following situations state how you might use such a graph as a model:

(a) You plan to construct a building. This involves doing many jobs, and each job may require that other jobs be completed first.
(b) You want to analyze a game in which you can move from one position to a certain number of other positions.
(c) You are doing a sociological study of likes and dislikes in a class and ask each person who his or her best friend is.
(d) In the crowded downtown section of a city, where there are many one-way streets, you want to know how to drive from one intersection to another.
(e) You want to describe the prey–predator relationships between various species; for example, a wolf may eat a bird, which eats insects, and so on.
(f) You want to use a family tree to represent the relationships between parents and children over several generations.
(g) You want to describe a chain of command in an army or a large corporation.

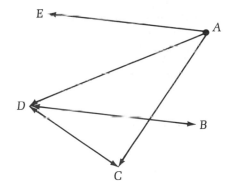

Figure 4

3 There are various times when you can avoid waiting by spending money or, conversely, save money by waiting longer. Explain how each of the following situations might be reflected by a single model. Imagine you are considering

(a) adding another telephone switchboard in an office.
(b) adding more tollbooths to a bridge.
(c) putting another airplane on call if demand warrants.
(d) hiring another person to sell soda at a baseball game.
(e) installing an extra elevator in a building.
(f) putting more trains in service during rush hour.

In each case imagine what the consequences are for each possible option.

Try to answer the questions below as best you can, and then use the hints suggested, if necessary.

1 The following problems may seem difficult at first glance, but they are easily solved if you look at them in the right way.

(a) A colony of microbes doubles its size every second. A single microbe is placed in a jar, and in an hour the jar is full. When was it half full?

(b) Two trains are approaching one another, each moving at 20 miles per hour. A bird perched on one of them flies directly to the other and upon reaching it goes back directly. If the bird flies at 60 miles per hour and continues to shuttle back and forth until the trains collide, how far will it have traveled if the trains were originally 120 miles apart?

(c) A square checkerboard with 8-inch sides is composed of 64 1-inch squares (see Figure 5). The two squares labeled *A* and *B*, at opposite sides of the main diagonal, are cut out. Is it possible to cover the remaining 62 squares with rectangular 2-inch by 1-inch squares?

Figure 5

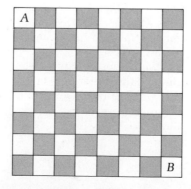

(d) Each of two players is given coins 1 inch in diameter. The players alternately place these coins on a table 20 inches in diameter so that no coin overlaps the edge of the table or any coins placed there earlier. The player who puts the last coin down wins. Which of the players can play in such a way as to guarantee that he wins? How can he do this?

(e) A red cup contains a pint of water, and a blue cup contains a pint of wine. A teaspoonful of water is taken from the red cup and placed in the blue cup, and the contents of the blue cup is mixed thoroughly. A teaspoonful of the mixture in the blue cup is returned to the red cup. Is there more wine in the red cup or water in the blue cup? Would it matter if there wasn't proper mixing?

(f) A prison has 1000 prisoners, and they wear numbers from 1 to 1000. Two prisoners occupy each cell. How can you assign them to cells so that the sum of all their numbers will be easy to calculate?

(g) Figure 6 shows a quarter-circle of radius 6 inches. Line segments $\overline{AE}$ and $\overline{AF}$ are radii. How long is the line segment $\overline{DB}$?

Figure 6

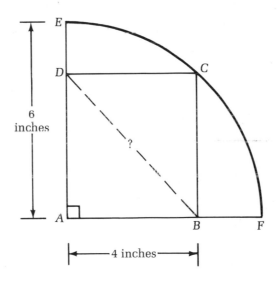

(h) In a subway train certain people are related to one another. (If A is related to B, then B is related to A, of course.) Show that there are at least two people in the train who have the same number of relations.

2   The following problems are a test of your intuition. Don't try to calculate an exact answer; estimate it and compare your estimate with the actual answer shown below.

(a) Assume a person is as likely to be born on one day of the year as another. If there are 60 people in a room, how likely is it that no two of them have the same birthday? Suppose there are N people in the room and the probability that two of them have the same birthday is about $\frac{1}{2}$; what is N?

(b) From 1,000,000 diamonds you want to select the largest one. The selection procedure is this: A diamond is picked at random from the million, and you must accept or reject it. If you accept it and it is not the largest diamond or reject it and it is the largest diamond, you fail. If you accept it and it is the largest diamond, you succeed. If you reject a diamond that's not the largest, you are shown a diamond selected randomly from the remaining 999,999 diamonds.

You make the same kind of decision again, except that now you know whether the second diamond is smaller or larger than the first one (you automatically reject a smaller diamond, of course). You accept this second diamond or reject it with the same consequences as before. You continue in this way, accepting or rejecting diamonds, making a mental note of the diamonds you've seen earlier. Without trying to pick a strategy, how likely is it that a clever chooser will pick the biggest diamond?

(c) There are three cards, of which one is black on both sides, another is red on both sides, and a third is black on one side and red on the other. You are shown one side of one of the cards, and it is red. What is the probability that the other side is also red? (Consider this argument: "Since I saw a red side, there are two possible cards I could have been shown, so the answer must be $\frac{1}{2}$.")

Hints to extras for experts

1 (a) Try working backwards—investigate what the population was just before the jar became full.

(b) Ignore the direction of the bird. Calculate the distance traveled (in either direction) by finding the time spent in the air and multiplying that by the speed of the bird.

(c) Observe that there are two kinds of squares: plain and shaded. Notice the squares that have been deleted and think about what kind of squares a tile covers.

(d) The first player to move always wins if he places his first coin in the center and plays symmetrically with respect to the center every time his opponent moves.

(e) Whether you mix well or not at all, there is as much wine in the red cup as there is water in the blue one. (Let x be the amount of wine in the red cup. Use the fact that you started and ended with a pint in each cup to show there must also be x amount of water in the blue cup.)

(f) Put prisoners 1 and 1000 in the first cell, 2 and 999 in the second cell, and so on. There are 500 occupied cells, and the sum of the numbers in each cell is 1001. The total sum is therefore 500,500.

(g) Line $\overline{AC}$ (not shown—that's the trick!) is a radius, and thus is 6 inches long. Line $\overline{BD}$ has the same length as $\overline{AC}$. The 4-inch length is window dressing.

(h) If there are N different people in the train, each person can have 0, 1, 2, . . . , $N - 1$ different relatives in the train: N different possibilities in all. If no two people have the same number of relatives, there must be someone with $N - 1$ relatives and someone with no relatives, which is impossible.

2   (a) If there are 60 people in a room, there is a probability of 0.006 that no two of them will have the same birthday. If there are 23 people in a room, the chances are about even that no two of them will have the same birthday.

(b) If you're clever enough, you can pick the largest diamond with a probability of better than $\frac{1}{3}$. The exact strategy that will accomplish this is a little complicated, but we'll describe a slightly inferior one: Let the first 500,000 diamonds go by and then pick the first diamond you see that's larger than all the diamonds you've seen earlier. If the largest diamond is in the second 500,000 diamonds you inspect and the second largest diamond is in the first 500,000, you'll succeed. (You'll also succeed in some other cases as well.) The probability of this occurring is about $\frac{1}{4}$.

(c) If you pick a red side, the chances are 2 in 3 (not 1 in 2) that the other side is red.

THE

FOUNDATION

SET THEORY

Robert Delaunay, Disques, *1938, 538 cm x 396 cm.*

1 Introduction

For most people who know little about it, mathematics seems to be an island of indisputable truth surrounded by the uncertainties of the ordinary world. People argue about whether one work of art is more beautiful than another or compare the relative merits of psychotherapy and long walks in the country, but they have no doubt at all about the validity of a mathematical theorem. Once a mathematical theorem is proved, it is true forever—at least according to popular belief.

But in fact the line between what has and what has not been proved is not absolutely clear, and mathematics has had its share of paradoxes and contradictions. A proof accepted by one generation may well be rejected by another. For example, after set theory was first introduced by Georg Cantor (1845–1918), a number of contradictions—the hobgoblins of mathematics—appeared, and some eminent mathematicians reacted by repudiating methods of proof that had been accepted for centuries. One of them, Herman Weyl, went so far as to say, ". . . that house [of mathematical analysis], in an essential part, is built on sand."*

Probably the most controversial area of mathematics is the one that treats of the infinite—both the infinitely large and the infinitely small. About 250 years ago the English philosopher George Berkeley expressed doubts about the development of the calculus. He sneered at the "ghosts of departed quantities," which were "neither finite quantities, nor quantities infinitely small, nor yet nothing." These parts of the theory were modified, but much later; the difficulties were avoided by restating the results in terms of finite quantities instead of the infinitely small.

To mathematicians the infinite has always been suspect. Galileo and Leibniz both tried to develop a theory for the infinitely large, and both decided the task was hopeless. The great mathematician Karl Friedrich Gauss asserted that the ". . . use of an infinite magnitude . . . is never permissible in mathematics. . . ." So for whatever reason, whether because it smacks too much of the mystical or because it is too much at odds with common sense, the infinite has generally been avoided; and when it has been used,

---

*For an account of the battle between the rebels (the intuitionists) and the traditionalists, see *The Development of Mathematics,* by Eric Temple Bell (New York: McGraw-Hill, 1945), pages 559–70.

mathematicians have remained uneasy until it has been replaced and the same results obtained by other means.

So it should not be too surprising that when Cantor confronted the analysis of the infinitely large, he received a reaction of hostility and abuse from many of his peers. A contemporary mathematician, Leopold Kronecker, was bitingly critical, and the great French mathematician Henri Poincaré stated that later generations would consider Cantor's work "disease from which one has recovered." But it wasn't too long before the tide turned; the great German mathematician David Hilbert, referring to Cantor's work on the infinite, asserted, "No one shall expel us from the paradise which Cantor has created for us." And at present, hardly 75 years later, Cantor's work is considered basic for graduate students in mathematics.

Since Cantor's work is fairly sophisticated, we will only sample a small part of it. What we will be primarily concerned with is set theory, a branch of mathematics originally fashioned by Cantor to study the magnitudes of infinite sets and ultimately used to lay the foundations of various other areas of mathematics.

To the student of mathematics one of the most puzzling things must surely be the way you begin. First you are given certain undefined terms, which must be accepted, at least formally, without any further explanation. You may be given an informal, intuitive explanation of what these terms "really" mean, but the picture you come away with may be different from mine. In addition to these undefined terms, you must accept certain basic axioms on which all your later conclusions will be based (if you are to play this mathematical game at all). And this is as it must be. To define one word you must use others, and these in turn must be defined in terms of still others, and so on until you come to words that aren't defined at all. And if you want to prove something, it must be done by using what you already know; if you start out knowing nothing, you can't get off the ground. In the usual tenth-grade geometry course, for example, you would accept the terms "point" and "line" as undefined and assume without proof that there is precisely one line through any two given points.

So mathematics, which is supposed to be so precise and reliable, turns out, ironically, to be based on words you can't define and assertions you can't prove. This is why Bertrand Russell once described mathematics as "the subject in which we never know what we are talking about, nor whether what we say is true." And yet mathematics has solved many problems in the ordinary world in which we live and has created deep and beautiful worlds of its own.

puzzles to ponder

1 Are there any differences between the "facts" of mathematics and the "facts" of physics? Do you think "A body falling in a vacuum accelerates at the rate of 32 feet per second per second" is a different kind of statement from "2 + 2 = 4"? Why?

2 Can you imagine a world in which the force of gravity makes falling bodies accelerate at 100 feet per second per second instead of at 32 feet per second per second? Can you imagine a world in which 2 + 2 = 5?

3 How did human beings learn about the facts governing the physical world? How did they learn the facts governing the mathematical world?

4 If mathematicians are free to choose any axioms they please, what is to prevent one mathematician from saying "X is true" in his system while another one says "X is false" in a different system? Would one or the other of them have to be wrong? (Look up the history of non-Euclidean geometry.)

5 In practice, mathematicians generally make the same assumptions. Why do you suppose this is true?

Sets

In set theory the terms "set" and "element" and the property of an element being *"in"* a set are all undefined. Still, we may explore the meanings of these terms informally. A **set** may be thought of as a collection of objects: We speak of the set of students in the honors history class, the set of whole numbers that are divisible by

Fred is not a member of the set of bearded men.

6, and the set of players on your college football team. An object contained in a set is called an **element** or a **member** of that set.

One way to describe a set completely is to list its members. You can do this directly by writing, for example, "$S = \{2, 4, 6, 8\}$," which means that the set S consists of the four elements 2, 4, 6, and 8. Or you can say the same thing indirectly by giving some property that is common to the members of S and *only* to the members of S. So an alternative definition of S might be, "S is the set of all even integers that are larger than 1 and less than 9." Whichever kind of definition you choose to use, the important thing is that you are always able to distinguish objects that are members of the set from those that are not and to distinguish one member of the set from another.

Universal, Null, and Complementary Sets

We will be talking about different kinds of sets at different times, but at any one time we will suppose there is a fixed universe of elements and will restrict our attention to it. This universe of elements will be called the **universal set** and will be denoted by E. Every other set we talk about will be assumed to consist of elements in E. At different times, the universal set may be the set of all American teenagers, the set of all people living and dead, the set of all IQ test scores, and so on.

At the other extreme we have the **null set**, denoted by $\varnothing$. The null set is defined as the set that contains no elements at all. The set of all living unicorns is the null set; the set of all square circles is also the null set.

For any set A we define the **complement** of A to be the set that contains precisely those elements in E that are not in A. We denote this set by A'. If E is the set of all married couples and A is the set of happily married couples, then A' is the set of unhappily married couples.

Relationships Between Sets

For clarity, we represent elements by small letters and sets by capital letters. To indicate that the element a is in the set B we write $a \in B$, which is read, "a is in B." Similarly, $a \notin B$ means that "a is not an element of B." If B is the set of mammals, x is a monkey, and y is a snake, then $x \in B$ and $y \notin B$.

When we write $A = B$, we mean that the sets A and B are really the same in that they have exactly the same elements. A single set may be defined in two different ways, and it may not be

Of course, the set of all unicorns represented by artists is not *the null set.*

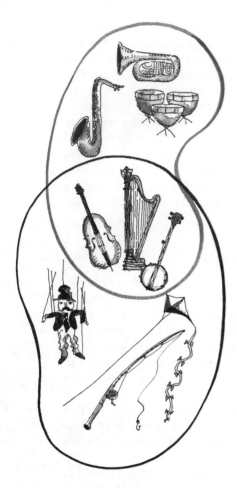

The set of objects with strings and the set of musical instruments are intersecting sets. Their intersection is the set of stringed instruments.

obvious that the sets determined by the two definitions are the same; consequently a set may go by many different names. For example, suppose A is the set of all years in which the Democrats win the Presidency in the last quarter of this century and $B = \{1976, 1984, 1992, 1996\}$. We don't know whether or not $A = B$. and we won't know with certainty for some time.

If every element of a set A is an element of some other set B, we say that A is a **subset** of B and denote this relationship by $A \subseteq B$.

If A is a subset of B but B is not a subset of A (that is, every element of A is in B but there is at least one element of B that is not in A), we say that A is a **proper subset** of B and denote this by $A \subset B$. The set of *happily* married couples is a subset of the set of all married couples, and to judge by the divorce courts, it's a proper subset.

New Sets from Old

If A and B are any two sets, then

1. The set consisting of precisely those elements which are in either A or B (or both) is called the **union of A and B** and is denoted by $A \cup B$.
2. The set consisting of precisely those elements that are in both A and B is called the **intersection of A and B** and is denoted by $A \cap B$. If E is the set of all students in a certain school, A is the set of all redheads in the school, and B is the set of all basketball players in the school, then $A \cup B$ is the set of all students who are either basketball players or have red hair (or both), and $A \cap B$ is the set of red-headed basketball players in the school.

The definitions of union and intersection may be extended to more than two sets:

$A \cup B \cup C \cup \cdots \cup Z$ consists of all the elements that are in at least one of the sets $A, B, \ldots, Z$, and $A \cap B \cap C \cap \cdots \cap Z$ consists of those elements that are in all the sets.

Two sets A and B are called **mutually exclusive** if they have no element in common, that is, if $A \cap B = \varnothing$. Any number of sets are called mutually exclusive if the intersection of *any two* of them is the null set. A shoe manufacturer might claim that the set of people using his product and the set of people with aching feet are mutually exclusive.

Venn Diagrams

It is often difficult to visualize the intersections and unions of various sets and their complements. It can be helpful to use a **Venn diagram**, which is a geometric expression of a set-theoretic statement. Figure 1 shows Venn diagrams in which the shaded portions represent a number of different sets.

Figure 1

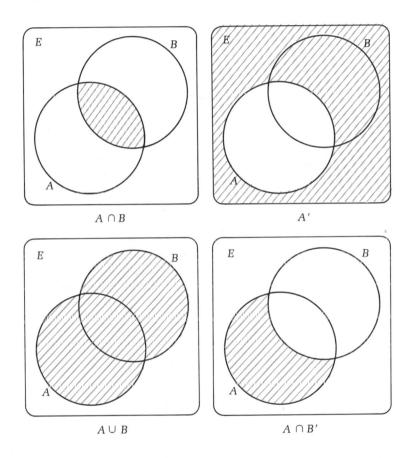

$A \cap B$ A'

$A \cup B$ $A \cap B'$

Figure 2

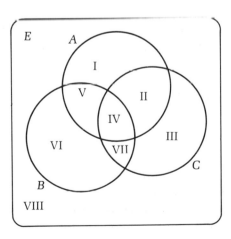

Where there are three intersecting sets, the diagram becomes a bit more complicated (see Figure 2). In all there are eight different subsets, which we may identify by Roman numerals. For example, the region designated by V consists of all elements that are in both A and B but are not in C, that is, $A \cap B \cap C'$. Region I consists of those elements that are in A but in neither B nor C, that is, $A \cap B' \cap C'$. Region VIII consists of those elements that are in none of the three sets A, B, and C.

Figure 3

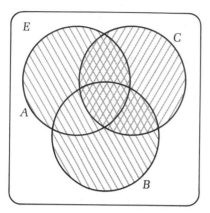

The region with the \\\ diagonal lines is $A \cup B$. The region with the /// diagonal lines is C. The region with *both* kinds of diagonal lines is $(A \cup B) \cap C$.

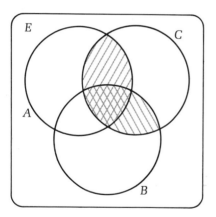

The region with the /// diagonal lines is $A \cap C$, and the region with the \\\ diagonal lines is $B \cap C$. The region that contains *either* of these types of diagonal lines is $(A \cap C) \cup (B \cap C)$.

By examining the diagram it is possible to confirm that certain sets are equal. For example, suppose we want to verify that

$$(A \cup B) \cap C = (A \cap C) \cup (B \cap C)$$

On the left side of the equation we notice that $A \cup B$ consists of the regions I, II, IV, V, VI, and VII, while C consists of II, III, IV, and VII. So $(A \cup B) \cap C$ consists of the regions common to both $A \cup B$ and C, or II, IV, and VII. On the right side we observe that $A \cap C$ consists of II and IV and $B \cap C$ consists of IV and VII. Thus $(A \cap C) \cup (B \cap C)$ consists of II, IV, and VII. This confirms the equality. (Another way to verify it is shown in Figure 3.)

If you want to show that two different sets are equal, you must show that any element of one set is also an element of the other. To show that $(A \cap B)' = A' \cup B'$, for example, takes two steps.

First, assume x is an element of $(A' \cup B')$. This means that x is either not in A or not in B, and therefore x is not in $(A \cap B)$. And this means x is in $(A \cap B)'$.

Now assume y is in $(A \cap B)'$. Since y is not in both A and B, either y is not in A or y is not in B. So either y is in A' or y is in B', and thus y is in $(A' \cup B')$.

The formula $(A \cap B)' = A' \cup B'$ can also be verified informally by using the Venn diagram in Figure 2. The set $(A \cap B)$ consists of the regions IV and V, so $(A \cap B)'$ consists of the other six regions. The set A' consists of the regions III, VI, VII, and VIII, while B' consists of the regions I, II, III, and VIII. Thus $(A' \cup B')$ consists of the same regions as $(A \cap B)'$: I, II, III, VI, VII, and VIII.

EXAMPLE 1

There are two honor societies in a certain college, Psi Phi Chi and Epsilon Upsilon. To be eligible for Psi Phi Chi you must *either* (1) be an athlete *or* (2) have at least a B average and play in the band. To be a member of Epsilon Upsilon you must (1) be in the band *and* (2) be an athlete or have at least a B average. Which of the two clubs is more exclusive?

Solution

Let P be the set of students that are in the band, Q the set of students that have at least a B average, and R the set of athletes. Then the set of students who are eligible for Psi Phi Chi is $R \cup (P \cap Q)$, and the set of those who qualify for Epsilon Upsilon

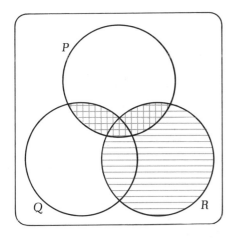

Figure 4

is $P \cap (Q \cup R)$. If student x is in the second set, then x is certainly in P and also in either Q or R. If x is in R, then x is in the first set, and if x is not in R, x is in $(P \cap Q)$ and again must be in the first set. However, if x is in R but not in P, x will not be in the second set but will be in the first one. So the second society is more exclusive.

This result can easily be shown with a Venn diagram, as in Figure 4. The qualifications for Psi Phi Chi are indicated by horizontal shading and those for Epsilon Upsilon by vertical shading.

EXAMPLE 2

Let E be the set of whole numbers from 1 to 12 inclusive. If A is the set of even numbers, B is the set of numbers greater than 8, and C is the set of numbers that are divisible by 3, then find

(a) $A \cap C$ (b) $B \cap C$

(c) $A \cup C$ (d) A'

(e) $B \cap C'$ (f) $(A \cup B)'$

Solution

The elements of sets A, B, and C are as follows: $A = \{2, 4, 6, 8, 10, 12\}$, $B = \{9, 10, 11, 12\}$, and $C = \{3, 6, 9, 12\}$. So

(a) $A \cap C = \{6, 12\}$ (b) $B \cap C = \{9, 12\}$

(c) $A \cup C = \{2, 3, 4, 6, 8, 9, 10, 12\}$ (d) $A' = \{1, 3, 5, 7, 9, 11\}$

(e) $B \cap C' = \{10, 11\}$ (f) $(A \cup B)' = \{1, 3, 5, 7\}$

EXAMPLE 3

In a certain college every freshman takes mathematics, English, and history. If M is the set of freshmen who failed mathematics, H is the set of freshmen who failed history, and E is the set of freshmen who failed English, justify the equation $(E \cap H \cap M)' = (E' \cup H' \cup M')$.

Solution

$E \cap H \cap M$ is the set of students who failed all three of these courses. Thus $(E \cap H \cap M)'$ is the set of students who passed at least one of them, that is, $E' \cup H' \cup M'$.

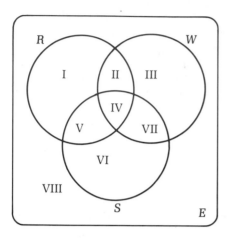

Figure 5

EXAMPLE 4

Suppose E is the set consisting of all the days in the previous year. Within this universal set let W be the set of windless days, R the set of rainy days, and S the set of snowy days.

Suppose you wore rubbers on the days when it either rained or snowed and only on those days, and you carried an umbrella when it both rained and it wasn't windy, and only then.

Express the following sets in terms of the sets C, R, and S: A, the set of days in which you both carried an umbrella and wore rubbers; B, the set of days in which you carried an umbrella and did not wear rubbers; C, the set of days in which you did not carry an umbrella but did wear rubbers; and D, the set of days in which you neither carried an umbrella nor wore rubbers. (See Figure 5.)

Solution

(a) $A = (R \cap W) \cap (R \cup S)$. You can use a Venn diagram or common sense to deduce that $A = R \cap W$. The reason is that any element of $R \cap W$ is also an element of $R \cup S$.

(b) $B = (R \cap W) \cap (R \cup S)'$. Since no element can be in $R \cap W$ and *not* be in $R \cup S$, the set B is actually $\varnothing$.

(c) $C = (R \cap W)' \cap (R \cup S)$. Using a Venn diagram, you can confirm that C is also $(R' \cap S) \cup (R \cap W')$.

(d) $D = (R \cap W)' \cap (R \cup S)'$. Using a Venn diagram you can confirm that $D = (R \cup S)'$.

exercises

1 (a) Define the union of two sets.
    (b) Define the intersection of two sets.
    (c) Define the complement of a set.
    (d) What is the complement of the universal set, E?
    (e) What is the complement of the null set, $\varnothing$?

2 Let E be the set of students in a certain French class, A the set of men students in the class, B the set of women students in the class, C the set of students in the class who are over 21, and D the set of students in the class who are under 21. Give a verbal description of each of the following sets:
    (a) $A \cup B$ (b) $A \cap C$ (c) $B \cup D$
    (d) $(A \cap C)'$ (e) $A \cup (B \cap D)$

3 Given each of the conditions below, state when, if ever, A is a subset of B, and when A is a proper subset of B.
    (a) $A = C \cap D$ and $B = C$
    (b) $A = C \cup D$ and $B = C$
    (c) $A = C$ and $B = C'$
    (d) $A = C$ and $B = (C')'$

4 Use a Venn diagram to confirm the following:
    (a) If every student with less than a B average is not on scholarship and there is at least one athlete on scholarship, then there is an athlete who has at least a B average.
    (b) If a certain company always employs high school graduates, then the set of people who do not work for the company is a subset of all people who are not high school graduates.
    (c) A talent scout for a baseball team accepts only players who have a batting average of more than .300 and either three good years of fielding experience or one excellent year of fielding experience. Another talent scout takes players who both bat .300 and have three good years at fielding or who both bat .300 and have one excellent year of fielding experience. Show that both scouts pick the same set of players.

5 Listed below are a number of different sets. In each case state which of the following are necessarily true: (i) $M \subseteq N$; (ii) $M = N$; (iii) $N \subseteq M$; (iv) $M \cap N = \varnothing$; (v) none of these. You may find a Venn diagram helpful.
    (a) $M = N \cap B$ (b) $M = N \cup B$
    (c) $M = \varnothing$ (d) $M = E$
    (e) $M = N'$ (f) $M = (N')'$
    (g) $M = (N' \cap A)'$ (h) $M = (N \cap B) \cup N$
    (i) $M = (N \cup B) \cap N$ (j) $M = (N \cup B) \cap B$
    (k) $M = (((N')')')'$

6 Listed below are a number of sets. Indicate any sets that are equal and indicate whether any sets are included in others; if the inclusion is proper, state that as well.

$A =$ the set of positive integers
$B =$ the set of nonnegative integers
$C =$ the set of prime numbers (that is, the positive integers that are divisible only by themselves and 1)
$D =$ the set of integers divisible by 4
$E =$ the set of positive even integers
$F = \{2, 4, 6, 8\}$
$G = \{1, 2, 3\}$

7 Each of the following statements follows directly from the definitions given earlier; convince yourself that they are true. (Use Venn diagrams to help.)

(a) The statements "$A = B$" and "$A \subseteq B$ and $B \subseteq A$" are equivalent; that is, if one is true, then so is the other.

(b) If $A \subseteq B$ and $B \subseteq C$, then $A \subseteq C$.

(c) $A \cap B = B \cap A$; $A \cup B = B \cup A$.

(d) For every set A, $\varnothing \subseteq A \subseteq E$.

(e) If $A \subseteq A'$, then $A = \varnothing$. If $A' \subseteq A$, then $A = E$.

(f) For any set A,
   (i) $A \cup \varnothing = A$ and $A \cap \varnothing = \varnothing$
   (ii) $A \cup A = A$ and $A \cap A = A$
   (iii) $A \cup A' = E$ and $A \cap A' = \varnothing$
   (iv) $A \cup E = E$ and $A \cap E = A$

(g) $A \subseteq A$

8 Corporation X plans to start manufacturing widgets and is seeking another corporation with which to merge. To make widgets you need a press and a die, and X is looking to its new partner for both machines. Let D, P, and C be the sets of companies that own dies, own presses, and have the cash to pay for both, respectively. X places the following advertisement in a local paper:

"Wanted for merger—a corporation with a press and a die or a million dollars."

Would it be reasonable to interpret this advertisement as addressed to a firm that is a member of the set $(P \cap D) \cup C$? To a firm that is a member of $P \cap (D \cup C)$? Which set does X have in mind? If the advertisement is misinterpreted, which will happen—will an ineligible firm mistakenly apply or will an eligible firm fail to apply because it feels it is ineligible?

9 Suppose E is the set of all the integers from 1 to 20 inclusive.

(a) List all the elements of each of the following sets:
   (i) S is the set consisting of the two largest elements in E.
   (ii) S is the set of all elements of E that are divisible by 3.
   (iii) S consists of the elements in E that have one digit.

(b) Define each of the following sets in a different way by giving a property that is common to the members (and only the members) of the set.
   (i) $S = \{2, 4, 6, 8, 10, 12, 14, 16, 18, 20\}$
   (ii) $S = \{1, 2, 3, 4, 17, 18, 19, 20\}$
   (iii) $S = \{1, 2, 3, 4\}$

10 Listed below are a number of sets, some of which contain others as subsets. Use the inclusion symbol, $\subseteq$, to express all of these relationships.
(a) Girls, boys, children, males, men, people, animals, mammals.
(b) Even numbers, odd numbers, numbers divisible by 8, numbers divisible by 3, numbers divisible by 6.

11 As in Illustrative Example 3, assume that E, M, and H are the sets of students who failed English, mathematics, and history, respectively. If every freshman takes these three courses, verify that $(E \cup H \cup M)' = (E' \cap H' \cap M')$.

12 We showed earlier that $(A \cap B)' = (A' \cup B')$. Verify that $(A \cup B)' = (A' \cap B')$ by using Venn diagrams or by showing that any element of either set must be in the other.

   These two formulas are known as De Morgan's Laws and they can be extended to any number of sets:

$$(A \cap B \cap C \cap D \cap \cdots)' = (A' \cup B' \cup C' \cup D' \cup \cdots),$$
$$(A \cup B \cup C \cup D \cup \cdots)' = (A' \cap B' \cap C' \cap D' \cap \cdots).$$

Can you see why these laws are true?

13 In ordinary arithmetic there is a law called the *distributive law*, which states that for any three numbers a, b, and c, $a(b + c) = ab + ac$. For example, $7(2 + 3) = (7)(2) + (7)(3)$. In set theory there are *two* such laws:

$A \cup (B \cap C) = (A \cup B) \cap (A \cup C)$ and
$A \cap (B \cup C) = (A \cap B) \cup (A \cap C)$

Verify these laws by any method you wish. (Notice that the other "distributive law" in arithmetic, which would state that $a + (bc) = (a + b)(a + c)$, is not true.)

14 The following quotation is from Lewis Carroll's *Alice in Wonderland:*
   " 'Then you should say what you mean' the March Hare went on.
      'I do' Alice hastily replied; 'At least—at least I mean what I
      say—that's the same thing you know.' "
Comment on Alice's logic.

15 To play baseball last year you either had to be off academic probation and not be in financial arrears or you had to have the Dean's permission. This year you can't play if you're both in arrears and on academic probation (whatever the Dean says). Were they stricter this year or last year or neither? Explain.

16 A job applicant must be approved by three interviewers before being offered a job. The first interviewer only approves applicants who are either over 40 or have a degree; the second interviewer only approves applicants who either are over 40 or have previous experience; and the third interviewer only approves applicants who are under 40. Show that the only applicants offered jobs will be under 40 with either a college degree or experience.

17 Assume that a job applicant need only be approved by one interviewer to receive an offer. One interviewer only approves applicants who are experienced, under 40, and not aggressive; the second interviewer approves applicants who are experienced, under 40, and aggressive; and the third interviewer approves applicants who are experienced, over 40, and aggressive. Show that the successful applicants will be experienced and either under 40 or aggressive.

18 By using the distributive laws or Venn diagrams, show that
(a) $(A \cup B) \cap B' = A \cap B'$
(b) $[A \cap (A \cap B)'] \cup B = A \cup B$
(c) $(A \cup B) \cap (A \cup B') = A$
(d) $(A \cup B) \cap A' = B \cap A'$
(e) $(A' \cap B')' \cup C = A \cup B \cup C$

19 There are four clubs in town. The first accepts any member who is either listed in *Who's Who* or a millionaire. The second accepts everyone who is either listed in *Who's Who* or has a graduate degree. The third takes anyone who is either a millionaire or has a graduate degree. And the fourth only takes those who are not millionaires. Show that anyone in all four clubs must be in *Who's Who,* must have a graduate degree, and must not be a millionaire.

20 Use a Venn diagram to show that
(a) $(B \cup C') \cap (B' \cup C) = (B \cap C) \cup (B' \cap C')$
(b) $(A \cap B') \cup (B \cap C') \cup (C \cap A') = (A \cup B \cup C) \cap (A' \cup B' \cup C')$
(c) $A \cup (B \cap A') \cup (C \cap A' \cap B') = A \cup B \cup C$

1 In a certain town the barber shaves those men, and only those men, who do not shave themselves. Does the barber shave himself or not? (Assume that he does and then assume that he doesn't, and see where each assumption leads.)

2 There are many ways you can describe a number. You can name one directly by saying "eighteen," you can define one arithmetically by saying "two times three times three," or you can define one in words by saying "the largest number that is divisible by six and is also less than twenty." Some fairly large numbers may have short descriptions ("the number of stars that are less than 1,000 light years away," for example), while others may not. Since there are only a finite number of English words, there can only be finitely many numbers that can be expressed in fifteen words or less.

Suppose X is *the smallest number that cannot be expressed in fifteen words or less*. But X has just been expressed in twelve words, so it can't be the number after all. But then it hasn't been expressed in twelve words, so we're back where we started from.

Infinite Sets

Cantor started his investigations into set theory to gain insight into the subject in which he was really interested, infinite sets. This theory of infinite sets, which provoked so much criticism by his contemporaries, is today considered his greatest contribution. Cantor's ideas are hard to follow at first, because they clash with our intuition. Our notion of what is "natural" and "proper" is strongly influenced by our experience with finite sets, and it is bewildering to find rules that have proved reliable breaking down when applied to infinite sets. Still, with a little patience, Cantor's assumptions begin to hang together with a logic of their own. We will state some of the problems that confronted Cantor as simply as possible and then indicate how he tried to solve them.

The infinite has always been a source of discomfort for mathematicians, especially the "absolute" infinite. A distinction is often made between the absolute and potential infinites. Suppose you are to receive $1 today, $2 tomorrow, $3 the day after, and so on. As more and more time passes, the amount of money you get may be said to "approach" infinity. Or, if $y = 1/x$ and x is a positive number getting very close to zero, we say y "becomes" infinite. (The value of x can never actually be zero, since division by zero is meaningless.) The idea of "becoming" or "approaching" infinity without quite getting there is sometimes called the *potential infinite*.

The absolute infinite—the quantity of integers, for example—is a very different thing. These infinite sets aren't "approaching" infinity—they've arrived. They must have their own rules, since the rules that work for finite sets break down, as we shall see. Still, a good way to learn about infinite sets is to start with what we know, finite sets, and then go on from there.

There is a basic problem concerning infinite sets which is easily resolved for finite sets: How do you determine whether two sets have the same number of elements?

Equivalent Sets

Two sets that have the same number of elements are called **equivalent**. (Remember, equivalence is not equality; for two sets to be equal they must have the same *elements,* not just the same *number* of elements.) Finite sets may be shown to be equivalent in more than one way. One obvious way is to count the elements in both sets. Another way is to set up a **1-to-1 correspondence**. If you want to find out whether there are as many chairs as people in a room, have everyone sit down; if there are no empty chairs and no one standing, the set of chairs and the set of people are equivalent. In general, if two sets are paired off so that every member of each set has exactly one mate in the other, there is a 1-to-1 correspondence between them, and the sets are equivalent. A roomful of married couples in which everyone has exactly one spouse determines an obvious 1-to-1 correspondence between the sexes. For finite sets, then, we have

Statement 1: If the elements of one set can be put into 1-to-1 correspondence with the elements of another set, then the two sets are equivalent, that is, they have the same number of elements.

You may not know how many dogs there are or how many dog walkers there are. But if every dog walker is holding exactly one *leash and every dog is being held by* exactly one *leash, then there is a 1-to-1 correspondence between the set of dogs and the set of dog walkers. This means that there are exactly as many dog walkers as dogs.*

Now suppose that in the room containing the people and chairs everyone sat down and some chairs remained vacant. It would then be obvious that there were more chairs than people. And in the same way we would conclude that there were more people than chairs if, after all the chairs were taken, there were some people left standing. So for finite sets we would also conclude

Statement 2: If all the elements of a set A can be put into 1-to-1 correspondence with the elements of a *proper* subset of B, then there are more elements in B than in A, and so A and B are not equivalent.

Everything we have said so far seems reasonable enough if you restrict your attention to finite sets. But problems arise when you try to extend these ideas to infinite sets. To see why, let's compare two simple infinite sets.

Suppose X is the set of all positive whole numbers and Y is the set of all positive even whole numbers:

$$X = \{1, 2, 3, \ldots\} \quad \text{and} \quad Y = \{2, 4, 6, \ldots\}$$

If Statement 2 were valid for infinite sets, it would follow that there must be more elements in X than there are in Y. It is easy to see why if you break X into two parts, one containing all the positive even integers and the other containing all the positive odd integers. Now you simply match the even integers in X with all the elements of Y, and the odd integers in X are left over.

On the other hand, if we use Statement 1 to determine whether X and Y are equivalent, we arrive at the opposite conclusion. Specifically, we can construct a 1-to-1 correspondence between the elements of X and the elements of Y:

$$\{1, 2, 3, \ldots, \quad i, \ldots, \quad j, \ldots\}$$
$$\updownarrow \updownarrow \updownarrow \qquad \updownarrow \qquad \updownarrow$$
$$\{2, 4, 6, \ldots, 2i, \ldots, 2j, \ldots\}$$

With each element, i, in X we associate the element in Y that is twice as big: $2i$. With each element, $2j$, in Y, we associate the element in X that is half as big: j. It is not difficult to confirm that each element in X corresponds to a unique element in Y and each element in Y corresponds to a unique element in X. So if we accept Statement 1, we must conclude that X and Y have the same number of elements.

Unless we are willing to assert that a set X can have the same number of elements as another set Y at the same time that we assert that the set X has a greater number of elements than the set Y, we will have to reject one statement or the other. This much was clear to Cantor, and after some consideration he decided it would be more fruitful if he dispensed with Statement 2.

So Cantor started by rejecting Statement 2 and assuming that Statement 1 was applicable not only to finite sets but to infinite sets as well. He defined two sets to be equivalent if they could be put into 1-to-1 correspondence.

A reader who sees this definition for the first time can easily be misled. If you can construct a 1-to-1 correspondence between set A and set B you will conclude that the two sets are equivalent—and so they are. But if you find a 1-to-1 correspondence between a proper subset of A and all of B you may be tempted to conclude that A is larger than B and therefore that the two sets cannot be equivalent—and in this you would be mistaken. As we observed earlier, if you look at the set of positive integers and the set of positive even integers in one way, they appear to match perfectly; if you look at them in another way, all the positive even integers and only part of the integers match and the odd integers are left over. No matter! It is possible to put the sets into 1-to-1 correspondence, and hence the sets are equivalent. We repeat:

If an entire set can be put into 1-to-1 correspondence with just a proper subset of another set, it does not necessarily follow that the two sets are not equivalent.

If you are only accustomed to dealing with finite sets, this will certainly seem to be a strange state of affairs. It is as though an audience entered a crowded auditorium and found that after all the seats were taken some people were left standing. And then, without adding any seats and without anyone leaving, the seated people arose and joined the standees, the audience milled about in some strange manner and then sat down again, but this time with no one left standing and with no more than one person to a seat—an apparently absurd situation.

There is an old saying that the whole is always greater than any of its parts. But in set theory this is not true, as we just observed. The set of *all* the positive integers is equivalent to the set of even positive integers. This attribute of a whole being equivalent to a proper part of itself is not just a curiosity; it is a property that distinguishes finite sets from infinite ones. In fact, an infinite set is sometimes defined as a set that is equivalent to a proper subset of itself.

Because of the way Cantor defined equivalence, sets that appear to have different magnitudes often do not. Consider the positive rational numbers, for example (these are just the ordinary fractions having integers as their numerators and denominators). When you compare the set of positive integers to the set of positive rational numbers, it seems that the integers are hopelessly outnumbered. Between any two integers—say 5 and 6—there are an infinite number of rationals: $5\frac{1}{2}$, $5\frac{1}{3}$, $5\frac{1}{4}$, 5.32, and so on, so that a 1-to-1 correspondence would seem to be impossible. And yet Cantor showed that the set of integers and the set of positive rational numbers *are* equivalent. Although the result is unexpected, the proof isn't hard, and we will outline it briefly.

To say that a set is in 1-to-1 correspondence with the positive integers means you can "count" it; that is, you can assign a positive integer to each element of the set so that each integer corresponds to one element and each element corresponds to one integer. The trick to counting the positive rational numbers is to avoid the temptation of considering them in order of size. Notice first that every positive rational number can be put in the form of a fraction in lowest terms with positive integers as numerator and denominator. Once this is done, the positive rationals can be ordered according to the *sum* of these two integers. There is only one positive rational number whose numerator and denominator sum to 2; that fraction is 1/1, and the positive integer 1 is assigned to it. There are two different rationals whose numerators and denominators sum to 3, namely, 2/1 and 1/2, and to these we assign the numbers 2 and 3. There are two more fractions whose numerators and denominators sum to 4, namely, 3/1 and 1/3, and these are assigned the integers 4 and 5. (We omit the fraction 2/2 because it is the same as 1/1, which we have counted already.) If we continue in this way, sooner or later every fraction will be assigned a number because every fraction's numerator and denominator have a finite sum and will therefore be preceded by only a finite number of other fractions. On the other hand there is no danger of running out of fractions, since there are an infinite number of them, and so every positive integer will be assigned to some mate.

There are many other cases in which sets that appear to be of different size turn out to be equivalent. For example, there is a 1-to-1 correspondence between the points on a one-inch line segment and the points on a line segment one million miles long, or on a line of infinite length, or in a two-dimensional, three-dimensional, or even an infinite-dimensional space. In fact, it might seem that *every* infinite set is equivalent to every other one, and this possibility occurred to Cantor. A measure that equates all infinite sets may be very democratic but wouldn't be very useful, and, not surprisingly, Cantor was very concerned. But he eventually dis-

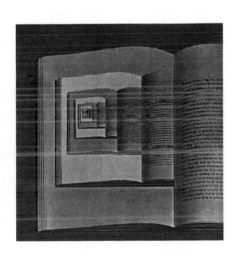

A representation suggesting an infinite set.

covered pairs of sets that were not equivalent. Cantor found that there are more points on a line than there are fractions, for example. Moreover, he found that there are an infinite number of different sizes that sets can have. He showed this by proving the following theorem:

> For any given set *A* there is a set *B* that is greater than *A*. (A set *B* is said to be *greater than* a set *A* if there is a 1-to-1 correspondence between the set *A* and some subset of *B* but there is no 1-to-1 correspondence between *B* and *any* subset of *A*.) In particular, the set consisting of all the subsets of *A* cannot be put into 1-to-1 correspondence with the elements of the set *A*.

The proof, which is quite ingenious, will be outlined in the exercises.

1   Describe how you would distinguish an infinite set from a finite one.

2   Give three examples of finite sets and three examples of infinite sets.

3   Suppose *E* is the set of positive integers, {1, 2, 3, . . .}, and *A* is a finite subset of *E* while *B* and *C* are infinite subsets of *E*. State whether each of the following sets is finite or infinite, or whether the set may be either finite or infinite. If the set may be either finite or infinite, give an example of each possibility.
(a) $A \cap B$
(b) $A \cap B'$
(c) $B \cap C$
(d) $A \cup B$
(e) $A \cup B'$
(f) $B \cap C'$

4   If every freshman has exactly one sophomore as a friend and every sophomore has exactly one freshman as a friend, a 1-to-1 correspondence between freshmen and sophomores exists. Suppose a similar 1-to-1 correspondence between freshmen and juniors also exists. How would you construct a 1-to-1 correspondence between sophomores and juniors?

5   If all freshmen can be put into a 1-to-1 correspondence with a *proper* subset of all sophomores and all sophomores can be put into a 1-to-1 correspondence with a subset of all juniors, show that all freshmen can be put into a 1-to-1 correspondence with a proper subset of all juniors.

6   Assume Eve had exactly one daughter, that daughter had exactly one daughter, that daughter had exactly one daughter, and so on ad infinitum. Show that the set of mothers can be put into 1-to-1 correspondence with the set of daughters. (Every woman is both a mother and daughter except Eve, who is only a mother.)

7   Construct a 1-to-1 correspondence between the set of positive whole numbers, $\{1, 2, 3, 4, 5, \ldots\}$, and the set of positive *and* negative whole numbers, $\{\ldots, -3, -2, -1, 1, 2, 3, \ldots\}$.

8   (a)   Show that there is a 1-to-1 correspondence between the set of all positive whole numbers that are divisible by 17 and the set of all positive whole numbers.

         (b)   Show that there is a 1-to-1 correspondence between the set of all perfect squares, $\{1, 4, 9, 16, \ldots\}$, and the set of all positive whole numbers.

1   When Cantor first defined equivalence for infinite sets, he found many pairs of sets that appeared to be very different in size but turned out to be equivalent. For a time it seemed that all infinite sets might be equivalent, but after some probing Cantor showed that if A is any set (finite or infinite) and B is the set of all subsets of A, then there are more elements in B than in A. To understand this better we will first show how it applies to finite sets and then outline the proof for the case in which A is the set of positive integers.

     In general, if there are n elements in A, there will be 2^n elements in B. If A is the null set, $\varnothing$, then B has one element: the null set itself. This is one more element than there is in A, since A has no elements at all. If A has one element, then B has two elements: the set A and the null set. If $A = \{p, q\}$, then $B = \{p, q, \{p, q\}, \varnothing\}$ has $4 = 2^2$ elements.

     Now suppose A is composed of all the positive integers, that is, $A = \{1, 2, 3, \ldots\}$. It is clear that a 1-to-1 correspondence exists between all the elements of A and those subsets of A which contain exactly one element: Each element i of A can be matched with $\{i\}$, the subset of A that contains i as its only element. The hard part of the proof, that there cannot be a 1-to-1 correspondence between the elements of A and the elements of B (consisting of all the subsets of A), remains to be shown.

     Let us suppose there was a 1-to-1 correspondence between the elements of A and the elements of B, that is, with each element of A we can associate one subset of A and with each subset of A we can associate an element of A. For each element of A determine whether that element is in its associated subset. Then form a *set T consisting of all those elements not contained in their associated subset.* Now *T* itself is a subset of *A*, so it must have its mate in *A* from the 1-to-1 correspondence; call its mate *t*. There are two possibilities: Either *t* is in *T* or it is not. We leave it as an exercise to show that each of the two assumptions leads to a contradiction.

This exercise (particularly the last part of the argument) makes rather rough going. If you can follow it—even after much time and effort— you're doing very well.

2   A police force has an infinite number of policemen, with badge numbers 1, 2, 3, 4, 5, ... However, they must cover an infinite number of street corners: 1st St. and 1st Ave., 1st St. and 2nd Ave., . . . , *i*th St. and *j*th Avenue where *i* and *j* are any two positive integers, etc. Is the police force big enough?

1   A very popular bar has a closing time of 2 A.M. The entrance or departure of patrons causes unpleasant drafts, so entrances and exits of patrons are restricted to certain times: an hour before closing, a half hour before closing, a quarter of an hour before closing, and so on, so that the successive waiting periods are halved as closing time approaches.

As time passes, the bar gets busier. At 1 A.M. patrons number 1 and 2 enter. At 1:30 A.M. patrons 3 and 4 enter and patron 1 leaves. At 1:45 A.M. patrons 5 and 6 enter and patron 2 leaves. At each break, two new patrons enter and one leaves. The patrons always leave the bar in the order in which they arrived.

The next morning you overhear a conversation between two customers. One says, "It must have been terribly crowded at closing time. During every break up until closing, two people entered while only one left, and with the breaks coming so frequently, especially toward the end, there could hardly be room to breathe."

"On the contrary," the other replies, "the place must have been empty, since everyone left before the bar was closed. You mention any patron (or at least the order of his arrival), and I'll tell you exactly when he left. And in every case it's before 2:00 A.M. (although in some cases it gets pretty close)."

Reread the problem, consider both comments, and then decide which point of view you accept.

2 There is an easily proved formula in set theory which says that the statements $A \subseteq B$ and $B' \subseteq A'$ are equivalent, that is, if either statement is true then the other must be true as well. The philosopher Carl Hempel formulated an amusing paradox from this simple relationship.

Suppose C is the set of all crows and B is the set of all things that are black. Suppose, further, that a scientist wants to determine whether $C \subseteq B$, that is, whether all crows are black. One approach is to go out into the fields where the crows are and look at as many crows as he can; each time he sees a crow that is black, the observation tends to confirm the theory. If all the crows he sees turn out to be black, he might tentatively conclude that all crows really are black (although he could never really be sure using this approach).

A lazy colleague with a bent for logic points out that the fields are cold and that he can save himself a great deal of trouble if he goes about his research a little differently. Since the statements $C \subseteq B$ and $B' \subseteq C'$ are equivalent, if you prove one, you have in effect proved the other. So why not try to confirm $B' \subseteq C'$?

To test $B' \subseteq C'$—that all nonblack things are not crows—you simply look for things that aren't black and check to see that they aren't crows. Since there are any number of things in the scientist's room that aren't black—paper clips, trash cans, particles of dust, a pet gerbil, etc.—he can make significant scientific progress without stirring from his office.

3 The following two paradoxes concerning the infinite are quoted in *Vicious Circles and Infinity*, by Patrick Hughes and George Brecht (New York: Doubleday, 1975).

(a) "Leinbach had discovered a proof that there really is no death. It is beyond question, he had declared, that not only at the moment of drowning, but at all the moments of death of any nature, one lives over again his past life with a rapidity inconceivable to others. This remembered life must also have a last moment, and this moment its own last moment, and so on, and hence, dying is itself eternity, and hence, in accordance with the theory of limits, one may approach death but can never reach it." (From Arthur Schnitzler, *Flight into Darkness*.)

(b) "Tristram Shandy, as we know, took two years writing the history of the first two days of his life, and lamented that, at this rate, material would accumulate faster than he could deal with it, so that he could never come to an end. Now I maintain that, if he had lived for ever, and not wearied of his task, then, even if his life had continued as eventfully as it began, no part of his biography would have remained unwritten." (From Bertrand Russell on the paradox of Tristram Shandy.)

4   A semicircle is drawn, as in Figure 6, with center P and with an arc in which the endpoints, A and B, are missing. Below the semicircle, parallel to the diameter $\overline{AB}$, an infinite line L is drawn. Observe that there is a simple 1-to-1 correspondence between the points of the infinite line L and the points of the arc $\overset{\frown}{AB}$. With any point R on the arc associate the point R', which is the intersection of the line L and the line through P and R; and with any point Q on the line associate Q', which is the intersection of the line L and the line through P and Q. Does the length of a line tell you anything about the quantity of points it contains?

Figure 6

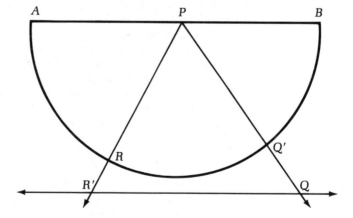

5   Suppose a generous bank offers its customers 100% interest; that means $1 deposited will grow to $2 a year later. A competing bank also offers 100% interest but compounds semiannually. That means you get 50% interest every 6 months, so that your $1 becomes $1.50 at the end of 6 months and then becomes $2.25 at the end of a year.

Not to be outdone, the original bank compounds quarterly, so that your money grows to $1.25, $1.5625, $1.953125, and $2.44140625 at the end of each of the four quarters.

If the banks continue to compete by making the interest period shorter and shorter, will you eventually get rich in a year? (*Hint:* If there are N periods in a year, your dollar becomes

$$\left(1 + \frac{1}{N}\right)^{N}$$

dollars in a year. Compute the value of this with a hand calculator for large values of N).

Summary

Set theory furnishes the basic rules that allow us to define and expand mathematical models. In this chapter we saw that the language of set theory is economical, in that a single statement may apply in many different contexts. For example, the statement $A' \cap B' = (A \cup B)'$ might be interpreted to mean that the set of sick adults is the same as the set of people who are neither children nor well; or it might be saying that those people who have bad credit ratings and are incompetent are precisely those people of whom it cannot be said that they either have good credit ratings or are competent.

The language of set theory is more precise than ordinary language. If an employment agency advertises that it is looking for someone who can type or take shorthand and do bookkeeping, will it accept an applicant who can type but who can't take shorthand and can't do bookkeeping? In set theory the agency would have to specify whether it wanted Typists ∪ (Shorthand ∩ Bookkeepers) or (Typists ∪ Shorthand) ∩ Bookkeepers. This extra clarity may have practical advantages. In an article in the *Yale Law Journal* Layman Allen pointed out various ways the language of set theory might be used to make legal documents unambiguous.*

Finally, set theory may be used to simplify. If you invite to a party (1) everyone who is in the chorus but not an athlete and (2) everyone who is an athlete but not in the chorus and (3) everyone who is neither in the chorus nor an athlete, then you have excluded those who are both athletes and in the chorus. In the language of sets, $(A \cap B') \cup (A' \cap B) \cup (A' \cap B') = (A \cap B)'$.

---

*Layman E. Allen, "Symbolic Logic: A Razor-Edged Tool for Drafting and Interpreting Legal Documents," *Yale Law Journal* 66 (May 1957), 833.

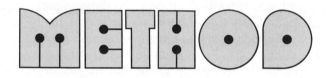

PROBABILITY

1 Introduction

In some ways probability theory is like any other branch of mathematics—you assume certain axioms and deduce theorems from them. But in one way it is different; the language of probability theory can often be understood by the mathematically unsophisticated. While other mathematicians talk about "groups," "fields," "functions," and "homeomorphisms," probability theorists use a vocabulary that includes many familiar terms. So when the evening newscaster puts the *probability* of rain at 40%, the local bookie gives two-to-one *odds* that the home team will win, or scientists tell us that our *chance* of dying from lung cancer will be *doubled* if we smoke a pack of cigarettes a day, we have a pretty good idea what is meant without having made any formal study of probability theory.

The reason people become familiar with the basics of probability theory is that they often apply it in everyday life. A motorist who knows that the meter maid appears at random an average of once in three hours can use probability theory to calculate whether it pays in the long run to park illegally for five minutes and risk a $10 fine, or to put a quarter in the meter. (One does better feeding the meter, but just barely.)

While this last problem is fairly easy, others can be more involved. For example, suppose the owner of a recently constructed ski resort wants to estimate the average number of customers attending each weekend so he can order food, set up a ski rental with sufficient skis, and prepare adequate parking. He knows from past experience that what a skier does during one weekend depends *only* on the skier's experience during the previous weekend. Specifically, he knows that someone who went skiing last week and had good, fair, or bad weather has a probability of skiing this week of .4, .3 and .1, respectively, and a skier who stayed home last week has a probability of .2 of skiing this week. In addition, he knows that 10%, 50%, and 40% of the weekends have good, fair, and poor weather for skiing, respectively, and that there are 2,000 potential customers who behave roughly as we have described. From this information how can the owner estimate his average weekend crowd? (As we will see in Chapter 7, on Markov chains, he can expect about 412 customers per weekend.)

Still another application of probability theory can be found in the stock market. Suppose a stock sells for $50 and someone offers you the following proposition: For a fee, you are given the right to buy 100 shares of the stock any time during the next six months for $50 a share, whatever the actual market value. Hence, if the stock price rises to $60, you can buy 100 shares at $50 and sell them at $60 for a profit of $1,000; if the stock does not rise, you lose nothing (except the fee). Such an option is referred to as a *call,* and calls are bought and sold on exchanges in the same way as stocks. If the call we just described is worth $X, what would a similar call be worth if the option could be exercised within a year instead of six months? You might think the lengthier call would be twice as valuable, but a probabilistic model suggests that it should be worth only about 40% more; actual market prices seem to confirm this model.

If you stop to think about it, there is something about the very existence of a theory of probability that almost smacks of alchemy—you start out ignorant and turn that ignorance into knowledge. Specifically, you don't know whether a stock will go up or down, you don't know how long a person will live, and you don't know how a die will turn up. And yet financial advisers select portfolios to minimize risk, insurance companies accurately predict death and accident rates, and gambling casinos flourish. In addition, probability is used in modern physics, it is used to predict how frequently airplanes will be overbooked and telephone switchboards overloaded, and it can be used to predict the fraction of blue-eyed people in future generations.

In this chapter we discuss the basics of probability theory. In Chapters 5 (on game theory) and 7 (on Markov chains) we will see how these basics can be applied.

puzzles to ponder

Suppose someone claims he can predict the winner of sporting events. When pressed, he says, "The better team will win"; and invariably, they do. After the fact he (very reasonably) points out that the better team *did* win—they would have lost if they were not the better team—and the original prediction is confirmed.

Not many people would buy this argument, and it's worth taking a moment to analyze why. The "prediction" is not a prediction at all because, whatever happens, it will be fulfilled. In other words, *there is nothing that can possibly happen to force you to conclude the prediction was wrong.*

Now consider the simple statement, "The probability of getting a head when I toss this coin is $\frac{1}{2}$." It can be argued that this straightforward probabilistic statement is of exactly the same type that we just discussed. What can you possibly observe in one toss, or in one million tosses, that would rule out the possibility that the chance of getting a head is $\frac{1}{2}$? Even if you observed a million heads in a row, the coin might still be unbiased. In fact, you can calculate how often you should expect such a lopsided result with a fair coin.

How would you respond to this argument?

Sample Spaces and Sample Points

Before he starts doing business for the day, a vendor who hawks refreshments at baseball games must pick one of several alternative options. He can choose to spend various amounts of money for food and beverages, and he can distribute the total amount of money he spends in various ways. Before making a decision he will normally try to determine what the consequences of his actions might be; he will want to know, for example, the profit on each item he sells, his loss if he fails to sell an item, which items have sold well before, and so on. But there are things that he can't know, factors that are determined by chance; he can't be sure of the team's popularity, the weather, or the particular tastes of that day's crowd, for example.

When you make decisions in everyday life, uncertainties are the rule and not the exception. You can't really be sure which life insurance policy is the best buy if you don't know how long you will live; you won't know if you should put another dime in the meter if you have no idea when the meter maid will turn up; and you can't be certain about whether to carry an umbrella if you're only guessing about whether it will rain. But you can make intelligent decisions even in the face of uncertainty if you use a probabilistic model. A simple example will show how you go about constructing one.

Assume A, B, and C are three horses competing in a race. In a three-horse race there are six possible finishing orders: *ABC*, *ACB*, *BAC*, *BCA*, *CAB*, and *CBA* (where *BCA*, for example, indicates that B finishes first, C second, and A last).

The race will be referred to as an *experiment;* "experiment" will be used as a general term in all probabilistic models. The individual possible outcomes are called *sample points;* in this experiment the sample points are the various finishing orders. And the set of all outcomes or sample points is called the *sample space.*

We just said that the set of all outcomes of an experiment makes up the sample space, but just what are the outcomes of an experiment? Returning to the three-horse race, let's suppose you bet $10 that horse *A* will win; in such a case you might view the race as having two possible outcomes: *A* wins or *A* doesn't win. If you bet $10 that *A* will place (finish first or second), you might once again describe the race as having two outcomes, but they would be different: *A* comes in third or *A* doesn't come in third. And if you bet $30 that *A* will win, $10 that *B* will place, and $5 that *C* will place, you would probably distinguish six different possible outcomes, since practically every finishing order would net you a different amount of money. The point is this: *For a single experiment you can construct many different sample spaces;* your choice of a sample space will generally be determined by your own interests and mathematical convenience. A little later we will see why one sample space may be easier to work with than another.

While there are many different possible sample spaces for a particular experiment, there is one condition that any sample space must satisfy: *Every possible outcome of an experiment must be associated with exactly one sample point; no more and no less.*

Suppose a coin is tossed twice. One possible sample space for this experiment consists of four sample points: (i) HH, (ii) HT, (iii) TH, and (iv) TT (where HT means you had a head and a tail in that order). Another possible sample space for the same experiment consists of three sample points: (i) HH, (ii) a head and a tail irrespective of order, and (iii) TT. If you use the second sample space, the outcomes HT and TH are indistinguishable, so your choice of sample space may depend on whether you are indifferent to these two outcomes. In any case, the important thing is that in both cases each possible outcome is included in exactly one sample point.

On the other hand, the three sample points (i) heads on the first toss, (ii) tails on the second toss, and (iii) TH do not constitute a sample space for the two coin tosses; neither do the two sample points (i) both tosses are the same and (ii) HT. In the first case the outcome HT is included both in sample point (i) and in sample point (ii), and in the second case the outcome TH is not included in either of the sample points.

(**EXAMPLE 1**)

Listed below are six different sets of sample points for the three-way horse race described earlier. Which of these sets of points make up a sample space?

(a)   (i)   *A* finishes second.
      (ii) *B* finishes second.
      (iii) *C* finishes second.

(b)   (i)   *A* finishes either first or second.
      (ii) *B* finishes either first or second.

(c)   (i)   *A* finishes before *B*.
      (ii) *B* finishes first.

(d)   (i)   *A* finishes before *B*.
      (ii) *C* finishes before *A*.
      (iii) The finishing order is *BAC*.

(e)   (i)   *A* finishes first.
      (ii) *B* finishes second.
      (iii) *C* finishes third.

(f)   (i)   *A* finishes before *B*.
      (ii) *B* finishes first.
      (iii) The finishing order is *CBA*.

Solution

Cases (a) and (f) are sample spaces; the others are not. Every possible outcome is included in (b), but *ABC* and *BAC* are included in both sample points (i) and (ii). In (c), *CBA* is included in neither sample point. All possible outcomes are covered by the three sample points in (d), but *CAB* is included in both (i) and (ii). In (e) the outcome *ABC* is included in all three sample points, and the outcomes *BCA* and *CAB* are included in none.

1   Construct a sample space for each of the following experiments:
   (a) A baseball fan picks one day of a given week to see his home team.
   (b) On a night out, a tourist must decide whether to see a musical or a drama and whether to eat at a French, Italian, or Greek restaurant.

    (c)   Two teams play a three-game match, and the first team to win two games is the winner.

    (d)   The three finalists in a spelling bee win the first, second, and third prizes.

2   Describe two different sample spaces that might be used for each of the following experiments:

    (a)   There are four foreign languages taught at Seward High School (French, Spanish, German, and Italian), and Arthur will study two of them. He will study one language for two years and the other for three years.

    (b)   Three chess players have a tournament in which each player meets every other player once.

    (c)   Fifty students from fifty different states compete in an essay-writing contest. There will be a first and a second prize.

    (d)   Ten thousand people buy lottery tickets for a lottery with 5 prizes.

3   Two dice are thrown. Why are the following point sets *not* sample spaces?

    (a)   (i)   The sum of the numbers on the faces of the dice is even.
          (ii) At least one number that turns up is odd.

    (b)   (i)   The number on the first die is greater than the one on the second.
          (ii) The number on the second die is greater than the one on the first.

    (c)   (i)   At least one of the numbers that turns up is greater than four.
          (ii) At least one of the numbers that turns up is less than three.
          (iii) Both numbers that turn up are three or four.

4   One of the families in the United States that have three children is selected at random. State which of the following sets of points constitute a sample space and why the others do not:

    (a)   (i)   The oldest child is a girl.
          (ii) The middle child is a girl.

    (b)   (i)   The oldest child is a boy.
          (ii) The oldest child is a girl, but there is at least one boy.
          (iii) There are three girls.

    (c)   (i)   There are more boys than girls.
          (ii) There are at least two girls.
          (iii) There are three girls.

    (d)   (i)   There is exactly one girl.
          (ii) There are more girls than boys.

5   A coin is tossed, and at the same time a die is thrown. Give two different possible sample spaces for this experiment. Under what conditions would you prefer each of them to the other?

6 At a certain race track there are five races, and in each of them three horses compete. List two different sample spaces that might be used for this experiment.

7 Four letters are put at random into four different envelopes. What sample space would you use if you were primarily concerned with whether your envelope contained your letter? Which one would you use if you were concerned about the disposition of all the letters?

8 Two high school students each apply for admission to five universities; list two possible sample spaces for the outcome. How many sample points would you use if you want to study the pattern of acceptances and rejections?

9 Three baseball games (A vs. B, C vs. D, and E vs. F) are to be played today. List a sample space that distinguishes all possible outcomes. (It consists of eight sample points.)

10 Three classes contain four, three, and two honor students, and one honor student is to be selected from each class. If each choice of three honor students constitutes a separate sample point, how many points will there be in the sample space?

11 Five senators (A, B, C, D, and E) want to form a committee of three. If each threesome of senators constitutes a sample point, how many sample points will there be in the sample space?

Probability Measures
of Sample Points and Events

After constructing a sample space for an experiment, the next step is to define a probability measure. A **probability measure*** assigns to each point x in the sample space a number denoted by $P(x)$. Intuitively, you can think of $P(x)$ as the fraction of times the sample point x would occur if the experiment were repeated a great many times. If x is the only possible outcome of the experiment, $P(x) = 1$; if x can never be an outcome of the experiment, $P(x) = 0$. If (as is most usual) x is sometimes an outcome of the

---

* Strictly speaking, we are defining a *discrete* probability measure. There are also *continuous* probability measures, but we will not be concerned with them here.

experiment and sometimes not, then $P(x)$ will be somewhere between 0 and 1. In short,

A probability measure assigns to every sample point x a number P(x) such that $0 \le P(x) \le 1$.

A natural extension of the concept of a sample point is that of an event. An **event** is a subset of the sample space. It may be the set of all sample points, in which case it would be the entire sample space E, or it may have no sample points at all, in which case it would be $\emptyset$, the null set. And an event can be any set in between; in particular, it can be a set consisting of only one sample point.

The **probability of an event** is the sum of the probabilities of all the sample points it contains. Intuitively, the probability of an event may be interpreted the same way as the probability of a sample point; it is the fraction of times you would expect the event to occur if the experiment were repeated many times.

(**EXAMPLE 1**)

If a fair die is tossed, what is the probability of each of the following events?
(a)   The number 4 turns up.
(b)   An even number turns up.
(c)   A number greater than 4 turns up.
(d)   Any number turns up.
(e)   No number turns up.

Solution

A *fair* die is one for which all numbers have an equal probability of coming up. Since there are six numbers, we can construct a sample space with six sample points, each with a probability of $\frac{1}{6}$.
(a)   The event "4 turns up" consists of one sample point, so its probability is $\frac{1}{6}$.
(b)   There are 3 even numbers, so the probability of this event is $\frac{3}{6} = \frac{1}{2}$.
(c)   There are 2 numbers greater than 4, so the probability is $\frac{2}{6} = \frac{1}{3}$
(d)   It is certain that one of the six numbers will turn up, so the probability of this event is 1.
(e)   The probability that no number will turn up is 0, since this event has no sample points. (We rule out the possibility that the die will balance on edge.)

⟨ **EXAMPLE 2** ⟩

A card is drawn at random from a deck of 52 cards. What is the probability of drawing

(a)   a spade?
(b)   an ace?
(c)   the ace of spades?

Solution

Each card represents a sample point with probability $\frac{1}{52}$.

(a)   Since there are 13 spades, the probability is $\frac{13}{52} = \frac{1}{4}$.
(b)   There are 4 aces, so the probability is $\frac{4}{52} = \frac{1}{13}$.
(c)   There is one ace of spades, so the probability is $\frac{1}{52}$.

For one cat to win the title "Best in Show" and for another cat to win the same title are mutually exclusive events.

If two events are mutually exclusive (that is, if they can't occur together), then the probability that one or the other of them will occur is obtained by adding their individual probabilities. If you pick a card out of a deck at random, the chance of getting an ace is $\frac{4}{52}$ (since there are four aces in the deck), and the chance of getting a king is $\frac{4}{52}$ (for the same reason). Since you can't draw an ace and a king at the same time, the probability of getting either an ace or a king is $\frac{4}{52} + \frac{4}{52} = \frac{8}{52}$. On the other hand, although the probability of getting a spade is $\frac{13}{52}$ and the probability of getting an ace is $\frac{4}{52}$, the probability of getting either an ace or a spade is *not* $\frac{13}{52} + \frac{4}{52}$ because you can get an ace and a spade on a single draw (by picking the ace of spades).

A similar formula holds for any number of events $A, B, C, \dots ,$ K. If any two of these events are mutually exclusive (that is, if they have no sample points in common), then the probability that one or the other of them will occur is the sum of their respective probabilities, that is,

$$P(A \cup B \cup C \cup \cdots \cup K) = P(A) + P(B) + P(C) + \cdots + P(K)$$

It is worth repeating that the events *must* be mutually exclusive. If the probability of my not being elected President is .99999 and the same probability holds for you, the probability that neither of us will be elected is *not* 1.99998. The formula fails because the events are not mutually exclusive; we both may fail to be elected.

The preceding comments are summarized below:

1. If ∅ is the null set, then $P(\varnothing) = 0$.
2. If E is the entire sample space, then $P(E) = 1$.
3. For any event S, $0 \leq P(S) \leq 1$.
4. If $A, B, C, \ldots, K$ are events and no two of these events have a sample point in common, then $P(A) + P(B) + P(C) + \cdots + P(K) = P(A \cup B \cup C \cup \cdots \cup K)$. In particular, if $A \cap B = \varnothing$, then $P(A) + P(B) = P(A \cup B)$.

But how do you actually assign probabilities to sample points? In practice you do it on the basis of experience and what you know about the world. An apparently symmetric coin is assumed to come up heads half the time, and a well-made die should show a 3 one sixth of the time. You may estimate the probability of rain to be $\frac{2}{3}$ because the sky is cloudy, on the basis of past experience. And an economist may decide the stock market will rise with a probability of .7 on the basis of subjective, but informed, judgment derived from a complex set of economic indicators, even without having experienced this exact situation before.

However, it may well happen that you don't have enough information to take a good guess at these probabilities. You may not know the probability that a long shot will win a horse race or the probability that General Motors stock will fall more than 10 points in the next three months. No matter; this problem, important though it is, will not concern us here. We will assume that in one way or another these probabilities have been assigned, and we will restrict ourselves to the problem of deducing what we can from them. In particular, we will assume that coins are as likely to turn up tails as heads, that all sides of a die have an equal chance of turning up, and so on.

Let's return to the three-way horse race and see how some of these ideas may be applied. If we believe all three horses have the same ability, then all possible finishing orders would be equally

This parimutuel board is showing a payoff of $325.60 on a $2 bet. We might interpret the payoff as meaning that the betting public assigned a probability of about $\frac{1}{160}$ to the event of horse number 8 winning this race—quite a long shot!

likely. So if we take as our sample space all six possible finishing orders, the probability of any particular finishing order would have to be $\frac{1}{6}$. Once the probabilities of all the sample points have been assigned, the probability of any event can also be calculated. To calculate the probability of the event "*A* wins," for example, you first observe that this event is composed of the two sample points *ABC* and *ACB*; since each of the finishing orders has a probability of $\frac{1}{6}$, together they have a total probability of $\frac{1}{3}$; thus $\frac{1}{3}$ is the probability that *A* wins. To find the probability that "*A* finishes before *B*" you first observe that this event is composed of three sample points (*ABC*, *ACB*, and *CAB*), and since each sample point has a probability of $\frac{1}{6}$, the probability that *A* will finish before *B* is $\frac{3}{6}$, or $\frac{1}{2}$.

We said earlier that there are many different sample spaces you can use for a given experiment. As a general rule it is convenient to use a sample space in which all sample points have the same probability. Suppose that in our three-horse race example you bet on *A* to win. It might seem that the sample space consisting of the two sample points (i) *A* wins and (ii) *A* doesn't win will serve best, since these are really the two outcomes that concern you. But if all the horses are of roughly the same ability and therefore all finishing orders have the same probability, it will be more convenient to use the six-point sample space and compute the probability of the event "*A* wins."

Similarly, suppose you toss a fair coin four times and bet $5 that you will get at least two heads. If you use the sample space composed of (i) more than one head tossed and (ii) fewer than two heads tossed, you will have a difficult time calculating the probabilities of the sample points. But if you take as your sample space all possible sequences of tosses (there are 16 in all, as you can see by counting them), you can count 11 of them with two or more heads. Then the chance of winning is easily calculated; it turns out to be $\frac{11}{16}$.

exercises

1 Next to each of the experiments described below are listed a number of events that are possible outcomes of the experiment. For each experiment construct a sample space in which all the sample points are equally likely. Then compute the probability of each event in the right-hand column by counting the number of sample points it contains.

	Experiment		Event
(a)	I pick a card at random from a deck of 52 cards.	(i)	I pick a diamond.
		(ii)	I pick an honor card (ten, jack, queen, king, or ace).
		(iii)	I pick a seven.
		(iv)	I pick a black card.
(b)	I throw a single die.	(i)	A six turns up.
		(ii)	An even number turns up.
		(iii)	A number greater than four turns up.
		(iv)	A number divisible by seven turns up.
(c)	Three candidates (Chiu, Lafarge, and Manson) run for political office. (You may assume all finishing orders are equally likely.)	(i)	Chiu finishes second or third.
		(ii)	Chiu or Lafarge finishes first.
		(iii)	Chiu, Lafarge, and Manson finish in that order.
(d)	I pick three out of five weekdays to go shopping. (Assume that I'm as likely to shop on one day as another.)	(i)	I don't go shopping on two consecutive days.
		(ii)	I shop on Friday.
		(iii)	I shop on Monday and Tuesday.
(e)	I pick two students at random from a class of three girls and two boys.	(i)	I pick two girls.
		(ii)	I pick a boy and a girl.
		(iii)	I pick two boys.

2 A card is selected at random from each of two decks.
   (a) Describe a sample space in which all sample points are equally likely.
   (b) How many points are there in this sample space?
   (c) Use the sample space to calculate the probability that both cards are the ace of spades.
   (d) Calculate the probability that both cards are (i) aces, (ii) black cards.

3 Two fair dice are thrown.
   (a) Describe a sample space in which all sample points are equally likely.
   (b) How many points are there in this sample space?
   (c) Calculate the probability that the sum of the numbers shown is five.
   (d) Calculate the probability that both dice show the same number.
   (e) Calculate the probability that the number shown on the first die is greater than the number shown on the second die.

4 A fair coin with a 1 marked on one face and a 5 on the other and a fair die are both tossed. Find the probability that the sum of the numbers that turn up is
(a) 3
(b) 7

5 Four horses are running at a race track, and you are unfamiliar with their past performance. You assume all finishing orders are equally likely. You bet that Tea Biscuit will win and Lucky will show (that is, come in first, second, or third). Find the probability that
(a) you win both bets.
(b) you lose both bets.
(c) you win exactly one bet.
(d) If you add the answers to parts (a), (b), and (c), you should get 1; could you have predicted this in advance?

6 In a certain restaurant you have a choice of (i) soup or salad, (ii) steak or fish, and (iii) pie, pudding, or sundae. Assuming that you make each choice at random, find the probability that your meal will consist of
(a) only things that begin with "s."
(b) one course that doesn't begin with "s" and two courses that do.

7 A president and vice-president are selected at random from four honor students in a class. Mary Jones and Arthur Smith are two of the honor students.
(a) Calculate the probability that Mary Jones will be selected for one of the offices by using (i) a sample space, (ii) common sense.
(b) Calculate the probability that Mary Jones and Arthur Smith will fill the two offices.

8 One hundred raffle tickets are sold, and you buy one of them. If there is one prize, what is the probability that you will win it?

9 One senator is picked at random from the hundred in the U.S. Senate. What is the probability that
(a) he or she is from Iowa?
(b) he or she is from Iowa or Kansas?

10 Each of two people decide to visit one of four museums on one of five weekdays. If each selects a museum and a day at random, what is the probability they pick the same museum and the same day?

11 On her vacation Elena visits four cities (A, B, C, and D) in a random order. What is the probability that she visits
(a) A before B?
(b) A before B and B before C?
(c) A first and B last?
(d) A either first or second?

12 There are three men and five women on the City Council. If a Council member is selected for a committee at random, how likely is it that it is a man?

13 Two dice are thrown. What is the probability that
(a) the difference between the numbers that come up is at least three?
(b) no odd numbers turn up?
(c) one odd number turns up?
(d) two odd numbers turn up?

14 When I leave work, I choose one of four elevators at random to reach the ground floor and then I choose one of three exits at random to leave the building. If a coworker does the same, what is the probability that we will take the same elevator and exit?

15 There are seven white keys and five black ones in a single octave on a piano. If I play a note at random, what is the probability that I strike a white key?

16 Three sales representatives pick three of five clients to visit at random. What is the probability that clients A and B will both be visited?

17 In order to select a student representative for an athletic conference, I first toss a coin to determine whether I will pick one from the basketball team (which has eight members) or the baseball team (with twelve members). After I decide which team to choose from, I pick a member of that team at random.
(a) What is the probability of picking a particular member of the baseball team?
(b) Would the answer to part (a) be any different if I combined all the members of both teams and selected one at random?

18 A fair coin is tossed four times, and you win a dollar for each head and lose a dollar for each tail that turns up.
(a) Construct a sample space with sixteen equally likely sample points.
(b) From this sample space calculate how many different amounts of money you can have after four tosses and the probability of having each of these amounts.

19 A committee of two is selected from two men and two women. What is the probability that the committee will have
(a) no men?   (b) one man?   (c) two men?

20 Bus drivers and police officers each work in three shifts (morning, evening, and "graveyard"), and a bus driver or police officer is as likely to be working one shift as another. By defining an appropriate sample space, calculate the probability that driver Larson and Officer Martez will be on the same shift.

21   To get to work you take two buses and a train, and each is late half the time. If the train or either of the buses is late, you will be late for work. Construct a sample space and calculate the probability that you will get to work on time.

In Chapter 1 we mentioned that it is impossible to pick a positive integer at random. We will (roughly) indicate why and leave the details for you.

If you want to pick a number at random between 1 and 10, then (a) the probability of picking any one number must be the same as picking any other number (that is what the word "random" means), and (b) since some number between 1 and 10 must be chosen, the sum of the probabilities, $P(1) + P(2) + P(3) + \cdots + P(10)$ must be 1.

If you extend this argument to the infinite set of all positive integers, then (a) $P(I) = P(J)$ for any two integers I and J, and (b) $P(1) + P(2) + \cdots = 1$. You can show that (a) and (b) can't both be satisfied by first assuming that the probability each number has of being chosen is zero and then assuming that this common probability is greater than zero. In each case you will be led to a contradiction.

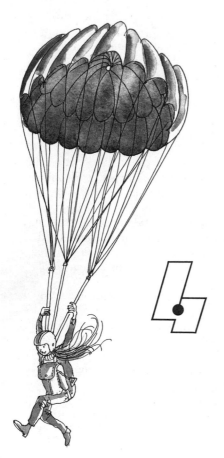

Probabilities of Compound Events

It often happens that you want to know the probability that either of two events will occur or, alternatively, the probability that both of them will. For example, if you use one parachute for jumping out of an airplane but also carry a spare one for emergencies, you will want to know the probability that at least one of them will work. And if a manufacturing process requires that each of two machines be in working order, it is important to know the probability that both of them will keep running.

To see how you calculate such probabilities, let's take a closer look at an example mentioned earlier: A card is drawn from a deck at random. What is the probability of getting either an ace or a spade?

It is easy to confirm that there are thirteen spades and four aces in a deck of 52 cards; $13 + 4 = 17$ favorable cards out of a possible 52 would seem to yield a probability of $\frac{17}{52}$. However, a simple count will show that there are not 17 but 16 cards that are either spades or aces. So the question becomes, What happened to the missing card?

The difficulty is most easily resolved by using a Venn diagram, as shown in Figure 1. In this diagram each of the events A and B are depicted as subsets of the sample space. The event $A \cup B$ is the subset of the sample space that consists of those sample points that are either in A or in B (or in both).

Suppose set A represents the event in which an ace is drawn and set B represents the event in which a spade is drawn. The compound event in which either an ace or a spade is drawn is represented by the set $A \cup B$. Since there are 52 points in the sample space, each representing the selection of a different card, and each card is equally likely to be drawn, $P(A \cup B)$ will be the number of sample points in $A \cup B$ divided by 52. But how many sample points are there in $A \cup B$? If we simply add the sample points in A to those in B, *we will be counting the sample points in $A \cap B$ twice!* The number of sample points in $A \cup B$ is equal to the sum of the numbers of sample points in A and B *less* those in $A \cap B$, that is,

$$P(A \cup B) = P(A) + P(B) - P(A \cap B)$$

In our original example the ace of spades was counted twice. When the 4 aces were added to the 13 spades, we obtained a total of 17, but we would have 16 if we counted the ace of spades only once. Using the formula, we obtain $\frac{16}{52} = \frac{4}{52} + \frac{13}{52} - \frac{1}{52}$.

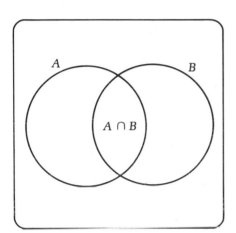

Figure 1

(**EXAMPLE 1**)

Two fair dice are thrown. What is the probability of getting a 1 on either (or both) of them?

Solution

Since the probability of getting a 1 on one die is $\frac{1}{6}$, it may be tempting to guess that the probability of getting a 1 on either of two dice is twice as great, or $\frac{2}{6}$. But if A is the event that you get a 1 on the first die and B is the event that you get a 1 on the second die, then the formula yields

$$P(A \cup B) = P(A) + P(B) - P(A \cap B) = \tfrac{1}{6} + \tfrac{1}{6} - \tfrac{1}{36} \text{ or } \tfrac{11}{36}$$

since the probability of getting a 1 on both the first and second throws is $\frac{1}{36}$. (There are 36 equally likely points in the sample space, and $(1, 1)$ is just one of them.)

EXAMPLE 2

In a mathematics course that is graded on the basis of two examinations, the probability of a randomly chosen student passing the first examination is .8 and the probability of passing the second examination is .7. The probability of passing at least one of them is .95. What is the probability of passing both?

Solution

If F and S are the events "pass the first test" and "pass the second test," respectively, then we need to find $P(S \cap F)$. We have $P(S \cup F) = P(S) + P(F) - P(S \cap F)$ or $.95 = .7 + .8 - P(S \cap F)$, so $P(S \cap F) = .55$.

1   If you toss a fair coin twice, what is the probability that you get at least one head? (The probability of getting two heads is $\frac{1}{4}$, as you should know.)

2   If you throw a die twice, what is the probability that you get a six on at least one throw?

3   There are six horses running in each of two different races, and you pick a horse to win at random. Your chance of guessing right in each race is $\frac{1}{6}$, and your chance of being right in both races is $\frac{1}{36}$. What is the chance of being right in at least one race?

4   Of all students in an English class, 70% were tutored, 60% studied at least an hour each day, and 50% were both tutored and studied an hour each day. If a student is selected at random, what is the probability that he or she either was tutored or studied an hour each day?

5 Before I go to work, I look out my window at the sky to see if I should carry an umbrella. It rains 40% of the time, I carry an umbrella 45% of the time, and 30% of the time I carry an umbrella in the rain.
   (a) What percentage of the time do I either carry an umbrella or find that it is raining?
   (b) How often do I get caught in the rain without an umbrella?
   (c) How often do I carry an umbrella in vain?

6 A tout at a race track approaches two different people and gives them each a tip in return for 10% of their winnings. If the tout has a probability of $\frac{1}{10}$ of being right in each case and if the probability of being right in both cases is $\frac{1}{100}$, what is the probability that the tout won't go home empty-handed?

7 A red, white, and a blue die are thrown, and you pick one of the dice at random. You win a dollar if either (i) you pick the red die or (ii) the face that turned up on the die you picked is a 4. What is your probability of winning?

8 At a county fair a customer pays a dollar and throws a die. If a six turns up, he gets 6 dollars (including the dollar he paid), and he gets nothing otherwise. After calculating a bit he concludes that the bet is fair, but he also concludes he shouldn't make two such bets at two different stands at the same time, since his chance of winning at least one of the bets is not $\frac{2}{6}$ (which is double $\frac{1}{6}$) but a little less. How would you reply?

9 On Monday you receive a dollar from every brunette in the class, on Tuesday you receive a dollar from every male in the class, on Wednesday you pay a dollar to everyone who is either a male or a brunette and on Thursday you pay a dollar to every male brunette in the class. What will you find when you count your money on Friday?

10 In a class, 40% of the students have special tutoring and 30% are athletes. If 10% of the class are athletes who have had special tutoring, what percentage of the class are either athletes or have had special tutoring?

11 You have eggs for breakfast 50% of the time and cereal 40% of the time. If you have one or the other 65% of the time, how often do you have both?

12 The probability that a student will pass the final examinations in both history and French is .6, and the probability of passing neither is .1. If the probability of passing the French examination is .8, what is the probability of passing the history examination?

13 The probability that an applicant will be admitted to his first choice college but not his second choice college is .1. The probability that the applicant will be admitted to his second choice college but not his first choice college is .4. If the probability of being admitted to both colleges is .2, what is the probability of being rejected by both colleges?

14 You are betting that the Lions and the Giants will win their respective baseball games. Let

Q be the probability that both teams win
R be the probability that at least one of the teams wins
S be the probability that the Lions win
T be the probability that the Giants win
U be the probability that the Lions win and the Giants lose
W be the probability that the Lions lose and the Giants win
V be the probability that both teams lose

Use a Venn diagram to show that

(a) $Q + R = S + T$
(b) $W + U + Q = R$
(c) $T + U + V = 1$
(d) $S = Q + U$

1 Consider the Venn diagram shown in Figure 2. Sets of sample points have been designated by letters r through x so that r is the set of sample points in the set $A \cap B' \cap C'$, s is the set of sample points in the set $A \cap B \cap C'$, and so on. Prove that

$$P(A \cup B \cup C) = P(A) + P(B) + P(C) - P(A \cap B) - P(A \cap C)$$
$$- P(B \cap C) + P(A \cap B \cap C)$$

by writing each of the expressions shown above in terms of the sets of sample points r through x, that is,

$$P(A \cup B \cup C) = P(r) + P(s) + P(t) + P(u) + P(v) + P(w) + P(x)$$

$$P(A) = P(r) + P(s) + P(v) + P(u)$$

$$P(A \cap B) = P(u) + P(s)$$

and so on.

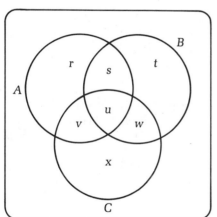

Figure 2

2 Three letters for Smith, Jones, and Green are dictated, and an envelope is addressed to each of them. The letters are inserted into the envelopes at random so that each envelope contains exactly one letter. Let S, J, and G be the events that Smith's, Jones's, and Green's letter was placed in the proper envelope, respectively.

Using the formula we derived for the probability of the union of events, find the probability that at least one letter is in its proper envelope; that is, find $P(S \cup J \cup G)$. Notice that $P(S) = P(J) = P(G) = \frac{1}{3}$, $P(S \cap G) = P(S \cap J) = P(J \cap G) = \frac{1}{6}$, and $P(S \cap J \cap G) = \frac{1}{6}$.

Suppose that after all attempts to persuade you otherwise you still believe that if you're given a card from each of two different decks you have a $\frac{2}{13}$ probability of getting at least one ace. (There's a $\frac{1}{13}$ probability of getting an ace from one deck, and you figure your chances are doubled with two decks.) What then would be the probability of getting at least one ace if you drew a card from three different decks? Use the same method to calculate the chance of getting at least one ace if you draw from 13 different decks and then 14 different decks, and then try to interpret your answer.

Permutations and Combinations

In many sample spaces all sample points have the same probability. In such spaces you can calculate the probability of an event by counting the number of sample points in the event and dividing this number by the total number of sample points in the space. This counting process may become cumbersome at times, and it will be useful to develop techniques to make it easier. To get an idea both of the difficulties involved and of the method of overcoming them, let's look at a simple example:

EXAMPLE 1

In a horse race in which four horses of equal ability (A, B, C, and D) are competing, what is the probability
(a) of A finishing first and B finishing second?
(b) of A finishing first?
(c) of A finishing before B?

Solution

If we take as our sample space all possible finishing orders, then there will be 24 different sample points in the sample space:

ABCD	*ABDC*	*ACBD*	*ACDB*	*ADBC*	*ADCB*
BACD	*BADC*	*BCAD*	*BCDA*	*BDAC*	*BDCA*
CABD	*CADB*	*CBAD*	*CBDA*	*CDAB*	*CDBA*
DABC	*DACB*	*DBAC*	*DBCA*	*DCAB*	*DCBA*

(a) The event "*A* finishes first and *B* finishes second" is composed of 2 sample points: *ABCD* and *ABDC*. The probability of *A* finishing first and *B* finishing second is $\frac{2}{24} = \frac{1}{12}$.

(b) There are 6 sample points in the event "*A* finishes first," namely, *ABCD*, *ABDC*, *ACBD*, *ACDB*, *ADBC*, and *ADCB*. So the probability of *A* finishing first is $\frac{6}{24} = \frac{1}{4}$.

(c) The event "*A* finishes before *B*" has 12 sample points (count them), so *A* has an even chance of finishing before *B*.

Figure 3 shows a *tree diagram* that illustrates the six possibilities if *A* wins. Each "path" from the top to the bottom of the tree represents a point in the sample space. For example, the point

Figure 3

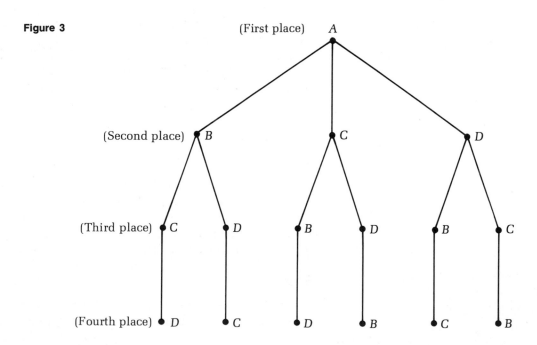

ACDB is represented by the path that takes the middle branch (to *C*) from *A*, and then the right-hand branch to *D* and *B*. There are also 18 more possibilities, which correspond to the cases in which *B*, *C*, and *D* win.

The direct calculation of the probability of an event is simple enough if the number of sample points is small, as in the last example, but it becomes impractical when the numbers are larger. If there were six horses running instead of four, there would be 720 different finishing orders; if 8 horses were running, there would be 40,320 orders.

Fortunately, you can calculate the number of finishing orders without actually listing them. In the 4-way horse race there are 4 possible winners. Once the winner is chosen there are 3 possible second-place finishers, so the first two positions can be filled in $(4)(3) = 12$ ways. After the first two positions are filled, there are two ways of filling the third position, and the fourth position is then determined. The total number of finishing orders, then, is $(4)(3)(2)(1) = 24$. If there were 6 horses, then there would be $(6)(5)(4)(3)(2)(1)$ finishing orders, and for 8 horses there would be $(8)(7)(6)(5)(4)(3)(2)(1)$ orders.

For convenience we will adopt a shorthand notation for this kind of product. For any positive whole number *N* we define *N* **factorial** to be $N(N - 1)(N - 2) \cdots (3)(2)(1)$ and write it as *N*!

It is also convenient to define 0! to be 1. This definition turns out to be useful and is suggested by the formula $N! = N(N - 1)!$, which holds for all positive integers *N* greater than 1 and which yields $1! = 1(0!)$, so $1 = 0!$ when *N* is set equal to 1.

(**EXAMPLE 2**)

Julia takes five one-hour courses from 10 A.M. to 3 P.M. How many different programs can she have? In how many different ways can she fill her first two hours? Her last 3 hours?

Solution

The first course can be chosen in 5 ways, the next course in 4 ways, the next in 3 ways, and the fourth course in 2 ways; the last course is then determined. Thus there are $(5)(4)(3)(2)(1) = 120$ ways to choose a program. Julia can fill her mornings in $(5)(4) = 20$ ways and her afternoons in $(5)(4)(3) = 60$ ways.

EXAMPLE 3

There are three prizes in a piano competition and 15 contestants. In how many ways can the prizes be won?

Solution $(15)(14)(13) = 2730$ ways.

EXAMPLE 4

In a class of 30 students, three class officers—a president, a vice-president, and a secretary—are to be picked at random. In how many ways can this be done? (Assume that no person can hold more than one office.)

Solution

The president can be picked in 30 ways. After the president is selected, there are 29 ways to pick the vice-president and then 28 ways to pick the secretary. So there are $(30)(29)(28)$ ways to pick the slate of class officers.

EXAMPLE 5

In a chess tournament with N contestants there will be a first prize, a second prize, . . . , and a Kth prize. How many different lists of prize winners can there be? (If A wins first prize and B wins second prize, a *different* listing would result from reversing their positions.)

Solution

There are N ways to award the first prize; after it is awarded, there are $(N - 1)$ ways to award the second prize, and so on until there are $N - K + 1$ ways to award the Kth prize (*not* $N - K$ ways—check it). So in all there are $N(N - 1)(N - 2) \cdots (N - K + 1) = N!/(N - K)!$ ways (why?).

In general, the rule is this:

The number of ways of choosing K objects from a total of N objects in a particular order is N!/(N − K)!. This number is defined to be the **permutation** *of N objects taken K at a time and is denoted by* P_K^N.

The phrase "in a particular order" is important; it means we are picking three class officers with different jobs and not just a set of three class officers. It makes a difference whether A is vice president and B is president or the other way around (as any vice-president will testify). By the same token we are concerned with which horse came in first, which came in second, and which came in third—not just a list of the horses that finished in the first three slots in arbitrary order. There are times, however, when we do want to know the number of different *sets* of class officers that can be chosen (without worrying about who holds what job) or the set of prize winners (without being concerned with who won what prize), and this is the problem we will discuss next.

EXAMPLE 6

In a class of 30 students a committe of 3 students is to be selected; in how many ways can this be done?

Solution

Again, it is important to distinguish this problem from the ones we did earlier. In this case we are only interested in *unordered* sets of 3 students (so that, for instance, ABC, BCA, and CBA all constitute the same committee).

The key to this new problem lies in the solution to the earlier one. Suppose we look at one particular committee of 3—the one composed of A, B, and C. From this single committee (which is unordered) six different permutations (orderings) can be formed: ABC, ACB, BAC, BCA, CAB, and CBA. This means that there are six times as many permutations of 30 people taken 3 at a time as there are committees. So the number of committees that can be formed will be the number of permutations divided by 6, or $P_3^{30}/6 = (30)(29)(28)/6$.

In general, the number of ways of picking an *unordered* committee of size K from a set of size N is the number of ways you pick an *ordered* committee of size K from a set of size N divided by K!, since a single unordered committee of size K can form K! different ordered committees. Therefore

The number of ways of choosing K unordered objects from a total of N objects is called the **combination** of N objects taken K at a time and is denoted by C_K^N.

$$C_K^N = \frac{P_K^N}{K!} = \frac{N!}{(K!)(N-K)!}$$

EXAMPLE 7

A, B, C, D, and E are five horses competing in a race.
(a) What is the probability that A, B, and C finish first, second, and third, respectively?
(b) What is the probability that A, B, and C are the first three horses to finish (in any order)?
(Assume that all finishing orders are equally likely.)

Solution

(a) If we consider the sample space consisting of all finishing orders in the first three places, we will have $P_3^5 = (5)(4)(3) = 60$ sample points, each with a probability of $\frac{1}{60}$. Since the event in (a) consists of only 1 sample point, its probability is $\frac{1}{60}$.
(b) If we use the sample space as we did in (a), the event "A, B, and C are the first three finishers" consists of $P_3^3 = 3! = 6$ sample points. (What are they?) So the solution to (b) is $(6)(\frac{1}{60}) = \frac{1}{10}$.

Alternatively, we might choose for (b) a sample space consisting of three-horse sets (those finishing in the first three positions). There would then be $C_3^5 = 5!/(3!)(2!) = 10$ equally likely sample points, of which the event defined in (b) would be one, so again the solution would be $\frac{1}{10}$.

EXAMPLE 8

In a certain lottery 100 tickets are sold and 3 equal prizes are awarded. What is the probability of not getting a prize if you buy
(a) 1 ticket?   (b) 3 tickets?   (c) 10 tickets?

Solution

The number of sample points is the number of ways 3 unordered tickets can be chosen from 100, which is $C_3^{100} = 100!/(3!)(97!)$. All sample points are assumed to be equally likely so the probability of each is $1 \div [100!/(3!)(97!)] = (3!)(97!)/100!$.

(a) If you missed all the prizes, all three winning tickets must have been chosen from the 99 you don't own. There are C_3^{99} ways this can happen, so the solution is $C_3^{99}/C_3^{100} = \frac{97}{100}$.

(b) The number of ways you can award 3 prizes without any of them being the 3 that you own is C_3^{97}, so the probability of not getting a prize is $C_3^{97}/C_3^{100} = (97)(96)(95)/(100)(99)(98) = \frac{7372}{8085}$.

(c) In this case we calculate the probability as $C_{10}^{90}/C_{10}^{100} = (90!)(90!)/(100!)(80!) = \frac{178}{245}$.

EXAMPLE 9

In a contest designed to discover musical talent there are to be two prize winners. Three of the eight entries come from the G-Clef Music School.

(a) How many different pairs of prize winners can be formed from students of the G-Clef School?

(b) How many different pairs of prize winners can be formed from those who are not students of the G-Clef School?

(c) How many possible pairs of prize winners can be formed from all eight of the entries?

(d) If the prize winners were picked at random, what is the probability that they would both be students at the G-Clef School?

(e) If the prize winners were picked at random, what is the probability that neither would be a student at the G-Clef School?

Solution

(a) $C_2^3 = 3$ (the number of ways a pair can be chosen from 3 students).

(b) $C_2^5 = 10$ (the number of ways a pair can be chosen from 5 students).

(c) $C_2^8 = 28$ (the number of ways a pair can be chosen from 8 students).

(d) $C_2^3/C_2^8 = \frac{3}{28}$.

(e) $C_2^5/C_2^8 = \frac{10}{28}$.

EXAMPLE 10

Thirteen cards are selected from a deck of 52. What is the probability of picking exactly 4 hearts?

Solution

Of the original 52 cards, 13 are hearts and 39 are not. There are in all C_{13}^{52} ways of choosing 13 cards from the original 52. We are interested in the probability of choosing 4 hearts and 9 nonhearts.

There are C_4^{13} ways of picking 4 hearts from 13 and C_9^{39} ways of picking 9 nonhearts from 39 nonhearts (Figure 4). Thus there are $C_4^{13}C_9^{39}$ different ways of picking 13 cards with 4 hearts. The probability is therefore $\dfrac{C_4^{13}C_9^{39}}{C_{13}^{52}}$.

Figure 4

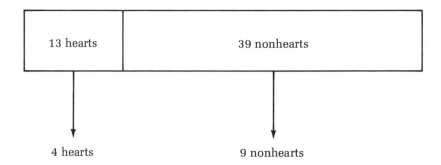

EXAMPLE 11

In a batch of 10 tomatoes, it is known that 3 are rotten. If I buy 4 tomatoes at random, what is the probability that I will pick (a) no, (b) one, (c) two, (d) three rotten tomatoes?

Solution

There are $C_4^{10} = 210$ ways of picking 4 tomatoes from 10. We are interested in the probability of picking a certain number, say i, of rotten tomatoes and the rest, $4 - i$, good ones. There are C_i^3 ways of picking i rotten tomatoes and C_{4-i}^7 ways of picking $4 - i$ good tomatoes. Thus the probability of picking i rotten tomatoes is

$$\frac{C_i^3 C_{4-i}^7}{C_4^{10}}$$

The answers then are

(a) $\dfrac{C_0^3 C_4^7}{210} = \dfrac{35}{210}$ (b) $\dfrac{C_1^3 C_3^7}{210} = \dfrac{105}{210}$

(c) $\dfrac{C_2^3 C_2^7}{210} = \dfrac{63}{210}$ (d) $\dfrac{C_3^3 C_1^7}{210} = \dfrac{7}{210}$

Notice that all four answers sum to 1, as they must.

1 Among six contenders in an Olympic event there will be a first, second, and third prize winner.
(a) In how many different ways can these prizes be awarded?
(b) How many different groups of three can win these prizes? (Disregarding who won what.)

2 How would your answers to Exercise 1 change if there were (a) five, (b) seven contenders rather than six?

3 Three rooms are to be painted, and different colors are to be used for the three rooms.
(a) If there are 5 colors in all, how many different color schemes can there be if you ignore which color was used for which room?
(b) Answer question (a) assuming that there are four colors.

4 If there are 25 members of a baseball squad, how many different teams of nine can be fielded? (Disregard which position a player takes.)

5 In how many different ways can a three-person committee be formed from the 100 members of the U.S. Senate? In how many ways can a two-person committee be formed?

6 In how many ways can a 97-person committee be formed from the 100 members of the U.S. Senate? In how many ways can a 98-person committee be formed? Compare your answers to this question with your answers to Exercise 5.

7 In how many ways can an $(N - 1)$-person committee be formed from N people?

8 You toss a fair coin N times. Calculate the probability of getting heads exactly half the time if N is
(a) 2   (b) 4
(c) 6   (d) 8

9 The police hear that there will be robberies in two of six high-crime neighborhoods, and they send reinforcements to two of the neighborhoods at random. What is the probability that reinforcements arrive at
(a) both robbery sites?
(b) exactly one of the robbery sites?
(c) neither of the robbery sites?

10 You receive 100 light bulbs from a manufacturer, and two of them are defective. If you test two of them, what is the probability that

(a) none of them are defective?
(b) one of them is defective?
(c) both of them are defective?
(d) Why do the answers to parts (a), (b), and (c) sum to 1?

11   There are fifty bags in a barrel, and ten of them contain prizes. If you pick five bags, how likely is it that you will get
(a) no prize?
(b) exactly one prize?
(c) exactly two prizes?

12   There are 12 U.S. senators from the New England states and 100 U.S. senators in all. A committee of six senators is picked at random. What is the probability that the committee contains
(a) no New England senator?
(b) exactly one New England senator?
(c) exactly two New England senators?

13   Out of 100 students, 20 are picked at random to form an experimental class and the remaining 80 form another, normal, class. If you and your friend are among the 100 students, what is the probability that
(a) you both enter the same class?
(b) you both enter the experimental class?

14   John Smith is in a class of 40 students.
(a) How many committees of 10 can you form including John Smith?
(b) How many committees of 10 can you form excluding John Smith?
(c) How many committees of 10 can you form in all?
(d) Use your answers to (a), (b), and (c) to show that $C^{39}_{10} + C^{39}_{9} = C^{40}_{10}$. Then verify this arithmetically.

15   If 13 cards are selected at random from a deck of 52 cards, what is the probability that exactly four spades are included?

16   In a class there are 8 boys and 12 girls. If 5 students are chosen at random, what would you guess is the most likely number of girls chosen? Confirm (or disprove) your guess.

17   A sales representative chooses three of fifteen accounts to visit on a given day
(a) How many different groups of three can he select?
(b) In how many different ways can he visit them if the order in which he visits them matters?

18   In a class of 10 students, 3 prizes will be awarded. If everyone in the class has the same ability, what is the probability that Joan and both of her two friends will take the three prizes?

19 In an 8-horse race you randomly pick 3 horses to finish first, second, and third. What is the probability that you picked
(a) the correct finishing order?
(b) the correct horses (though not necessarily in the correct order)?

20 What would the answer to Exercise 19 be if
(a) 6 horses were running?
(b) 4 horses were running?

21 A broadcasting company has 3 slots available for commercial advertising. There are 7 companies interested in buying a slot, and 3 of them are tobacco companies. If the spots are filled at random (with no more than one spot to a customer), what is the probability that none of these slots will be obtained by a tobacco company?

1 In a class there are 5 boys and 4 girls.
(a) If 3 children are chosen at random show that the number of ways of picking i boys and $3 - i$ girls is $C_i^5 C_{3-i}^4$.
(b) Deduce from this that $C_0^5 C_3^4 + C_1^5 C_2^4 + C_2^5 C_1^4 + C_3^5 C_0^4 = C_3^9$.
(c) Generalize this result by showing that $C_0^M C_K^N + C_1^M C_{K-1}^N + \cdots + C_K^M C_0^N = C_K^{M+N}$.

2 Show that there are C_K^N subsets of size K of a set of size N. From this and the fact that a set of N elements has 2^N subsets (which will be proven later), deduce that $C_0^N + C_1^N + \cdots + C_N^N = 2^N$.

3 Show that $(C_0^N)^2 + (C_1^N)^2 + \cdots + (C_N^N)^2 = C_N^{2N}$. (*Hint:* Observe that $C_i^N = C_{N-i}^N$ and use the answer to Question 1.)

4 Show that if N and K are positive whole numbers and $N > K$,
$$\frac{N(N-1)(N-2) \cdots (N-K)}{(K+1)!}$$
is a whole number.

5 (a) In a class of 3 students, in how many different ways can you select a committee of (i) 0? (ii) 1? (iii) 2? (iv) 3?
(b) By adding these answers calculate the number of committees of any size that can be formed.

6 Calculate the number of committees (of any size) that can be formed if the class size is (a) 2, (b) 4, (c) 5.

The number of ways of picking K students out of a class of N students is

$$C_K^N = \frac{N!}{K!(N-K)!}$$

and the number of ways of picking $(N - K)$ students out of this class is the same. Can you see intuitively why this must be so? (*Hint:* When you pick a set of students out, you simultaneously leave a set of students in. How large are these respective sets? Why must there be the same number of each?)

Conditional Probability and Independence

A detective's attention may be drawn from some suspects to others when a new clue turns up.

It sometimes happens that a jury's verdict is reversed because new evidence has appeared. Initially, perhaps, the jury decides that the defendant is probably guilty, but on the basis of new information an appellate court may estimate the probability of guilt to be less. A detective's attention may be drawn from some suspects to others when a new clue turns up. If your new car breaks down frequently, you change your opinion about its reliability. After observing the effects of certain food and drug additives, a government agency may decide to ban these substances, which previously were considered harmless.

In each of the situations we just described, someone was faced with this problem: How does new evidence modify your estimate of the probability of an event? Before trying to answer the general problem let's look at a few particular examples.

EXAMPLE 1

Two fair dice are thrown.
(a) What is the probability that the sum of the numbers that turn up on the dice is 8?
(b) How will this probability be modified if we know that the two numbers that turned up are different?

Solution

Suppose we let A be the event "the sum of the faces is 8" and B the event "the numbers on the dice are different." For our sample space we will take the 36 sample points consisting of pairs (i, j) where i and j are the numbers on the first and second die, respectively. Since the dice are fair, all 36 sample points have the same probability of occurring, and five of them—$(2, 6)$, $(3, 5)$, $(4, 4)$, $(5, 3)$, and $(6, 2)$—have a sum of 8. So $P(A) = \frac{5}{36}$.

The Original Sample Space

(1, 1)	(2, 1)	(3, 1)	(4, 1)	(5, 1)	(6, 1)
(1, 2)	(2, 2)	(3, 2)	(4, 2)	(5, 2)	(6, 2)
(1, 3)	(2, 3)	(3, 3)	(4, 3)	(5, 3)	(6, 3)
(1, 4)	(2, 4)	(3, 4)	(4, 4)	(5, 4)	(6, 4)
(1, 5)	(2, 5)	(3, 5)	(4, 5)	(5, 5)	(6, 5)
(1, 6)	(2, 6)	(3, 6)	(4, 6)	(5, 6)	(6, 6)

└──── The favorable sample points, A

In part (b), however, we are given new information: The numbers on the dice are different. So we delete the six sample points $(1, 1)$, $(2, 2)$, $(3, 3)$, $(4, 4)$, $(5, 5)$, and $(6, 6)$ from the sample space, because we *know* these didn't occur. Now there are only 4 sample points that are favorable and 30 sample points in all. Hence the probability of A, after we learn B has occurred, is $\frac{4}{30}$.

The Restricted Sample Space

	(2, 1)	(3, 1)	(4, 1)	(5, 1)	(6, 1)
(1, 2)		(3, 2)	(4, 2)	(5, 2)	(6, 2)
(1, 3)	(2, 3)		(4, 3)	(5, 3)	(6, 3)
(1, 4)	(2, 4)	(3, 4)		(5, 4)	(6, 4)
(1, 5)	(2, 5)	(3, 5)	(4, 5)		(6, 5)
(1, 6)	(2, 6)	(3, 6)	(4, 6)	(5, 6)	

└──── The favorable sample points, $A \cap B$

Figure 5

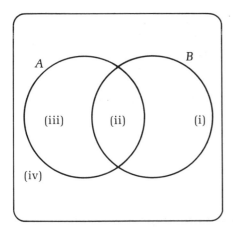

Region (i) = $B \cap A'$
Region (ii) = $B \cap A$
Region (iii) = $B' \cap A$
Region (iv) = $B' \cap A'$

Table 1

Grade	Men	Women
A	5	12
B	2	8
C	3	10

It is useful to express this *conditional probability* in terms of our original probability distribution. The favorable sample points in the restricted sample space make up the event $A \cap B$ and the entire restricted sample space is B, so *the probability of A after B is known to have occurred, which is written* $P(A|B)$, turns out to be $\dfrac{P(A \cap B)}{P(B)}$. In this case $P(A \cap B) = \frac{4}{36}$, $P(B) = \frac{30}{36}$, so $P(A|B) = \frac{4}{30}$.

The same point may be made geometrically with a Venn diagram, as in Figure 5. Initially, $P(A)$ is the sum of the probabilities of all the sample points in regions (ii) and (iii). But once we learn that B has occurred, we may eliminate the sample points in (iii) and (iv) because we know the outcome must be in (i) or (ii). So $P(A|B)$, the probability of A given B, is

$$\frac{\text{Sum of probabilities of sample points in (ii)}}{\text{Sum of probabilities of sample points in (i) and (ii)}} = \frac{P(A \cap B)}{P(B)}$$

EXAMPLE 2

A class of 30 women and 10 men received grades as shown in Table 1.
(a) What is the probability that a student picked at random from the class received a B?
(b) If you are told the student picked at random from the class is a woman, what is the probability that she received a B?

Solution

(a) Since 10 of the 40 students received a B, the answer to part (a) is $\frac{10}{40}$ or $\frac{1}{4}$.
(b) Once we are told that a woman is selected, all information about male students can be discarded; the conditional probability will depend only on how the grades of women are distributed. There are a total of 30 women students, of whom 8 received a B, so the conditional probability is $\frac{8}{30}$. If we rewrite this as $\left(\frac{8}{40}\right)/\left(\frac{30}{40}\right)$, this is

$$\frac{P(\text{woman} \cap \text{B grade})}{P(\text{woman})}$$

The formula $P(A|B) = P(A \cap B)/P(B)$ can also be written and interpreted differently, as $P(A \cap B) = P(B)P(A|B)$. In this form the equation states that the probability of A and B is the product of the probability of B and the conditional probability that A will occur once we know that B has occurred. Applying this formula to the last example, we conclude that the probability of picking a woman student who received an A is $\frac{12}{40}$; this is the product of the probability of picking a woman $\left(\frac{30}{40}\right)$ and the probability of the student having received an A once you learn the student you picked was a woman $\left(\frac{12}{30}\right)$, or $\frac{12}{40}$.

Independence

Suppose a student who gets 100 on one third of his daily examinations lives in a climate where it rains a quarter of the time. What is the probability that on a particular examination day it will rain *and* he will get 100?

A plausible analysis might go something like this: Of any 12 examination days, 3 might be expected to be rainy; and on one of these 3 days you might expect the student's examination grade to be 100. So of the original 12 days, one should bring both rain and a grade of 100, on the average. That is, the probability is $\frac{1}{12}$.

This argument suggests that for any two events A and B, $P(A \cap B) = P(A)P(B)$. In fact, this is sometimes the case but not always. If it rains $\frac{1}{3}$ of the time and I carry an umbrella $\frac{1}{3}$ of the time, the probability of my carrying an umbrella in the rain is not likely to be $\frac{1}{9}$. In fact, it would be $\frac{1}{3}$ if I carried an umbrella only when it rained. When you toss a coin, you have a probability of $\frac{1}{2}$ of getting a head and the same probability of getting a tail, but your chance of getting both is not $\frac{1}{4}$ but 0. And in our original example, if the student finds rain depressing, he may never get 100 when it rains.

In all these cases we must consider the possibility that the occurrence of event A *changes* the probability that B will occur. If the formula $P(A \cap B) = P(A)P(B)$ is to hold, the events A and B must in some sense be unrelated. Intuitively, you can look at the problem this way: If A and B are independent, then a person betting on A should be indifferent to whether B occurred, that is, $P(A) = P(A|B) = P(A \cap B)/P(B)$ or $P(A \cap B) = P(A)P(B)$. So we make the following definition:

Two events A and B are **independent** if and only if $P(A \cap B) = P(A)P(B)$.

(In general, any number of events are said to be independent if the product of the probabilities of any subset of them is equal to the probability that all the events in the subset will occur simultaneously.*)

EXAMPLE 3

Assume that the probability of having naturally curly hair is $\frac{1}{10}$, that the probability of being taller than 6 feet is $\frac{1}{5}$, and that hair type and height are independent. Find the probability of being taller than 6 feet and also having naturally curly hair.

Solution

Since the two events are independent, we multiply their probabilities, obtaining $\frac{1}{50}$.

EXAMPLE 4

Of 40 students, 10 are boys and 30 are girls. Two boys and 9 girls are in the band. A student is picked at random. If G denotes the event that the student picked is a girl and B denotes the event that the student is a member of the band, find
(a) $P(G)$
(b) $P(G|B)$
(c) $P(B|G)$

Solution

(a) Since there are 30 girls in a class of 40, $P(G) = \frac{30}{40}$.

(b) $P(G|B) = \dfrac{P(G \cap B)}{P(B)} = \dfrac{\frac{9}{40}}{\frac{11}{40}} = \dfrac{9}{11}$

(c) $P(B|G) = \dfrac{P(B \cap G)}{P(G)} = \dfrac{\frac{9}{40}}{\frac{30}{40}} = \dfrac{9}{30}$

---

*The terms "mutually exclusive" and "independent" are sometimes confused, but they are entirely different. If events A and B are mutually exclusive and A happens, you know B will not happen. If A and B are independent and A happens, you have no new information about B.

EXAMPLE 5

In a group of 35 people there are 15 smokers, 20 nonsmokers, 14 tall people, and 21 short ones. Suppose a person is to be selected at random from this group, and let A be the event "the person selected is short" and B the event "the person selected smokes." If A and B are independent, how many short smokers are there?

Solution

If A and B are independent, $P(A \cap B) = P(A)P(B) = \left(\frac{15}{35}\right)\left(\frac{21}{35}\right) = \frac{9}{35}$. If the fraction of short smokers is $\frac{9}{35}$, there must be 9 short smokers.   .

   If there were more than 9 short smokers, smokers would tend to be short and short people would tend to be smokers. If there were fewer than 9 short smokers, smokers would tend to be tall; in either case smoking and height would be dependent.

EXAMPLE 6

What is the probability that everyone in a room has a different birthday if there are (a) 2, (b) 3, (c) N people in the room? (Assume there are 365 days in the year and one day is as likely to be a birthday as another.)

Solution

(a)   Whatever the first person's birthday, there are 364 days that are different from it. So the probability that the second birthday will be different from the first is $\frac{364}{365}$.

(b)   If A is the event "the first two people have different birthdays" and B is the event "the third person's birthday differs from the first two," then $P(A \cap B) = P(A)P(B|A) = \left(\frac{364}{365}\right)\left(\frac{363}{365}\right)$.

(c)   Let A be the event "the first $N - 1$ people have different birthdays" and B be the event "the Nth person has a birthday different from everyone else." Then $P(A \cap B) = P(A)(366 - N)/365$. (There are $365 - (N - 1)$ days he can't use.) Using parts (a) and (b), we calculate $P(A)$ to be $\left(\frac{364}{365}\right)\left(\frac{363}{365}\right)$

when $N = 4$ and $P(A) = \left(\frac{364}{365}\right)\left(\frac{363}{365}\right) \ldots (367 - N)/(365)$ for any N. So

$$P(A \cap B) = \left(\frac{364}{365}\right)\left(\frac{363}{365}\right) \cdots \left[\frac{(366 - N)}{365}\right]$$

The arithmetical calculations are cumbersome, but it turns out that when $N = 23$, $P(A \cap B)$ is about $\frac{1}{2}$, and when $N = 100$, it is almost certain that everyone will have different birthdays. (See the discussion of the birthday problem in Chapter 1.)

1   Of 3 families, one has 3 girls and a boy and the other two have only boys. If a child is selected at random from a family selected at random, what is the probability that the child will be a girl?

2   At two of the three theaters in town, each customer is given a lottery ticket that gives him a $\frac{1}{200}$ chance of winning a prize. If you select a theater at random, what is your chance of winning a prize?

3   Corporations A and B are bidding on a contract, and each has an even chance of having its bid accepted. If A wins the bid, you will become the subcontractor with probability $\frac{1}{3}$. What is the probability that you will become the subcontractor?

4   Two examinations are given in a course, and to pass the course a student must pass both of them. The performance on one examination is independent of the performance on the other.
(a)   If the student has a probability of .8 of passing the first examination and .7 of passing the second, what is his chance of passing?
(b)   What is his chance of passing if you know he did *not* fail both examinations?

5   If the chance of getting a hit is .4 whenever a player comes to bat, what is the probability of failing to get a hit three times in a row?

6   My probability of winning the first of two games of tennis is .6. If I win the first game, I have a .7 probability of winning the second game; if I lose the first game, I have a .4 probability of winning the second game.
(a)   What is the probability of winning both games?
(b)   What is the probability of winning exactly one of the two games?

7 If two evenly matched teams play a series of 3 games, what is the probability that
(a) a team will win at least 2 games if it wins the first game?
(b) a team will win all three games if it wins the first game?

8 If the chance of a horse winning a race in the rain is .4 and the chance of rain is .6, what is the probability that it will rain *and* the horse will win?

9 If you and I independently pick a number from 1 to 10, what is the probability that
(a) both numbers are even?
(b) one is even and the other is odd?
(c) both numbers are the same?

10 If two fair dice are thrown, what is the probability that at least one die will show a 1 if you know the sum of the faces is 5? What is the probability that the sum of the faces is 5 if you know at least one of the dice shows a 1?

11 A die has the numbers 3, 4, 5, and 6 painted green and 1 and 2 painted red. If G is the event that a green side turns up and O is the event that a 1 turns up, calculate
(a) $P(G|O)$ (b) $P(O|G)$
(c) $P(O)$ (d) $P(G)$
(e) $P(O \cup G)$ (f) $P(O \cap G)$

12 A number is chosen at random from 1 to 24. What is the probability that
(a) it is divisible by 8?
(b) it is divisible by 2?
(c) it is divisible by 8 if you know it is divisible by 2?
(d) it is divisible by 3 if you know it is divisible by 8?
(e) it is divisible by 8 if you know it is divisible by 3?

13 A card is selected at random from a deck of cards. What is the probability that the card is
(a) a spade?
(b) a ten?
(c) a spade given that it is a ten?
(d) a ten given that it is a spade?

14 Two dice are tossed. What is the probability that the sum of the two numbers that turn up is 9
(a) if no other information is available?
(b) if both numbers are greater than 3?

15   Four horses of equal ability, A, B, C, and D, are entered in a race. Find the probability that A will win if you know that
    (a) B finished third.
    (b) B did not finish third.

16   In a class of 40 students, 10 are in the glee club but not the chess club, 12 are in the chess club but not the glee club, and 5 are in both clubs. A student is selected at random.
    (a) What is the probability that the student will belong to the chess club?
    (b) What is the probability that the student will belong to the chess club if
        (i) the student is known to belong to the glee club?
        (ii) the student is known not to be a member of the glee club?

17   Letters are sent to the four prize winners in a contest. The envelopes are addressed appropriately, but the letters are placed in the envelopes at random.
    (a) What is the probability that the first-prize winner received his own letter?
    (b) What is the answer to (a) if you are told in advance that the second-prize winner received his own letter?

18   A student is selected at random from a class. Let M be the event that the student is male and A the event that the student received an A in the course. If 10 female students received A, 15 male students did not get an A, and 30 male students did get A, how many female students did not get A? Assume that sex and grades are independent.

19   You pick a card from a deck; A is the event that you pick a spade, B the event that you pick an ace, and C the event that you pick a king. Which two of the pairs of events are independent?

20   Two fair dice are thrown. What is the probability that
    (a) at least one of the faces is a 6?
    (b) at least one of the faces is a 6 given that the sum of the faces on the two dice is (i) 8? (ii) 10?

21   You toss a coin twice. Let A be the event that the first toss is a head, B the event that the second toss is a head, and C the event that there is an uneven number of heads.
    (a) Are each of the three pairs of events independent?
    (b) Are the three events independent?

22   Show that if the events A and B are mutually exclusive and $P(A)$ and $P(B)$ are not zero, then they cannot be independent.

23   Of families with two children, $\frac{1}{4}$ have two boys, $\frac{1}{4}$ have two girls, and $\frac{1}{2}$ have a boy and a girl. If you pick a family with two children at random, what is the probability that it will have two boys if you know it has at least one?

24   There is a pseudolaw known as the law of "maturity of chances," which states that when a succession of coin tosses are all heads the next toss is more likely to be a tail (assuming an unbiased coin), since there must be the same number of tails as heads in the long run.

       Let A_k be the event that the first k tosses are heads, and assume that the tosses are independent, so that $P(A_k) = 1/2^k$. Compute $P(A_{k+1}|A_k)$. What are your conclusions concerning this law?

25   Show that $P(A \cap B|B) = P(A|B)$.

26   Show that $P(A \cup B|B) = 1$.

27   The first team to win two games is declared the winner of a three-game series. There is a probability of p that the favorite will defeat the underdog in each game, and the outcomes of the games are independent.
(a)   If $p = .6$ and the underdog wins the first game, what is the probability that the favorite will win the series?
(b)   If, after the underdog wins the first game, each team has an even chance of winning the series, what must be the value of p?

1   Suppose that $P(A|B) = P(A)$. You may interpret this to mean that a person betting on A would find information about B irrelevant to the bet. Show that
(a)   A person betting on B would find information about A irrelevant to the bet as well, that is, $P(B) = P(B|A)$.
(b)   If the bettor is unconcerned that B occurred, he will also be unconcerned if B does not occur, that is, $P(A|B') = P(A)$. [*Hint:* $P(A \cap B') = P(A) - P(A \cap B)$ and $P(B') = 1 - P(B)$.]

2   Someone throws two fair dice, and you win \$1 if a 4 turns up on at least one of them. You are told the sum on the faces of the two dice; what sum would maximize your chance of winning the dollar?

3   Two baseball teams, A and B are so well matched that there is an even chance that one will beat the other in every game (the games are independent). If A beats B 3 games to 1 in a five-game match, what is the probability that B won the first game?

1 (a) Two of the 100 members of the U.S. Senate are picked at random and asked, "Are either of you from Wyoming?" When you receive an affirmative answer you ask the senator from Wyoming to step forward. (If both are from Wyoming, the senior senator steps forward.) What is the probability that the senator who did *not* step forward is from Wyoming?

(b) Once again you pick two senators from the Senate at random, but now you ask just one of them if he's from Wyoming, and he says yes. What is the probability that the other senator you selected is also from Wyoming?

　　　The interesting point here is that the answers to parts (a) and (b) are not the same; in fact, the answer to part (b) is almost double the answer to part (a). Before you try to work out the exact answers, try to see why the answers might be different.

2　Suppose you were told that the probability of team A beating team B is $\frac{1}{3}$. Suppose you were also told that the probability of team A beating team B is greater than $\frac{1}{3}$ if it rains and also greater than $\frac{1}{3}$ if it doesn't rain. These three statements seem inconsistent, and indeed they are. Can you show that if $P(A|B) > P(A)$, then $P(A|B') < P(A)$? [*Hint:* $P(A \cap B) = P(A) - P(A \cap B')$.]

Bayes's Theorem

Whenever I catch a cold, I pick a bottle at random from the medicine chest and swallow one of the pills. The medicine chest contains two bottles of aspirin, three bottles of cold capsules, and five bottles of vitamin C. Aspirin cures my cold $\frac{1}{4}$ of the time, cold capsules cure it $\frac{1}{2}$ the time, and vitamin C cures it $\frac{5}{6}$ of the time. If I catch a cold, what is the probability of my being cured?

The tree diagram in Figure 6 is helpful in analyzing what is happening. It shows the six possible outcomes and their respective probabilities; I take an aspirin but am not cured $\frac{9}{60}$ of the time, I take vitamin C and am cured $\frac{25}{60}$ of the time, and so on. The chance of being cured is $P(\text{vitamin C} \cap \text{cure}) + P(\text{cold capsule} \cap \text{cure}) + P(\text{aspirin} \cap \text{cure}) = \frac{3}{60} + \frac{9}{60} + \frac{25}{60} = \frac{37}{60}$. (We can add the probabilities because the events do not overlap.)

This much is straightforward. The real problem arises when we turn the problem around: If I was cured, what is the probability that the pill I took was an aspirin?

To find $P(\text{aspirin}\,|\,\text{cure})$ we proceed in the same way as when we discussed conditional probability. Since we know there was a cure, we can eliminate the three outcomes in which there was no cure. By the same informal argument we used earlier, we would

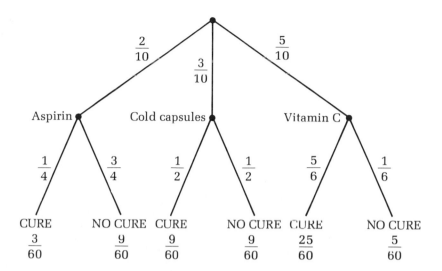

Figure 6

expect that on 60 trials, I would take aspirin and be cured on 3 days, would take a cold capsule and be cured on 9 days, and would take vitamin C and be cured on 25 days. (On the other days I would not be cured, and this possibility has already been eliminated.) Of the 37 days I was cured I took aspirin on 3, so

$$P(\text{aspirin} \mid \text{cure}) = \frac{P(\text{aspirin} \cap \text{cure})}{P(\text{aspirin} \cap \text{cure}) + P(\text{cold capsule} \cap \text{cure}) + P(\text{vitamin C} \cap \text{cure})}$$

$$= \frac{\frac{3}{60}}{\frac{3}{60} + \frac{9}{60} + \frac{25}{60}} = \frac{3}{37}$$

EXAMPLE 1

The commanding general of an army defending an island is preparing for an invasion. On the basis of intelligence reports he estimates that the probability of an invasion on July 8 is $\frac{1}{3}$. From experience he knows that $\frac{4}{5}$ of the time invasions are preceded by a high level of radio traffic on the previous day, while ordinary days are preceded by a high level of radio traffic only $\frac{1}{4}$ of the time. A high level of radio traffic is observed on July 7; how should the general modify his initial probability of an invasion?

Figure 7

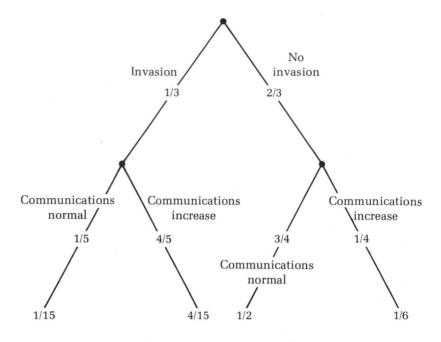

Solution

See Figure 7. There are four possibilities initially, but two are eliminated when an increase in communications is observed. The probability that there will be an invasion, given an increase in communications, is

P(invasion | communications increase)

$$= \frac{P(\text{invasion} \cap \text{communications increase})}{P(\text{invasion} \cap \text{communications increase}) + P(\text{no invasion} \cap \text{communications increase})}$$

$$= \frac{\frac{4}{15}}{\frac{4}{15} + \frac{1}{6}} = \frac{8}{13}$$

(**EXAMPLE 2**)

A diagnostic X-ray machine will show positive 98% of the time when a person with tuberculosis is tested but will do so only 1% of the time if the person is healthy. If the machine shows positive, what is the probability that the person tested has tuberculosis?

Solution

This is an illustration of a pitfall to avoid. Although the question seems to be properly posed, vital information is missing; to solve this problem you must first be told what proportion of the population has tuberculosis. To show this, we will consider two cases: (a) 10% of the population has tuberculosis and (b) $\frac{1}{10}$ of 1% of the population has tuberculosis. We will abbreviate a positive showing of the X-ray machine by P, a negative showing by P', actually having tuberculosis by TB, and not having tuberculosis (being healthy) by H. The tree diagram for case (a) is shown in Figure 8. We have

$$P(TB|P) = \frac{P(TB \cap P)}{P(TB \cap P) + P(H \cap P)} = \frac{.098}{.098 + .009} = \frac{98}{107}.$$

In case (b), as shown in Figure 9, we have

$$P(TB \cap P) = \frac{P(TB \cap P)}{P(TB \cap P) + P(H \cap P)} = \frac{.00098}{.00098 + .00999} = \frac{98}{1,097}.$$

So despite the positive indication, the probability is less than .1. If someone shows a positive reading, the odds are better than 10 to 1 he's been misdiagnosed!

Figure 8

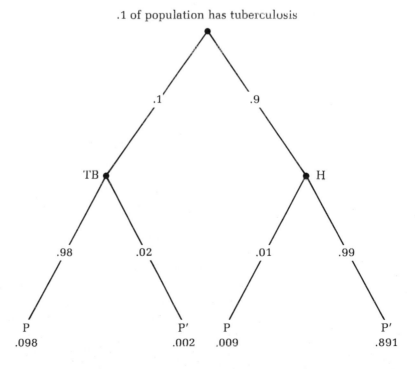

.1 of population has tuberculosis

Figure 9

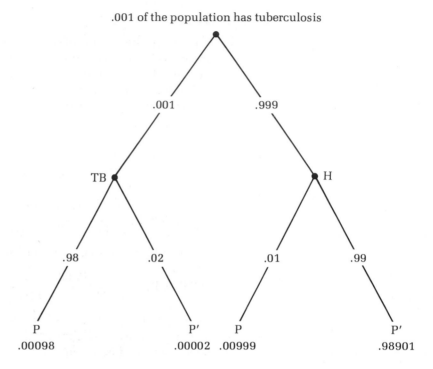

.001 of the population has tuberculosis

Statement of Bayes's Theorem *(Optional)*

Although we have not explicitly mentioned Bayes's Theorem, the examples we have worked out actually illustrate several applications of it. We have deliberately refrained from spelling out just what Bayes's Theorem states, because the statement looks rather forbidding in its most general form. But with a little thought it can easily be applied in a particular case, as we did in Illustrative Examples 1 and 2. Still, it is appropriate to state the theorem generally and precisely, although we recommend that in practice you reason through a problem rather than memorize the theorem and plug in the appropriate numbers.

Suppose there are n events, $A_1, A_2, \ldots, A_n$, with respective probabilities $P(A_1), P(A_2), \ldots, P(A_n)$. After one of the events A_i has occurred, one of m events, $B_1, B_2, \ldots, B_m$, will occur. As before, $P(B_j | A_i)$ is the probability that B_j will occur when you know that A_i has already occurred.

Suppose that some A_k occurs (but you don't know which A_k) and then you observe that B_j has occurred. With this information,

Figure 10

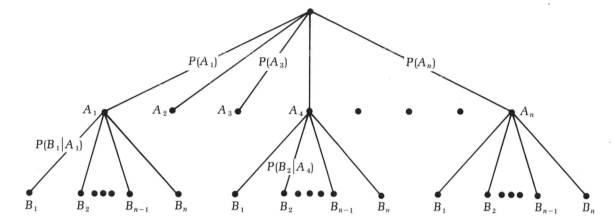

how do you calculate the probability that it was one particular A_i that occurred earlier?

Let's draw a diagram, as in Figure 10. Once we are told B_j has occurred, we eliminate all sample points in which B_j does not occur. By the same argument we used in our earlier examples, we deduce the following formula, which is known as Bayes's Theorem:

$$P(A_i|B_j) = \frac{P(A_i)P(B_j|A_i)}{P(A_1)P(B_j|A_1) + P(A_2)P(B_j|A_2) + \cdots + P(A_n)P(B_j|A_n)}$$

EXAMPLE 3

Assume that boys and girls are equally likely to be born and the sexes of newborn children are independent. A child is chosen at random from a family with 3 children, and the child turns out to be a girl. What is the probability that all 3 children in the family are girls?

Solution

Let A_1 be the event that the family has i girls. It turns out that $P(A_0) = P(A_3) = \frac{1}{8}$ and $P(A_1) = P(A_2) = \frac{3}{8}$.*

---

* If you set up a sample space in which you have all orders of birth [(boy, boy, girl), (girl, boy, girl), . . .], you obtain 8 sample points. Then you just count the sample points in each event and multiply by $\frac{1}{8}$ to find the probabilities.

Let B_0 be the event that a boy is picked from the family and B_1 the event that a girl is picked. Then

$$P(B_1|A_0) = 0, \qquad P(B_1|A_1) = \frac{1}{3}, \qquad P(B_1|A_2) = \frac{2}{3}, \quad \text{and} \qquad P(B_1|A_3) = 1$$

Therefore

$$P(A_3|B_1) = \frac{P(A_3)P(B_1|A_3)}{P(A_0)P(B_1|A_0) + P(A_1)P(B_1|A_1) + P(A_2)P(B_1|A_2) + P(A_3)P(B_1|A_3)}$$

$$= \frac{(\frac{1}{8})(1)}{(\frac{1}{8})(0) + (\frac{3}{8})(\frac{1}{3}) + (\frac{3}{8})(\frac{2}{3}) + (\frac{1}{8})(1)} = \frac{\frac{1}{8}}{\frac{4}{8}} = \frac{1}{4}$$

A similar calculation will show that

$$P(A_2 \mid B_1) = \frac{2}{4}, \qquad P(A_1 \mid B_1) = \frac{1}{4}, \text{ and } P(A_0 \mid B_1) = 0$$

1 Of the students in a graduating class, 10% are in the Honor Society. It is known that 40% of the Honor Society students win scholarships, and only 20% of the other students do. A student selected at random from the graduating class turns out to have a scholarship. What is the probability that this student is a member of the Honor Society?

2 In mathematics there are five regular classes and one honor class, and all classes have the same number of students. In the honor class 50% of the students received an A, and in the regular classes only 15% did. A mathematics student is selected at random and is found to have received an A. What is the probability that the student was from the honor class?

3 There are three cards; one of them has both sides black, another has both sides white, and the third has black on one side and white on the other. A person picks a card at random and then he selects one of the sides at random. He finds that it is white. What is the probability that the other side is white as well?

   (Comment on this argument, which suggests that the answer should be $\frac{1}{2}$: "Since black–black has been eliminated, there are only two possible cards you could have chosen—white–white and white–black. Since both cards were equally likely to be chosen, the answer is $\frac{1}{2}$.")

4   Jar I contains 3 black balls and 2 white balls, and jars II and III each have 3 white balls and 2 black balls. If I pick a jar at random and then pick a ball at random, what is the probability that the ball came from jar I if the ball I drew was
(a) white?   (b) black?

5   One quarter of the freshman students taking French come from high school A and the other three-quarters come from high school B. Students from A pass the course 90% of the time, and students from B pass it 80% of the time. If I pick a student at random and find he or she has passed the course, what is the probability that he or she came from high school A?

6   You believe that an arson suspect set a fire with probability .01. The suspect, who claims he's innocent, is identified as a liar by a lie detector. A lie detector identifies a liar as such 90% of the time and mistakenly labels the truth as a lie 5% of the time. What is the probability that he is a liar?

7   I employ 10 salesmen who make a sale 10% of the time, 5 who make a sale 20% of the time, and 1 who sells 50% of the time. I observe a salesman successfully making a sale. Calculate the probability that he is each of these three types.

8   In a certain class 10% of the students cheat on examinations. Successful cheaters always get 100% and a noncheater has a 40% chance of getting 100%. If Janet gets 100% on an examination, what is the probability that she cheated?

9   A handicapper who predicts the outcome of sporting events is right 80% of the time when the event isn't fixed. (Whenever it is fixed, he's invariably wrong.) Assume that 5% of sporting events are fixed. If the handicapper is wrong about a particular prediction, what is the probability that the event was fixed?

10   An IRS auditor knows from past experience that 10% of taxpayers actually donate more than $500 to charity. Everyone who donates this money deducts it, but 15% of the population who do not make such a donation deduct it as well. If a certain taxpayer deducts this contribution, what is the probability that the deduction is legitimate?

11   Everyone who trains as a pilot takes an aptitude test. It has been found that (i) 10% of those who apply for training make successful pilots; (ii) applicants who eventually become pilots pass the examination 80% of the time, and those who fail to become pilots pass the examination 20% of the time. What is the probability of
(a) a candidate who passes the examination becoming a successful pilot?
(b) a candidate who fails the examination becoming a successful pilot?

12   One coin has two heads, and another has a head and a tail. I choose one of the coins at random, flip it twice, and get two heads. What is the probability that it was a two-headed coin?

13   A firm buys $\frac{4}{5}$ of the parts it uses from supplier A and the remaining $\frac{1}{5}$ from supplier B. Experience has shown that 5% of the parts A sends are defective and 15% of the parts B sends are defective. What is the probability that a part chosen at random comes from A if
(a)   it is defective?
(b)   it is not defective?

14   Each of 2 bottles have 3 black and 2 white balls and a third has 1 black ball and 4 white balls. A bottle is chosen at random, and a ball is taken from the bottle at random. What is the probability that the ball chosen comes from the third bottle if the ball is
(a)   white?
(b)   black?

15   In a race, 5% of the competitors take drugs. Of those who take drugs, 20% complete the mile in less than 4 minutes; and of those who do not, 5% complete the mile in less than 4 minutes. What is the probability that a competitor takes drugs if he (a) does, (b) does not complete the mile in 4 minutes?

1   Three prisoners—A, B, and C—are condemned to die but later learn that one of them has been pardoned. When A inquires, the jailer says he can't tell the lucky prisoner the news until he is informed officially. Prisoner A says, "At least one of the two other prisoners wasn't pardoned, possibly both; can you tell me a prisoner who wasn't pardoned?" The jailer tells A that B wasn't pardoned. Now A feels he is better off, since he knows that either he or C is the lucky one, so he has a $\frac{1}{2}$ chance rather than a $\frac{1}{3}$ chance of survival. What would Bayes say to that?

2   It is known that 5% of those who pass through a certain airport are smugglers. At a certain time there are known to be 5 smugglers at the airport and N ordinary travelers, but N is not known. At this point someone whose status is unknown enters the airport; a person is chosen randomly from everyone present, and that person turns out to be a smuggler. What is the probability that the person who entered last is a smuggler? (The answer does *not* depend on N.)

The Binomial Distribution

An airline regularly overbooks its flights because it has found from experience that 10% of its passengers are "no shows." If it sells 52 tickets for an airplane that has only 50 seats, what is the probability that an excess of passengers will turn up for the flight?

This is a problem involving the **binomial distribution**. The binomial distribution* has two basic properties:

1. *There is one basic experiment, which has exactly two possible outcomes, called* **success** *and* **failure**. *The probabilities of success and failure are denoted by p and q, respectively. Since there are only two outcomes, p + q = 1.*
2. *The basic experiment is repeated N times, and the outcomes of the experiments are independent—the probability of success is always p and the probability of failure is always q.*

---

*The term "binomial distribution" comes from the Binomial Theorem, which you may have studied in your algebra courses. If there are N trials and p is the probability of success and q the probability of failure on each trial, then the probability of exactly K successes is calculated by expanding the binomial $(p + q)^N$ and taking the term containing $p^K q^{N-K}$.

In the airline example the basic experiment has two possible outcomes: the appearance or nonappearance of a customer who has purchased a ticket. The probability p of success (the customer's appearance) is .9, and the probability q of failure is .1. The number of times the experiment is repeated, N, is 52—the number of customers who bought tickets for the flight.

Other experiments that can be regarded as having only two possible outcomes are (a) coin tossing (you get either a head or a tail), (b) taking an examination (you either pass or fail), and (c) applying to college (you either get in or don't).

The basic question we will try to answer is this: If conditions (1) and (2) hold, what is the probability of having exactly K successes in N experiments?

Before we try to solve the general problem, we'll start with a simpler one:

A die is thrown 4 times. If a 1 or 2 turns up, it will be considered a success, a 3, 4, 5, or 6 is to be considered a failure. If the die is fair, what is the probability of exactly 2 successes in 4 throws?

Let's first list all the sample points—all possible sequences of success and failure:

(i)	SSSS	(ii)	SSSF	(iii)	SSFS	(iv)	SSFF
(v)	SFSS	(vi)	SFFS	(vii)	SFSF	(viii)	SFFF
(ix)	FSSS	(x)	FSSF	(xi)	FSFS	(xii)	FSFF
(xiii)	FFSS	(xiv)	FFFS	(xv)	FFSF	(xvi)	FFFF

The key to solving this problem lies in two observations:

1. All sample points in which there are 2 successes and 2 failures have the same probability of occurring. Because the tosses are independent we can calculate the probability of each sample point by multiplying the probabilities of the individual tosses, and in each case we get $\left(\frac{2}{3}\right)^2\left(\frac{1}{3}\right)^2$. The order in which the successes and failures occur is irrelevant.

2. The next observation is based on a simple matter of counting; we must determine how many sample points there are with 2 successes and 2 failures. Since there are 4 tosses and the successes may occur on any 2 of them, the number of different sample points will be $C_2^4 = 6$. In our example the six sample points turn out to be (iv), (vi), (vii), (x), (xi), and (xiii).

Now all that's left is to put the pieces together: If you multiply the number of sample points by the common probability of each, you get $C_2^4(\frac{2}{3})^2(\frac{1}{3})^2$.

In general, suppose there is a sequence of n experiments and we want to know the probability of k successes (and $n - k$ failures). If the probability of success is p and that of failure is q, the probability of any single sample point with k successes and $n - k$ failures is $p^k q^{n-k}$. The number of such sample points is C_k^n. So the probability of k successes is given by the formula

$$P(k \text{ successes}) = C_k^n p^k q^{n-k}$$

Now let's return to the problem the airline faced. Since the appearance of a passenger is termed a success, we must calculate the probability that there are an excessive number of passengers, that is, the probability of either 51 or 52 successes.

In this case $n = 52$, $p = .9$, and $q = .1$, so the probability of an excessive number of passengers is $(.9)^{52} + C_1^{52}(.9)^{51}(.1)$, which turns out to be about $.004 + .021 = .025$. Thus they will overbook about $2\frac{1}{2}\%$ of the time.

(EXAMPLE 1)

If 12 dice are thrown, what is the most likely number of dice that will show a 1?

Solution

The probability of k dice showing a 1 is given by the formula

$$P(k) = C_k^{12}\left(\frac{1}{6}\right)^k\left(\frac{5}{6}\right)^{12-k}$$

so

$$P(0) = \left(\frac{5}{6}\right)^{12}$$

$$P(1) = 12\left(\frac{5}{6}\right)^{11}\left(\frac{1}{6}\right)^1 = \left(\frac{12}{5}\right)P(0)$$

$$P(2) = 66\left(\frac{5}{6}\right)^{10}\left(\frac{1}{6}\right)^2 = \left(\frac{66}{25}\right)P(0)$$

$$P(3) = 220\left(\frac{5}{6}\right)^{9}\left(\frac{1}{6}\right)^3 = \left(\frac{44}{25}\right)P(0)$$

At this point the probabilities begin to drop (can you prove this?), so the answer is 2. Since the probability of getting a 1 is $\frac{1}{6}$, it is not too surprising that in 12 tosses the most likely result is $\frac{1}{6}$ of 12; still, this is not a proof.

1   In a psychological experiment a mouse is repeatedly put in a position in which it must turn either right or left at a certain point in a maze. In one direction it finds food, and in the other it receives an electric shock. Experience proves that a mouse is as likely to go one way as the other the first time around, but as it gains experience it tends to turn toward the food. If you wanted to calculate the probability that the mouse will go left 10 times in succession, why would the binomial distribution be inappropriate?

2   There are 4 people in a car pool. Each person has a 90% chance of showing up on any given day. If 3 people must be in the car to be eligible for a reduced bridge toll, what is the probability that a full toll will have to be paid on any given day? (Assume that the probability of nonappearance of the different commuters is independent.)

3   A family has *n* children. What is the probability of having the same number of children of each sex if the births are independent and one sex is as likely as the other when *n* is equal to
(a)   2?   (b)   4?
(c)   6?   (d)   8?

4   Company *A* is owed money by 4 other companies, each of which is in financial trouble. The probability of bankruptcy for each company (independently of one another) is .4. If 3 or more of the companies go under, *A* will go under too. What is the probability that Company *A* will survive?

5   If the probability of beating my opponent is $\frac{2}{3}$ in each game of a series of 6, what is the probability that I'll win at least 4 of the 6 games?

6   If I toss a fair coin 7 times, what is the probability that I get
(a)   3 heads?
(b)   4 heads?
(c)   5 heads?

7   I toss a fair coin N times. Am I more likely to get an even or an odd number of heads if N is
(a)   2? (b)   3? (c)   4? (d)   5? (e)   6?
Can you guess the general rule?

8   Each of 2 women in a class has a 50% chance of getting an A, and each of four men has a 25% chance of getting an A (independently). What is the probability that exactly 2 of the men and women get an A?

9   What is the probability of throwing a die three times and getting
(a)   all 6's?
(b)   no 6's?
(c)   two numbers greater than 4 and one number less than or equal to four?
(d)   all even numbers?

10   If 10% of a company's output is defective, how likely is it that a sample of 3 items will include (a) 0, (b) 1, (c) 2, (d) 3 defective ones?

11   Whenever a player comes up to bat, he has a probability of $\frac{1}{4}$ of getting a hit. What is the probability of coming up to bat 4 times and getting (a) 2, (b) 3, (c) 0 hits?

Rod Carew of the California Angels.

12   In a family with 5 children what is the probability of having
    (a)   two children of one sex and three of the other?
    (b)   one child of one sex and four of the other?
    (c)   all children of one sex?
    You may assume that boys and girls are equally likely to be born and that births of different children are independent.

13   Three dice are tossed.
    (a)   How often would you expect at least two of them to show numbers greater than or equal to 5?
    (b)   How often would you expect all of them to show numbers less than 6?

14   A manufacturer assures you that only 5% of the parts he sends you are defective. You test 3 parts and find 2 of them defective. How often would you expect to find that many parts (or more) defective if he were telling you the truth?

1   If a coin is tossed N times, the probability of getting k heads is $C_k^N(\frac{1}{2})^k(\frac{1}{2})^{N-k} = C_k^N(\frac{1}{2})^N$. Use this fact to prove what we proved earlier in another way: $C_0^N + C_1^N + C_2^N + \cdots + C_N^N = 2^N$.

2   If the probability of success of an event is p and it is repeated N times (each trial being independent of the others), show that the most likely number of successes is the whole number k such that

$$p(N + 1) - 1 \leq k \leq p(N + 1)$$

[*Hint:* Show that

$$C_k^N p^k q^{N-k} \geq C_{k+1}^N p^{k+1} q^{N-k-1} \text{ if and only if } k \geq p(N + 1) - 1$$

and

$$C_k^N p^k q^{N-k} \geq C_{k-1}^N p^{k-1} q^{N-k+1} \text{ if and only if } k \leq p(N + 1).]$$

3   Use the result of Question 2 to find the most likely number of successes if an experiment is repeated 5 times and p is
    (a) $\frac{1}{2}$ (b) $\frac{1}{3}$
    (c) $\frac{1}{4}$ (d) $\frac{1}{5}$

Summary

Probability theory has been applied to the real world in many different ways, but in one respect all these applications are alike. You first assume something about the world, such as that a coin is fair, or that boys are as likely to be born as girls, that intensive tutoring increases any student's chance of passing a course from .6 to .8. From these simple assumptions you derive more complicated conclusions, such as the probability of getting three heads in five tosses, the probability that a family with seven children will have no boys, or the chance that a class of ten students who are intensively tutored will have fewer students who passed than another class of ten students in which no one is tutored.

This same procedure can be followed but in reverse: If we observe that a coin turned up heads ten times in twelve tosses, is it likely that the coin is fair? How plausible is it that brand X bulbs last longer than brand Y bulbs on the average if the brand Y bulbs we actually tried lasted longer? These problems (when posed more precisely) are the subject matter of statistics, and this is what we will discuss next.

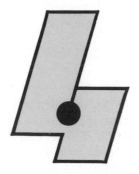

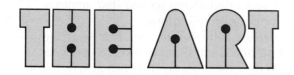

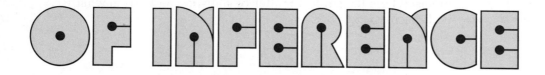

STATISTICS

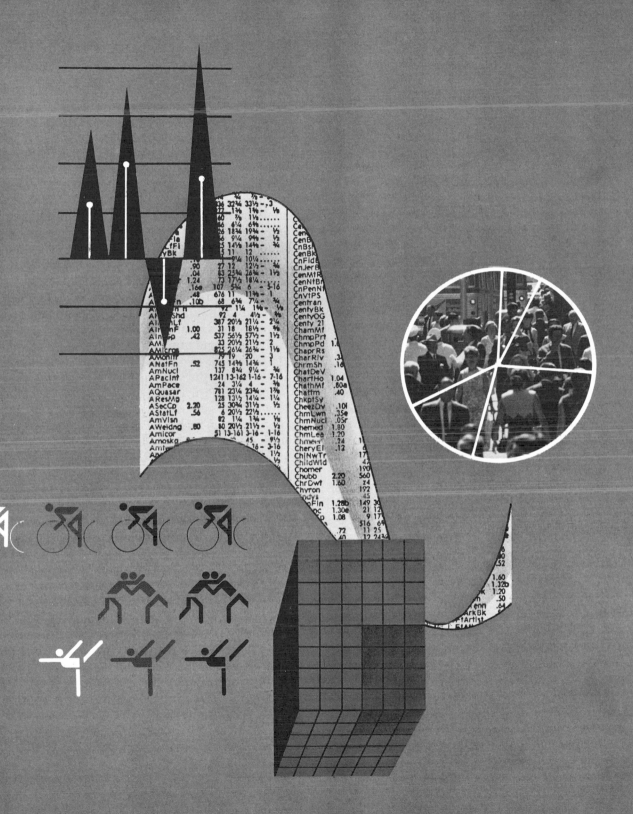

Expected Value

Suppose you were studying the history of medical care in the United States and wanted to compare the longevity of people living today with those who were living fifty years ago. What sort of number or numbers might you use?

Or imagine that a retailing chain is thinking about opening a new store in a city of 100,000 people. The company wants to know the income of its potential future customers, so it hires a management consultant to find out what the residents are earning. What kind of information should the consultant obtain, and in what form should it be given to the client?

It is easy enough to compare the life span of one person with another—just give the ages of both at death. And a person's salary can be described by a single number. But the problem is that we are dealing not with one or two people but with large numbers of them. How do you determine which of two groups live longer when there is a great deal of variation in longevity within each group? And how do you describe what people are earning in a town when no two people may be earning the same amount?

There is a straightforward way of attacking both problems: You might list every person who died this year along with his or her age at the time of death and come up with a statement such as "2% of the people died before they reached the age of one year; 1.5% died in their second year; and so on." You would then go through the same process for people who died fifty years ago. In a similar way, you might list the various salaries earned by the people in the town and the fraction of people earning them.

But the fact is that this straightforward method of compiling exhaustive master lists is almost worthless. For one thing, it is very expensive to spell out in detail when every person died or what every single wage earner makes. And even more important, if you did spell it out, the mass of information you would have would be virtually unmanageable. What we need instead is some sort of summary that reduces a massive list of numbers to a single number or a few key numbers and that can be easily interpreted and applied (so you would be able to compare salaries in two different towns, for example). Of course, if you use a few key numbers to describe a large set of data you must inevitably discard some information, and often the information discarded is useful; so it is important to proceed with care.

Suppose, to take the extreme case, the retailing chain wants just a single number that "best" describes the income of its future customers. There is more than one way of selecting such a number; different people may have different ideas about which number they consider reasonable, and the same person may use different approaches in different situations.

A possible choice for representing all salaries is the **median*** salary. A salary is a median salary if no more than 50% of the other salaries are higher than it and no more than 50% of the other salaries are lower. To put this another way, if the potential customers are listed in order of their income, the person halfway down the list is earning the median income.

The **mode*** is another single number that is sometimes used to describe a set of numbers. The mode is the number that occurs most frequently; in terms of our earlier examples it would be the age at which more people die than any other or the salary earned by more people than any other.

EXAMPLE 1

Find the median and mode heights in a class in which five people are 64 inches tall, six people are 66 inches, eight people are 68 inches, and ten people are 70 inches.

Solution

Since there are more people (10) with a height of 70 inches than any other height, 70 inches is the mode. If the 29 people were ordered by height, the fifteenth person in order—that is, the middle person—would be 68 inches, so that would be the median. Ten people are taller than 68 inches and 11 people are shorter, and each of these numbers is less than half the total population.

A third possibility for a number that describes a group of numbers, the one that is used most often by far, is the **expected value** (also called the *mean* or the *average*). Before formally

---

* The median and the mode need not be unique. It is possible that instead of having one single number in the middle of the range there may be two candidates for the median. For the set of four numbers {1, 2, 3, 4}, both 2 and 3 satisfy the definition of the median. In such a case some take the number halfway between them, $2\frac{1}{2}$. Similarly, there may be more than one number that occurs most frequently.

A Table of thermometrical observation made at Monticello from Jan. 1. 1810. to Dec. 31. 1816.

It is a common opinion that the climates of the several states of our union have undergone a sensible change since the dates of their first settlements; that the degrees both of cold & heat are moderated. the same opinion prevails as to Europe: & facts gleaned from history give reason to believe that, since the times of Augustus Caesar, the climate of Italy, for example, has changed regularly at the rate of 1.° of Fahrenheit's thermometer for every century. may we not hope that the methods invented in latter times for measuring with accuracy the degrees of heat and cold, and the observations which have been & will be made and preserved, will at length ascertain this curious fact in physical history?

At his home in Monticello, Thomas Jefferson kept this weather record showing the maximum, minimum, and mean temperatures for each month during the years 1810–1816. He also calculated the mean annual temperature for each of these years (last row) and the mean of each month over the seven-year period (last column). Presumably through an oversight, Jefferson interchanged the maximum and minimum listings.

defining the expected value we will digress a little to explain why it is defined as it is.

Suppose that in a certain city, $\frac{1}{10}$, $\frac{2}{10}$, $\frac{3}{10}$, and $\frac{4}{10}$ of the population earn $25,000, $20,000, $15,000, and $10,000 per year, respectively. If you took a random sample from this population, you could reasonably expect the representation of each salary group within your sample to reflect the numbers in the population as a whole. For example, if you choose a sample of 10 people, you might expect that roughly 4, 3, 2, and 1 of them would be earning $10,000, $15,000, $20,000, and $25,000, respectively. The sum of all their earnings would be $(4)(\$10,000) + (3)(\$15,000) + (2)(\$20,000) + (1)(\$25,000) = \$150,000$. If this is the amount earned by all the members in the sample, a "typical" member of the sample would get $\frac{1}{10}$ of that amount, or

$$\left(\frac{4}{10}\right)(\$10,000) + \left(\frac{3}{10}\right)(\$15,000) + \left(\frac{2}{10}\right)(\$20,000)$$
$$+ \left(\frac{1}{10}\right)(\$25,000) = \$15,000$$

The expected value is defined formally as follows:

Assume that an experiment has as its possible outcomes the n different numbers $Y_1, Y_2, \ldots, Y_n$ and that these numbers have respective probabilities $P(Y_1), P(Y_2), \ldots, P(Y_n)$ of occurring. The **expected value** is then defined to be

$$Y_1[P(Y_1)] + Y_2[P(Y_2)] + \cdots + Y_n[P(Y_n)]$$

In our example the Y_i values were the salaries of the members of the population and the $P(Y_i)$ values were the fraction of the population earning these respective salaries; $15,000 turned out to be the expected value.

The mean, mode, and median (and there are other possibilities as well) may each be considered a number that in some sense typifies the population; so it is not too surprising that these numbers are often very close to one another. But it can also happen that they vary considerably, as we will see in the following examples.

(EXAMPLE 2)

In a certain population the probability of dying at age 55, 60, 65, and 70 is .1, .45, .25, and .2, respectively. Find the median, mode, and expected value of the population's life span.

Solution

The median is 60; $\frac{1}{10}$ of the people die before 60 and $\frac{45}{100}$ of them die after 60, and neither of these numbers is greater than $\frac{1}{2}$. The mode is also 60, since more people die at 60 than at any other age. The expected value of a person's age at death is

$$(.1)(55) + (.45)(60) + (.25)(65) + (.2)(70) = 62.75$$

(**EXAMPLE 3**)

In a certain city a person selected at random has a probability of .3, .1, .2, .2, and .2 of earning $5000, $6000, $10,000, $30,000, and $50,000, respectively. Find the median, mode, and expected value of the salaries earned in that city.

Solution

The median is $10,000; 40% of the people earn less than that and 40% earn more than that, and both figures are less than 50%. More people earn $5000 than any other salary, so that is the mode. And the expected value is

$$(.3)(\$5000) + (.1)(\$6000) + (.2)(\$10,000) + (.2)(\$30,000)$$
$$+ (.2)(\$50,000) = \$20,100$$

Notice that the values are considerably different and the largest is more than 4 times as great as the smallest.

     While the term "expected value" is in general use, it is a little misleading; the "expected value" may actually be quite unexpected. If a fair coin is tossed once, the expected number of heads is $\frac{1}{2}$, but half a head is clearly *not* one of the possible outcomes of a coin toss. This is not to say that the expected value is of no practical use; a casino is very much concerned about the expected payoffs of its gambling games, and life insurance companies devote much effort to determining their expected payout on the life insurance contracts they write. The basis of their concern lies in what mathematicians call the *law of large numbers*. Although an experimental outcome may vary considerably from the expected value when the experiment is repeated a few times, the law of large numbers asserts that the expected value will almost certainly be an excellent predictor of the outcome when an experiment is repeated often (that is, when many bets are taken or many life insurance contracts written).

Since the mean, mode, and expected value may vary considerably from one another (witness our previous example), which of them presents the "truest" picture of the actual population? In fact, there really is no one right number; each is useful insofar as it reflects one attribute of the real population, and each can be misleading as well. Consider a few examples to illustrate this point:

1. In a certain population 1% of the people earn $1,000,000 and the rest earn $1000; the expected value of the earnings is $10,990. This is almost 11 times the amount that 99% of the population earns, and some people would consider this figure misleading. Many would regard the mode and the median, both of which have the value $1000, as much more typical of what people earn. There is really no decisive way of determining which number is best, of course; which one you use depends on your own purposes (as we will see in the following problems) and intuition.

2. The towns of Pleasantville and Laurelton have very similar populations. In Pleasantville 49% of the people earn $5000, 48% earn $20,000, 1% earn $25,000, and 2% earn $5001. So the median is $5001, the mode is $5000, and the expected value is $12,400.02.

   In Laurelton, as in Pleasantville, 49% of the people earn $5000, 48% earn $20,000, and 1% earn $25,000; but the remaining 2% earn $19,999 (instead of the $5001 they earned in Pleasantville). In this case the median is $19,999, the mode is $5000, and the expected value is $12,699.98.

The law of large numbers asserts that the expected value will almost certainly be an excellent predictor of the outcome when an experiment is repeated often. However, it does not guarantee that an experimental outcome will never deviate much from the expected value.

If you compare these two populations, the median does not seem to be a plausible "typical" salary. When only 2% of the population have their salary changed from $5001 to $19,999, the median for the entire population increases by the same amount. This seems to be a radical increase for a relatively modest change.

The effect of this change on the mode is also somewhat surprising, but in a different way. It does not change radically—in fact, it doesn't change at all. Only the expected value seems persuasive here; it shows a modest change reflecting a modest change in the population.

EXAMPLE 4

Sherman's Corner Grocery makes a $100 profit on 40% of the days, a $50 profit on 30% of the days, a $10 profit on 10% of the days, and takes a $20 loss on 20% of the days. Calculate the store's median, mode, and expected daily profit.

Solution

The median profit is $50, since the store makes more than that less than 50% of the time and less than that less than 50% of the time. The mode is a $100 profit, since the store makes that more than any other amount. And the expected profit is

$$(.4)(\$100) + (.3)(\$50) + (.1)(\$10) + (.2)(-\$20) = \$52$$

(The minus sign in $-\$20$ reflects the fact that the $20 quantity is a loss rather than a profit.)

1 Find the mean, mode, and median of each of the following sets of numbers:
   (a) 1, 2, 4, 5, 6, 8, 8, 9, 9, 9, 16
   (b) 0, 7, 7, 7, 8, 11, 12, 14, 15
   (c) 1, 1, 2, 2, 7, 7, 7, 7, 29
   (d) −23, 7, 7, 7, 7, 13, 14, 15, 16

2   For each of the five classes listed below, calculate the mean, mode, and median grade:

		Grade received on the examination						
		100	90	80	70	60	50	40
Percentage of the class	(a)	10	15	15	10	20	15	15
who received that grade	(b)	2	5	20	13	40	20	0
	(c)	5	10	15	20	25	10	15
	(d)	0	0	20	50	30	0	0
	(e)	13	14	15	16	15	14	13

3   In a tournament in which there are 100 entrants, five people win cash prizes. The winner receives $10,000, two people who tie for second and third place each receive $3000 and the fourth and fifth place winners receive $2000 and $1000, respectively. Calculate the mean, median, and mode of the winnings of all the entrants.

4   In a lottery there is a $100,000 first prize winner, one hundred people who win $1000, and one thousand people who win $100. One million people each pay a dollar for a lottery ticket.
(a)   What is the average amount of money won?
(b)   What is the average gain or loss of each person buying a ticket?

5   The members of the mathematics department contribute to the entertainment fund according to their rank: The four professors each pay $25, the 10 associate professors each pay $15, the 15 assistant professors each pay $10, and the 20 lecturers each pay $5. Find the mean, mode, and median amount contributed.

6   In a certain college 20% of the freshmen are 17 years old, 40% are 18, 20% are 19, and 20% are 20. Find the expected value, mode, and median of the ages in this class.

7   Of the trips made by the planes of an airline, 25% are 1000 miles long, 20% are 1200 miles, 15% are 2000 miles, and the remaining 40% are 800 miles long. Find the expected value, mode, and median length of each trip.

8   In a lottery in which 100,000 tickets are sold, you have a 1% chance of winning $10, a $\frac{1}{1000}$ probability of winning $50, and a $\frac{1}{10,000}$ probability of winning $1000.
(a)   What are the expected winnings of a lottery ticket buyer?
(b)   If lottery tickets cost $1, what is the expected return to the buyer of a lottery ticket?
(c)   What is the expected profit of the lottery?

9   You toss a coin four times and receive $1 for every head that turns up. What is your expected payment?

10 The East–West streets of a large city are numbered consecutively, and the distance between any two adjacent streets is always the same. If you choose a morning commuter at random as he enters the subway, he has a probability of .3, .15, .2, .1, and .25, respectively, of getting off at the 13th, 20th, 30th, 36th, and 42nd Street stations.

 If you wanted to pick a single street that was in some sense the "center" of the morning commuter traffic, you might use the median, mode, or expected value of the street numbers at which people arrive. Calculate each of these. If the person interested in this information is planning to open a newspaper stand and it is known that commuters buy their newspapers at the station at which they arrive, which of these measures would he use?

11 The only issue in the local election is the education budget. There are five proposals being considered: a budget of (in thousands of dollars) 30, 35, 40, 50, or 90. The percentage of the population supporting each of these is 10%, 10%, 35%, 5%, and 40%, respectively. Assume that a voter will vote for the candidate who supports a proposal that is closest to his own preference. Find the median, mode, and expected value of the amount people are willing to spend. If you were one of only two candidates and were interested only in winning the election (and not in the amount of the budget), what budget size would you support? In general, would a candidate who only wants to win the election be interested in the mean, mode, or expected value?

12 Twenty-five identical coins are placed on a narrow metal strip, and the distance of each coin from one end of the strip is noted. A coin is selected at random.
 (a) If a knife edge were placed at a certain location under the metal strip, it would be perfectly balanced. Is this location at the median, mode, or expected value of the distances from the end? (You may assume the strip itself is weightless.)
 (b) If the coins are in stacks, would the highest stack be located at the median, mode, or expected value of the distances from the end?
 (c) If a particular coin has as many coins on one side as it has on the other, is it located at the median, mode, or expected value of the distances?
 (d) Suppose a single coin close to one end is moved further away from the center. State whether each of the following may be changed, must be changed, or cannot be changed: (i) median, (ii) mode (iii) expected value.

13 After examining the scores of a college's admission examinations, an admissions officer decides that the grade required for admission that year should be the highest possible score that will allow at least 50% of the applicants to enter. Will that score be the median, mode, or expected value of the scores?

14 Every evening a theater conducts a lottery in which it passes out numbers from 1 to 100 to its customers, then picks one such number at random and pays the lucky customer that amount in dollars. If the theater is planning its annual budget and wants to estimate a typical evening's lottery payment, should it calculate the median, mode, or expected value of the numbers drawn?

15 The East–West streets of a city are arranged as described in Exercise 10. A messenger service is planning to open a branch office in that city and will have as its customers seven hotels that all lie on the same North–South street. The locations of these hotels are 5th, 10th, 15th, 40th, 45th, 55th, and 60th Streets. The messages are to be delivered one at a time, and each hotel gets approximately the same number of messages.
(a) What is the best location for the new branch office (that is, which street location would minimize the traveling time for the messengers in the long run)?
(b) What would be the answer to (a) if a new hotel were built on 80th Street?
(c) Are the answers to (a) and (b) based on the median, the mode, or the expected value of the distances between the hotels and the office?

16 In a certain carnival game a wheel is spun and the house wins or loses money as follows: The house loses \$18 with a probability of $\frac{1}{6}$, wins \$2 with a probability of $\frac{3}{4}$, and wins \$1 with a probability of $\frac{1}{12}$. What are the median, mode, and expected value of the house's winnings? If you were running this gambling game, which of these figures would interest you?

17 When a fair die is thrown, each of the numbers from 1 to 6 has a $\frac{1}{6}$ probability of turning up. What is the expected value of the number that turns up?

18 If a fair coin is tossed three times, what is the expected number of heads? What would the expected number of heads be if the coin were tossed twice? Four times?

19 Two cards are selected from a 52-card deck containing 4 aces. Find the expected number of aces selected
(a) if the first card drawn is replaced in the deck before the second card is drawn.
(b) if the first card drawn is not replaced before the second card is drawn.

20   In a certain company 10%, 25%, and 65% of the employees have a salary of $30,000, $20,000, and $10,000, respectively. What would be the effect on the median, mode, and expected value of the employees' salaries if

(a) everyone received a $3000 raise?

(b) everyone's salary was tripled?

Can you make a general statement on the basis of your answer?

21   In a given year an automobile driver has a probability of .8, .07, .06, .05, and .02 of causing $0, $100, $200, $300, and $600 worth of damage, respectively. What should an insurance company charge for a one-year policy if it wants to get three dollars more than it pays out, on the average, per customer?

22   A driver estimates that he has a .05, .06, .04, .03, and .02 probability of causing $200, $100, $75, $50, and $25 worth of damage, respectively, in a given year. The driver can either pay $35 and be covered for all damage or buy a $100-deductible policy for $18. (A $100-deductible policy requires the holder to pay the first $100 in damages himself.) Which policy will give him a lower expected loss?

23   Jay, Leslie, and Fran share an office, and each adjusts the thermostat to his or her own taste. Jay is in the room $\frac{5}{12}$ of the time and prefers a temperature of 72°, Leslie is in the room $\frac{3}{12}$ of the time and prefers 76°, and Fran is there $\frac{4}{12}$ of the time and prefers 75°. You would like to calculate the anticipated fuel bills, which requires finding the expected value of the room's temperature. What is this value? (No two of the occupants are ever in the office at the same time, so the office is always occupied.)

Suppose that on a certain examination 9 students receive a grade of 60, 24 get 70, 36 get 80, and 6 get 100.

(a) Calculate the median, mode, and expected value of the grades.

(b) Suppose that some students appeal their grades, and as a result 15 who formerly had grades of 80 have them raised to 100. Show that there would be no effect on the median. Show that the mode actually drops.

(c) Show that if a single person does better or worse, it will have no effect on either the median or the mode.

(d) Show that if every grade of 60 or 70 is raised to 80, the median and mode remain unchanged.

(e) Show that for *any* distribution of grades, if even one grade is modified (either higher or lower), the expected value of the grades will become higher or lower as well.

Variance

If you want a bird's-eye view of some quantitative aspect of a population—earned salaries or shirt sizes, perhaps—the expected value is generally a very useful number. But if you want to dig a little deeper, the expected value by itself may be inadequate. A house in which half the rooms are at a freezing 32° and the remaining rooms are kept at a sweltering 104° has a deceptively comfortable average (expected) temperature of 68°. A store ordering size 10 shoes will not be happy to find that half of their order is composed of size 11 shoes; and they will not consider it much of a compensation if the rest of the shoes are size 9, although this would make the "average" size of the shoes they receive exactly what they ordered. You might be willing to risk $50 to gain $100 but unwilling to risk $10,050 to gain $10,100; yet your expected gain in both cases is exactly the same: $25.

In each of these examples more information than just the expected value is required. What is needed is some indicator of how much the experimental outcomes deviate from the expected outcome, that is, a measure of "spread"; and this is what we will discuss next. To illustrate some of the ideas we will use a spatial analogy that we used earlier: that of a town built along a road with streets numbered in the usual order and at some unvarying distance apart.

A house in which half the rooms are kept at a freezing 32° and the remaining rooms are kept at a sweltering 104° has a deceptively comfortable average (expected) temperature of 68°.

Let's suppose there are three large hotels (*A*, *B*, and *C*) situated along the main road at 12th, 18th, and 22nd Streets with 30, 50, and 20 stories, respectively. (See Figure 1.) A fire station is to be built somewhere along this road, in the most strategic spot. From past experience it is known that the further the fire station is from a burning house, the longer it takes for the fire engine to arrive and the greater the damage to the house on the average. Specifically, if the distance between the house and the fire station is *d*, the expected value of the damage done is d^2. Let's assume that because of the relative sizes of the hotels, if there is a fire there is a probability of .3, .5, and .2 that it will be at hotel *A*, *B*, and *C*, respectively. Where should the fire station be located to minimize the expected financial loss? What will the expected loss be if the fire station is at this ideal location?

The expected loss, if the fire station is at street x, is given by the formula

$$\text{Loss} = (.3)(x - 12)^2 + (.5)(x - 18)^2 + (.2)(x - 22)^2$$

After a little calculation the expected loss can be put into a more convenient form:

$$\text{Loss} = (x - 17)^2 + 13$$

Since both of the terms are always at least zero, the smallest value of the expected loss is attained when x = 17. So the fire station

Figure 1

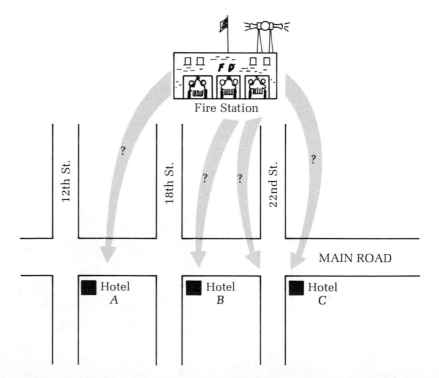

Fire Station

12th St.

18th St.

22nd St.

?   ?   ?   ?

MAIN ROAD

Hotel
A

Hotel
B

Hotel
C

should be built on 17th Street; if it is, the expected loss will be 13.

If we calculate the expected value of the street where a fire breaks out, we get $(.3)(12) + (.5)(18) + (.2)(22) = 17$. This is also the location of the ideal site, and it is not a coincidence that they are the same.

The **variance** is generally used as a measure of the "spread," or deviation from the expected value, in a group of numbers. Suppose $X_1, X_2, \ldots, X_n$ are all the possible outcomes of an experiment and the expected outcome is $\bar{X}$. Then the variance is defined to be

$$\text{Variance} = P(X_1)(X_1 - \bar{X})^2 + P(X_2)(X_2 - \bar{X})^2 + \cdots$$
$$+ P(X_n)(X_n - \bar{X})^2$$

That is, the variance is the expected value of the square of the difference between $\bar{X}$ and the outcome of the experiment.

EXAMPLE 1

A television quiz program is conducted in the following way: A contestant is given 4 questions to answer, and experience has shown that there is a 25% chance that a contestant will answer any particular question correctly. (We will assume that answering one question correctly is independent of answering another one correctly.) Another contestant is given $16,000 to start with and must guess how many questions the first contestant will answer correctly The difference between the number of questions actually answered correctly and the second contestant's estimate is squared and forfeited (in dollars) by the second contestant.

(a)   What guess should the second contestant make to minimize the expected value of his forfeiture?

(b)   If he makes the most sensible guess, what is the expected value of the second contestant's forfeiture?

Solution

(a)   The best guess for the second contestant is the expected number of right answers, which can be computed by using the binomial distribution:

$$(0)\left(\frac{3}{4}\right)^4 + (1)C_1^4\left(\frac{3}{4}\right)^3\left(\frac{1}{4}\right)^1 + (2)C_2^4\left(\frac{3}{4}\right)^2\left(\frac{1}{4}\right)^2 + (3)C_3^4\left(\frac{3}{4}\right)\left(\frac{1}{4}\right)^3 +$$

$$(4)\left(\frac{1}{4}\right)^4 = 0 + \frac{27}{64} + \frac{27}{64} + \frac{9}{64} + \frac{1}{64} = 1$$

(b)   The expected forfeiture is

$$(0 - 1)^2 \left(\frac{3}{4}\right)^4 + (1 - 1)^2 C_1^4 \left(\frac{3}{4}\right)^3 \left(\frac{1}{4}\right) + (2 - 1)^2 C_2^4 \left(\frac{3}{4}\right)^2 \left(\frac{1}{4}\right)^2 +$$

$$(3 - 1)^2 C_3^4 \left(\frac{3}{4}\right) \left(\frac{1}{4}\right)^3 + (4 - 1)^2 \left(\frac{1}{4}\right)^4 = \frac{81}{256} + 0 + \frac{54}{256} +$$

$$\frac{48}{256} + \frac{9}{256} = \frac{3}{4}$$

So by guessing that the first contestant will answer one question right, the second contestant can expect to lose $\frac{3}{4}$ of a dollar on the average.

Just as the average, or expected value, is used as a measure of the "center" of a population, the variance is used as a measure of its "spread," that is, of the extent to which experimental outcomes differ from one another. But why should we be concerned with the *square* of the differences between the average and the experimental outcomes rather than the differences themselves?

There are really two reasons for squaring the deviations; in some applications it may be more realistic. For some purposes an error of 10 may be much more than 10 times as bad as an error of 1. A traffic officer may feel that a motorist's inconvenience is not proportional to the length of time he is delayed, for example; it may be considered preferable to have 100 motorists each delayed 5 minutes than to have a single motorist delayed 3 hours, although the sum of the waiting times is actually longer in the first case. The effect of using the errors squared rather than the errors themselves is to emphasize large deviations disproportionately. For much the same reason elevators are often controlled by a computer that staggers their starting times. While this may increase the total waiting time, it minimizes the number of long delays. And finally, if you wanted to measure the accuracy of a thermostat that determines the temperature of your room, squared deviations are more appropriate. It's better to have the actual temperature differ from the ideal by 1° a thousand times than to have it differ by 200° even once.

But the main reason for using squared deviations as a measure of spread is mathematical convenience. It turns out that theorems become simpler and calculations more tractable if this definition is used. The following example will illustrate some of the effects of using these two definitions of "spread."

Suppose that

$$X = 100 \text{ with probability } \frac{1}{2000}$$

$$= -100 \text{ with probability } \frac{1}{2000}$$

$$= 0 \text{ with probability } \frac{999}{1000}$$

and

$$Y = 1 \text{ with probability } \frac{1}{2}$$

$$= -1 \text{ with probability } \frac{1}{2}$$

The average value of X and Y is 0. The variance of X, which we will denote by V_X, is

$$V_X = \frac{(100 - 0)^2}{2000} + \frac{(999)(0 - 0)^2}{1000} + \frac{(-100 - 0)^2}{2000} = 10$$

while the expected deviation of X from its average is given by

$$D_X = \frac{(100 - 0)}{2000} + \frac{(999)(0 - 0)}{1000} + \frac{0 - (-100)}{2000} = \frac{1}{10}$$

A similar calculation yields $V_Y = D_Y = 1$. So the variance of X is much larger than its expected deviation, while the variance and expected deviation for Y are identical. Why should this be the case?

Notice that X is almost always equal to its average value, 0, and only rarely is it a substantial distance away. This large but rare deviation has a substantial effect on the variance but much less of an effect on the deviations themselves.

On the other hand, Y always deviates from its average but never by very much. The variance of Y is only a tenth of the variance of X, but its distance deviations are ten times as large.

In general, experimental outcomes that happen rarely and differ markedly from the average have a much greater effect on the variance, which is a sum of squared deviations, than on the simple deviations themselves.

1   For the sets of data in Exercises 1–9 at the end of Section 1 (pages 114–16), find the variance in each case.

2   On a nationwide examination $\frac{1}{6}$, $\frac{1}{3}$, $\frac{1}{20}$, $\frac{1}{5}$, and $\frac{1}{4}$ of the grades are 72, 60, 80, 65, and 92, respectively. Find
   (a)   the average grade.
   (b)   the variance.
   (c)   the expected value of the magnitude of the difference between the average examination grade and the examination grades themselves.

3   Incoming planes of an airline arrive on time, arrive 10 minutes late, arrive 30 minutes late, and arrive 12 minutes early $\frac{1}{3}$, $\frac{2}{5}$, $\frac{1}{10}$, and $\frac{1}{6}$ of the time, respectively.
   (a)   How late are the planes on the average?
   (b)   What is the variance of the arrival times?
   (c)   What is the average deviation from the expected arrival time?

4   A manufacturer orders shipments of 100 parts. The shipments contain one item too many 20% of the time, one too few 10% of the time, and the correct quantity 70% of the time. If X is the number of parts over or under the order, find the expected value and variance of X.

5   I pick a number from 1 to N at random. Find the expected value and variance of the number I choose if N is (a) 2, (b) 4, (c) 7.

6   The examination grades in a class of 27 are four 90's, eight 80's, six 70's, two 50's, and seven 40's. Find the expected value and variance of the grades.

7   You toss a coin (which may be biased) once. If the probability of getting a head is p, what are the expected value and variance of the number of heads that turn up? How would your answer change if the probability of heads were $1 - p$?

8   Calculate the expected value and variance of X if
   (a) $P(X = 0) = \frac{1}{6}$, $P(X = 12) = \frac{1}{3}$, and $P(X = 24) = \frac{1}{2}$.
   (b) $P(X = 14) = \frac{1}{7}$, $P(X = 28) = \frac{2}{7}$, and $P(X = 35) = \frac{4}{7}$.
   (c) $P(X = 3) = \frac{1}{5}$, $P(X = 5) = \frac{3}{10}$, and $P(X = 7) = \frac{1}{2}$.

9   In a certain population 20% of the people earn $10,000, 45% earn $16,000 and 35% earn $8000. The salary of a person selected at random from the population is denoted by $X.
   (a)   What is the expected value of $X?

(b) What is the variance of X?

(c) How would the answers to parts (a) and (b) be affected if everyone's salary were doubled?

(d) How would the answers to parts (a) and (b) be affected if everyone received a $2000 raise?

10 In each of the three parts of Exercise 8:

(a) Double the values of X but leave the probabilities unchanged. How does this affect the expected value and the variance?

(b) Add 1 to the values of X but leave the probabilities unchanged. How does this affect the expected value and the variance?

11 A manufacturer shipping goods by railroad arranges for a truck to pick up the load at the station and deliver it locally. There are four trains that may be carrying the load, and they arrive at 2 A.M., 6 A.M., 2 P.M., and 10 P.M.; the shipper has no idea which train they'll be on. The shipper must pay a penalty in dollars equal to the square of the difference between the truck's and the train's delivery times (in hours). When should the truck arrive, and what should he expect to pay?

12 In a certain class $\frac{1}{4}$ of the students receive each of the following grades: 80, 68, 52, and 40. After rechecking, the ones who received 80 and 52 have their grades raised 8 points and those who received 40 and 68 have their grades lowered by 8 points. So the final grade of $\frac{1}{4}$ of the students is 88, another $\frac{1}{4}$ get 32, and the remaining $\frac{1}{2}$ get 60.

(a) Find the expected value of the grades before and after the grade change.

(b) Find the variance before and after the grade change.

13 (a) You choose at random an integer from 2 to 10. What is the expected value and variance of the number you chose?

(b) You pick an even number from 2 to 10. What is the expected value and variance of the number you chose?

The class average on a final examination is 50%. Later 20% of the class have their grades changed. The results of these changes are that 5% of the class who originally received 70% have their grades reduced by x%, 5% of the class who received 30% have their grades raised by x%, 5% who received 50% have their grades raised by x%, and another 5% of the class who received 50% have their grades reduced by x%. If x is between 0 and 20, show that the class average remains 50% and the variance of the grades is decreased.

Testing Hypotheses

A company manufactures two grades (A and B) of a certain part, and these are identical except in one respect: Grade A parts are satisfactory 60% of the time and grade B parts are satisfactory only 30% of the time.

A retailer orders three grade A parts, receives them, and observes their performance. Presumably, all three parts are of the same grade. How can the retailer determine whether the parts are grade A, as ordered, or of the inferior, grade B type?

This is a problem in **hypothesis testing**. (A *hypothesis* is a tentative assumption constructed to explain observed facts.) For problems of this type you start by

1. devising a test
2. listing all possible outcomes of the test
3. listing all possible actions the experimenter can take (or, alternatively, all possible conclusions that can be drawn)

The decision the retailer must make is how to translate each outcome of the test on the first list into an action (or conclusion) on the second list. The rule by which this is done is called a **decision rule**.

In the particular example we started with, there are four possible outcomes to the test: There can be 0, 1, 2, or 3 satisfactory parts. Also, there are two different conclusions the retailer can

draw (and two possible actions): He can decide that he was sent inferior, grade B parts and take his future business to a more reliable (but possibly more expensive) competitor, or he can conclude that he received the parts he ordered and stay with this manufacturer.

Whatever strategy the retailer adopts, he must accept the possibility of error. As a matter of fact, there will generally be a possibility of two different kinds of error. To clarify this point, let's examine what happens when the retailer adopts a specific strategy.

Let's suppose the retailer adopts the following plan: If at least 2 of the 3 parts prove satisfactory, he will stick to his old manufacturer; if fewer than 2 of them are satisfactory, he will look for another supplier. Consider what can possibly go wrong.

Case 1: The retailer has actually received grade A parts as ordered. In this case the danger is that the retailer will mistakenly conclude that he received grade B parts; this will happen if either one or none of the parts he receives are satisfactory. We are dealing with a binomial distribution in which the chance of a single success, p, is .6 (so that $q = .4$) and $n = 3$. Thus the chance of 1 or fewer successes is $(.4)^3 + (3)(.4)^2(.6) = .352$. In short, if the retailer adopts this plan and he is sent grade A parts, a little more than 35% of the time he will mistakenly conclude that he has been cheated and sent grade B parts.

Case 2: The manufacturer has sent grade B parts instead of the grade A parts that were ordered. Now the danger is that the inferior parts will perform well enough to be mistakenly accepted as grade A parts. The probability of having either 2 or 3 successes in 3 trials in a binomial distribution in which $p = .3$ is $(.3)^3 + (3)(.3)^2(.7) = .216$. So there's about a 20% chance that the grade B parts will be accepted as grade A parts.

In general, any strategy of this type, which concludes that either grade A or grade B parts were sent on the basis of the performance of a sample, will be vulnerable to one or the other of these errors and generally to both. You can always change strategies so that one kind of error becomes less likely, but the other kind of error must then become more likely. So in certain applications, where you are particularly concerned about keeping the likelihood of one error very low, you must expect to see the likelihood of the other error correspondingly high.

There are times when it may be considered very important to avoid one kind of error; this would be the case if the issue were one of life and death, for example. A jury in a criminal trial must decide how convincing the evidence will have to be before it decides to convict. The stricter the standard, the less likely it will be that an innocent person will be convicted, but the greater the likelihood that a guilty person will go free. It is even possible to avoid one of the two potential errors completely; you could set a policy in which there is *no* chance of convicting an innocent person. Unfortunately, in this imperfect world the only way to guarantee that no innocent person is convicted is to make sure that *no one at all* is convicted. This would mean that a guilty person would necessarily go free. In numerical terms, the chance of an innocent person being convicted would be 0%, but the chance of a guilty person going free would be 100%.

Let's go back to our original problem and see what other options the retailer might have. Since the grade A parts are more reliable than the grade B parts, the retailer's strategy should look something like this:

If *at least i* parts work, conclude that *grade A* parts were received; if *fewer than i parts* work, conclude that *grade B* parts were received.

There are basically five values that can be used for i, and each represents a different strategy for the retailer. If $i = 0$, for example, the retailer always concludes that he received grade A parts; and if $i = 4$, he always concludes he received grade B parts, whatever he observes. For i between 1 and 3 his conclusion will depend on his experience. Table 1 lists five different values of i along with the probabilities of each of the two kinds of error that can be made. It is clear from the table that as one kind of error increases, the other

Table 1

i	Probability of concluding that you have received grade A parts when you really have received grade B parts	Probability of concluding that you have received grade B parts when you really have received grade A parts
0	1.0	0.0
1	.657	.064
2	.216	.352
3	.027	.784
4	0.0	1.0

decreases. The only way to be sure of avoiding one kind of error is to lay yourself wide open to the other. In such a situation you choose a decision rule by balancing the kind of harm the two kinds of errors may cause.

EXAMPLE 1

Suppose the retailer of our previous example receives 6 parts instead of 3. List the various strategies he can use to determine whether he received the parts he ordered, and state the kind of errors he can make and the probability of making them for each of his plausible strategies.

Solution

For $i = 0, 1, 2, \ldots, 7$ the dealer can decide the parts are grade B if fewer than i parts work and grade A if at least i parts work. Table 2 lists the probabilities of the two types of error.

Table 2

i	Probability of concluding that you have received grade A parts when you really have received grade B parts	Probability of concluding that you have received grade B parts when you really have received grade A parts
0	1.0	0.0
1	.882	.004
2	.580	.041
3	.256	.179
4	.070	.456
5	.011	.767
6	.001	.953
7	0.0	1.0

So once again there is a kind of conservation law at work here. You can lower the probability of either kind of error or even eliminate it entirely, but you must expect the other error to become correspondingly more likely. This much is clear from our two examples. Still, there is a way of decreasing both kinds of errors at the same time—simply test more parts.

Suppose the retailer has had a good relationship with his present supplier and is reluctant to change. He decides to try someone new only if none of the three parts tested works. If he has been sent inferior parts, he has a .657 probability of still staying with his present supplier; and if he has been sent grade A parts, he has a .064 probability of mistakenly leaving his supplier.

If he tests 6 parts, he has much more flexibility. Suppose he decides to stick to his present supplier if 2 or more of the 6 parts tested work and go elsewhere only if all or all but one fail. Notice that the probability of both kinds of error go down. If he is sent grade A goods, he will shift suppliers with a probability of .041, which is lower than the probability of .064 in the other case; and if he is sent grade B goods, he will stay with his old supplier with a probability of .580 rather than .657.

As you increase the number of parts you use for testing, both kinds of errors tend to decrease. But for any fixed number of test items the tradeoff is always there—you raise or lower one error by doing the reverse to the other. Continuing with the same example (in which grade A parts are successful 60% of the time and grade B parts are successful 30% of the time), we list in Table 3 some of the errors when 25 parts are tested. (As before, the retailer infers that the parts are grade B if fewer than i of them are satisfactory.) Only values of i from 11 to 18 have been tabulated.

It is clear that the errors you can make are substantially less likely if you test 25 rather than 3 or 6 parts. Whatever strategy you choose, there is at least a 35% chance of making one error or the

Table 3

i	Probability of concluding that you have received grade A parts when you really have received grade B parts	Probability of concluding that you have received grade B parts when you really have received grade A parts
11	.098	.034
12	.044	.078
13	.017	.154
14	.006	.268
15	.002	.414
16	.000+	.575
17	.000+	.726
18	.000+	.846

other if you test 3 parts, and this chance can be lowered to about 25% if you test 6 parts. However, by setting $i = 12$, and testing 25 parts you make your chance of misdiagnosing less than 8% if you're given type A parts and even smaller if you're given type B parts.

If large samples make hypothesis testing more reliable, why should you ever be content with small ones? Generally, it's because large samples are expensive, although there may be other reasons as well. If you want to test the relative significance of environment and heredity on human beings, you may be hard put to find identical twins who have been reared in different environments. If you want to evaluate the success of heart transplants, it would be ideal to find two people in similar circumstances one of whom had surgery while the other did not, but such pairs are hard to come by. And even in the more run-of-the-mill applications of statistics in which you test a new product or poll voters about their likes and dislikes, large samples are expensive. So the hypothesis tester not only must balance the sizes of the two kinds of statistical errors but also must weigh them against the cost of obtaining a sample.

EXAMPLE 2

Machines A and B produce sewing needles that apparently are identical. However, the needles produced by machine A are defective 20% of the time, while those from machine B are defective 75% of the time. You are given a trial machine and must reject or accept it; you will test n needles (the number n to be determined by you). You want to be sure that, if you were given machine A, you will have at least an 80% chance of accepting it, and that if you were given machine B, you will have at least a 90% chance of rejecting it. What is the smallest number of needles you can test, and what criterion for acceptance will you use?

Solution

First, we'll show that testing 3 needles is not enough. If you accept when i or more needles are not defective and reject otherwise, there are five possible strategies corresponding to five possible different values of i:

Strategies for different values of i	Probability of accepting if you are given machine B	Probability of rejecting if you are given machine A
0	1	0
1	$\frac{37}{64}$	$\frac{1}{125}$
2	$\frac{10}{64}$	$\frac{13}{125}$
3	$\frac{1}{64}$	$\frac{61}{125}$
4	0	1

If you want to be sure that when you're given A you accept it at least 80% of the time, i cannot be greater than 2. If you want to reject 90% of the time when you're given B, then i must be 3 or more. So both conditions cannot be satisfied.

Four inspections are sufficient, however. Your strategy should be to reject a machine if there is more than one defective needle and accept it otherwise. If you actually get A, the chance of having more than one defective needle is $6(\frac{4}{5})^2(\frac{1}{5})^2 + 4(\frac{4}{5})(\frac{1}{5})^3 + (\frac{1}{5})^4 = .1808$. This means you will reach the right conclusion with probability .8192 when you get machine A, which is better than 80% of the time. If you get B, you will reject it with probability $6(\frac{1}{4})^2(\frac{3}{4})^2 + 4(\frac{1}{4})(\frac{3}{4})^3 + (\frac{3}{4})^4 = .9492$, which is more than 90% of the time.

EXAMPLE 3

An insurance company must decide whether it wants to continue with an accident policy for someone who appears to be a bad risk. For simplicity, assume the insurance company considers someone a good risk if they have an 80% or greater chance of avoiding an accident during any given year. The insurance company is reluctant to lose a customer, but it will cancel a policy if it finds that a policy holder has accidents during any 3 years of a 5-year period. What is the probability that a customer who is really a good risk will have a policy canceled because of bad luck? (Assume that the probability of having an accident is the same every year for a particular motorist.)

Solution

We can assume that accidents occur independently. Since the probability of not having an accident in a given year is .8, the probability of having accidents during 5, 4, and 3 years is .00032, .0064, and .0512, respectively. Thus the probability of having accidents during at least 3 years is .05792. So persons who have accidents only 20% of the time will have their policies canceled less than 6% of the time. For those who have accidents less frequently than that, the chance of cancellation will be even smaller.

There is a companion question, which we have not considered but which would also be of interest to the insurance company: What is the probability that a bad risk will have a policy for 5 years and still remain eligible to continue? Here the answer is more complicated, since it depends on how bad a risk the person is. For example, those who have a 90% probability of having an accident in any given year would not be renewed after 5 years 94.2% of the time; those who only have accidents during 30% of the years will be barred from renewing 16.3% of the time.

1   There are three possible outcomes on an aptitude test: Your score can indicate that you are an outstanding, fair, or poor prospect. Of those who ultimately perform the job successfully, 50%, 30%, and 20% receive respective scores of outstanding, fair, and poor; of those who fail to do the job, 10%, 40%, and 50% receive scores of outstanding, fair, and poor. If you hire only those who get an outstanding score, calculate the probability that
(a) someone who is potentially successful is not hired.
(b) someone who will fail is hired.

2   You know a certain coin is either a fair coin or two-headed. To test it you flip it three times. If it always shows a head, you conclude it's two-headed; otherwise, you conclude it's a fair coin. What is the probability that you conclude
(a) it is a fair coin when it is really two-headed?
(b) it is two-headed when it is really a fair coin?

3   You throw two dice three times. Either both dice have 4 green faces and 2 red ones or both have 5 red faces and 1 green one. If at least 1 green face turns up, you assume both dice have 4 green faces; if no green face turns up, you assume both dice have 5 red faces.
(a) State two different kinds of errors you can make and their respective probabilities.
(b) State the probabilities of the errors mentioned in (a) if you decide a die has 5 red faces if at least 1 die turns up red.

4   A cigarette manufacturer advertises that 75% of the people who sample its product switch to it permanently. You choose 5 people at random and decide that if 3 or more of them don't switch to the new product, you will discount the advertisement as puffery. What is the probability that the claim is true but you reject it nonetheless? Can you determine the probability that the claim is false but you fail to dismiss it?

5   Pauline claims that she can predict how a fair coin will land. You toss a coin 10 times and decide to dismiss her claim if she predicts fewer than 9 of the tosses correctly. What is the probability that she is really a random guesser but is sufficiently lucky to get by? Can you estimate the probability that her predictions are not just guesses and yet you dismiss her claim?

6   Students have passed a certain examination 40% of the time during the last decade. A tutor claims the chance of passing the examination will go up to 70% if you make use of his services. To test the claim you and 3 other students go to the tutor, but you decide to demand that he refund the fee if fewer than 3 of you pass the examination. Assuming that he is either a complete sham or that he does exactly what he claims, find the probability that he is telling the truth and yet you demand a refund. What is the probability that he is a sham and you don't demand one?

7   You are given a die, which may be either loaded or fair. You are told that the probability of a 1 turning up when you throw the die is $\frac{1}{6}$ if the die is fair and $\frac{1}{2}$ if it is loaded. You throw the die twice and decide the die is fair if and only if a 1 comes up on neither throw. What are the two kinds of errors you can make, and what are their probabilities?

8   Tina and Ken are typists. Tina's letters have at least one error 50% of the time, but only 25% of Ken's letters do. You receive 3 letters, all typed by the same person. You decide the letter was typed by Ken if there were fewer than k letters with errors. Describe the two kinds of errors you can make and your probability of making them if k is
(a) 1   (b) 2
(c) 3   (d) 4

9   Pete knows that Jane has a Saturday night date at least 4 days in advance 40% of the time. He calls Jane on six different Tuesday nights, and she says she is busy every time. He wonders whether he's had bad luck or whether she just doesn't want to go out with him. How likely is it, if she was willing to go out with him, that he would find her busy six times in succession?

10   On your first visit to a race track, with four races still remaining to be run, you are approached by a tipster. "I see that you're betting," he says. "For a small consideration I can see to it that you also win." To forestall your obviously rising suspicions he adds, "Don't pay now; I'll tell you the names of the winners of the next 4 races. If I'm right on at least 3 of them, you pay me 10% of your net winnings; otherwise, we'll forget we ever spoke." If there are 5 horses in each race and he picks his predicted winner at random, what is the probability that he will be successful in guessing 3 winners? What is his chance of making some money if he approaches 40 other people with the same proposition?

11 Duncan has been selling insurance, with reasonable success, for several years. But one day his tailor remarks, "Clothes make the man, you know. With a whole new wardrobe you could have better results in selling." In the past Duncan found that he was successful in making a sale with 10% of his prospects. To test his tailor's theory he orders a small number of new outfits. He decides he'll invest in a complete wardrobe if, with the new outfits, he makes sales to 2 or more of his next 4 prospects. What is the probability that he will be lucky enough to make 2 or more sales out of the next 4 opportunities if clothes have nothing to do with his ability to sell?

12 An airline claims that its planes arrive on time on 4 out of 5 trips, but its competitor claims they arrive late half the time. You assume that one or the other claim is correct. You observe 5 landings at random and decide to believe the airline if N or more flights are on time and to believe the competitor otherwise. What is the probability of (i) believing the competitor when the airline's claim is true, (ii) believing the airline when the competitor's claim is true, if N is
(a) 2?
(b) 3?
(c) 4?

13 A certain type of tire lasts 50,000 miles 40% of the time. The company that manufactures the tire claims that it has been improved so that 70% of the tires last 50,000 miles, but you suspect there has been no change at all. You observe 4 tires and decide to accept the claim if N or more of the tires last 50,000 miles.
(a) What is the highest value of N you can use if your chance of rejecting an improved tire is to be no more than 35%?
(b) What is the chance of accepting a false claim for this value of N?

14 During a certain war, when a suspected spy is captured, he is asked 5 standard questions. (Who won last year's World Series? What picture did _____ star in?, And so on.) Experience has shown that spies answer each question wrong $\frac{2}{3}$ of the time and nonspies answer them wrong only $\frac{1}{3}$ of the time. If you want to be sure that spies are not released more than 5% of the time, how many correct answers should you insist on? If you do insist on them, how often will a nonspy fail to be released? (Assume the probabilities of answering different questions correctly are independent.)

1 Many people who own stocks want to know whether the value of their holdings will rise or fall in the future. Suppose a firm that is in the business of counseling stockholders agrees to the following test: A pair of leading companies is selected from each of 8 separate industries, and the counseling firm selects the one it considers most attractive. A year later prices are compared to see if the firm's choice was correct.

(a) If the firm made its recommendations by flipping a coin, how likely is it that it would be correct on each of the 8 predictions? If you took a trial subscription to a newsletter that gave financial advice and it predicted the better company 8 out of 8 times, would you find that convincing?

(b) Now suppose there isn't just one company that gives advice about the market but 250. If the company with the best record of these 250 is examined (on the same test described above) and it is found to have made 8 correct choices out of 8 tries, would you be as confident as you were in part (a)?

2 Vera, a psychology student, has some doubts about the existence of extrasensory perception (ESP). Since this talent is alleged to vary from person to person, she arranges to test 5000 people by flipping a coin 12 times and having them guess the outcome. When a subject guesses all 12 outcomes, she feels ESP has been proven. Someone suggests that now that the "best" performer has been identified, he should be retested. But Vera responds, "If the subject were to get 12 out of 12 now, I would be convinced. But he's already done it—why do it again?" Can you see why? Can you see why people doing research are reluctant to draw conclusions from an experiment unless they spell out in advance exactly what and whom they are testing?

Puts and Calls

When your income or savings exceed your immediate needs, you may want to look around for ways to use this extra money. There are a number of possible alternatives you may want to consider: You can invest in stocks or in real estate, speculate on the commodity exchange, or gamble in Las Vegas. What you finally do will probably be determined by such factors as your attitude toward risk and your own previous experience, but whatever you decide, a knowledge of probability should prove useful. In this section we will show how probability theory can be applied to certain spec- ulative devices in the stock market, called *puts* and *calls*.

A **call** is the right to buy 100 shares of a stock at some agreed price, called the **strike price**, at any time during some specified period.

For instance, the owner of a 6-month call in U.S. Steel with a strike price of 55 would have the right to buy 100 shares of U.S. Steel at $55 per share at any time during the next 6 months, whatever the actual market price of the stock.

A **put** is the right to sell 100 shares of a stock at some fixed price, called the **strike price**, at any time during some specified period.

The owner of a 6-month put in U.S. Steel with a strike price of 55 would have the right to sell 100 shares of U.S. Steel at $55 per share at any time during the next 6 months.

If the market price of a stock drops below the strike price of a call, the call would simply not be exercised; if the market price rises above the strike price, a put would not be exercised.

(EXAMPLE 1)

A stock is currently selling for $85. An investor buys a put with strike price $70 and a call with strike price $50, each of which is to expire in six months. How much money can the investor make if, at the end of that period, the price of the stock is
(a) $50?   (b) $75?   (c) $60?   (d) $65?

Solution

(a) In this case, as in the ones that follow, the current price of the stock is irrelevant. The call is worthless, since the investor can buy the stock for $50 on the open market. But the put is worth $2000, since the investor can sell 100 shares at $70 per share after buying them on the open market at $50 per share.

(b) At $75 the put is worthless, but the call is worth $2500, since the investor can buy 100 shares at $50 each and sell at $75 per share.

(c) The put is worth the same as the call ($1000), so together their value is $2000.

(d) The put is worth $500 and the call is worth $1500, so together they are worth $2000.

EXAMPLE 2

A stock is currently selling for $45. An investor buys a call and plans to hold it until it expires 6 months from now. At expiration there is a 30%, 15%, 20%, and 35% chance that the stock will be selling at 45, 50, 55, and 60, respectively. What is the expected value of the call if the strike price is
(a)   45?   (b)   50?

Solution

(a)   On the expiration date, if the stock is selling for $45 or less, the option is worthless. If the stock is selling for $50, each one of the 100 shares that the call entitles the investor to purchase may be bought for $45 and sold for $50 on the open market, so the value of the option is $500. If the stock is selling for $55,

the option is worth $1000; and if the stock is selling for $60, the option is worth $1500. The expected value of the option is therefore

$$(.15)(\$500) + (.20)(\$1000) + (.35)(\$1500) = \$800$$

(b) At a strike price of $50, the option is worthless if the stock isn't selling for more than that; if the stock is selling for more than $50, the value of the option is $500 less than it was in part (a), so the expected value of an option at 50 is

$$(.20)(\$500) + (.35)(\$1000) = \$450$$

A fundamental problem that faces everyone who is involved in the business of buying and selling puts and calls is determining what a "fair price" should be. For this purpose the following coin-tossing problem will be very helpful:

EXAMPLE 3

A fair coin is tossed N times; If H heads and T tails turn up, you are paid $\$(H - T)$ if there are more heads than tails and nothing otherwise. Calculate your expected winnings if
(a) $N = 2$ (b) $N = 4$ (c) $N = 8$

Solution

(a) The chance of getting 2 heads is $\frac{1}{4}$, and your profit in that case is $2. If you don't get two heads, you win nothing. Your expected winnings, then, are $(\frac{1}{4})(\$2) = \$.50$.

(b) There are two ways of winning something if you toss four times: You could get four heads and win $4 with probability $\frac{1}{16}$ or get three heads and a tail and win $2 with probability $\frac{1}{4}$. Your expected winnings are $(\$4)(\frac{1}{16}) + (\$2)(\frac{1}{4}) = \$.75$.

(c) There are four ways of winning when the coin is tossed eight times: if there are eight, seven, six, or five heads. If you calculate the probability of each of these cases and the winnings obtained, you find the expected winnings to be

$$(\$8)\left(\frac{1}{256}\right) + (\$6)\left(\frac{1}{32}\right) + (\$4)\left(\frac{7}{64}\right) + (\$2)\left(\frac{7}{32}\right) = \$\left(\frac{35}{32}\right)$$

The technique used to set the price of a put or a call is by no means an exact science. The precise alchemy that determines the price varies from person to person, but it is generally agreed that such factors as the price and volatility of the underlying stock, the outlook for the economy as a whole, and interest rates play a role.

A particularly interesting factor in the price-setting business is the length of the period during which you can exercise your option. Specifically, suppose a stock is selling for 50, the strike price on a call is also 50, and, by one method or another, someone sets the price of a call with a 6-month period. How do you fix the price of a call that is identical except for having a different period—3 months, 12 months, or whatever? (We assume that the price of the original call was set "correctly.")

The factors we mentioned earlier—volatility, economic conditions, etc.—are all relevant, of course, but these have already been taken into account in setting the price of the 6-month call. The only difference is the length of the period, and this alone will account for the difference between the two prices. At first glance it might be supposed that the prices should vary according to the length of the period: The 3-month call would be half the price of the 6-month call and the 12-month call double. In fact, a very persuasive case can be made to the effect that this is not the relationship at all, and for this purpose the coin tossing experiment is very useful.

The first question you may be tempted to ask is, What has the price of a call to do with the tossing of a coin? The answer is quite a bit—at least in the opinion of some economists. They believe that the day-to-day changes in a stock's price are what mathematicians call a *random walk*. This would mean that whether a stock's price rises or falls tomorrow is independent of its past history (see Chapter 7, on Markov chains).

A bird's idea of a random walk.

Suppose you assume the following:

1. On a given day the market price of a stock will either rise a dollar or fall a dollar, and it has a probability of $\frac{1}{2}$ of doing one or the other.
2. A call is written on the stock with a strike price equal to the current market price.
3. The call has a period of 2N days.
4. The stock's performance on any given day is independent of its performance on other days.

Then we really have the coin-tossing problem only slightly disguised. The stock price at the end of 2N days corresponds to $H - T$ in Illustrative Example 3. (In real life a stock's movements are more complicated, of course, so we have simplified the problem. But in principle the analogy is reasonable.)

It is often believed on Wall Street that the value of a call on a stock that satisfies (1) to (4) above is not proportional to the time period itself but to its square root. Thus, if the stock price and the strike price were the same, a 4-week call would be 2 times as expensive as a 1-week call (because the square root of 4 is 2) and a 36-week call would be 3 times as expensive as a 4-week call (because 36 ÷ 4 = 9, and the square root of 9 is 3).

In Table 4 the first column lists the period of the call in days or the number of coin tosses, depending on which model is being used. In the second column we have calculated the exact expected

Table 4

Length of period (in days)	Exact value of the call	Approximate value of a call obtained by using the square root rule	Percentage difference between columns 2 and 3
2	.5	.557	11.4
4	.75	.788	5.07
6	.938	.965	2.88
8	1.094	1.111	1.55
10	1.230	1.246	1.30
12	1.354	1.365	.81
14	1.466	1.474	.55
16	1.571	1.576	.32
18	1.669	1.672	.14
20	1.762	1.762	.0

value of a call with a period corresponding to the value in the first column. In the third column the value of the 20-day call is calculated exactly as in the second column; the values of all the other calls in the third column are approximated by taking the value of the 20-day call and applying the "square root rule":

> If the market price and the strike price of a call are the same, the value of a call is proportional to the square root of the length of the period.

The last column lists the percentage difference between the exact values in the second column and the approximate values in the third column.

To see how the numbers in Table 4 are calculated, look at the entries corresponding to a 6-day call. The various things that can happen are shown in Table 5. Note that if there are fewer than 4 up days, the profit is zero (since you can never lose), so the total expected profit is $\$\left(\frac{60}{64}\right) = \$.9375$.

Table 5

Number of up days	Number of down days	Profit (up days minus down days)	Probability of occurrence	Expected profit
6	0	$6	$\frac{1}{64}$	$\$\left(\frac{6}{64}\right)$
5	1	$4	$\frac{6}{64}$	$\$\left(\frac{24}{64}\right)$
4	2	$2	$\frac{15}{64}$	$\$\left(\frac{30}{64}\right)$

To calculate the entry in the second column of Table 4 we observe that 1.762 is the value of a 20-day call, so the value of a 6-day call is $(1.762)\left(\sqrt{\frac{6}{20}}\right) = .965$. The entry in the last column is found by dividing the difference between the numbers in columns 2 and 3 by the exact value.

(**EXAMPLE 4**)

A stock now selling for 100 has an even chance of going up 5 points or dropping 5 points in any given month. (We assume that what a stock does in one month is independent of what happened in

previous months.) What is the expected value of a call with a strike price of 100 if the period is
(a) 9 months?   (b) 4 months?

Solution

(a) We can use the binomial distribution to calculate the probability that the price of the stock at the end of the 9-month period will be 105, 115, 125, 135, and 145. The respective probabilities are $\frac{126}{512}, \frac{84}{512}, \frac{36}{512}, \frac{9}{512}$, and $\frac{1}{512}$. So the value in dollars of the 9-month call is

$$(500)\left(\frac{126}{512}\right) + (1500)\left(\frac{84}{512}\right) + (2500)\left(\frac{36}{512}\right) + (3500)\left(\frac{9}{512}\right) +$$

$$(4500)\left(\frac{1}{512}\right) = 615.23$$

(b) With a 4-month call you will gain 1000 one fourth of the time and 2000 one sixteenth of the time, so the expected value of the call (in dollars) is

$$\left(\frac{4}{16}\right)(1000) + \left(\frac{1}{16}\right)(2000) = 375$$

exercises

1   A stock presently at 50 has a probability of $\frac{1}{2}$ of moving up $2\frac{1}{2}$ points and the same probability of moving down $2\frac{1}{2}$ points in any given month. Show that the value of a 6-month call with strike price
(a) 50 is $234.38
(b) 45 is $562.50
(c) 55 is $62.50

2   A company is involved in a major sales campaign. There's a 60% chance that the campaign will succeed, in which case the price of the stock will be $80 in a month, and a 40% chance it will fail, in which case it will go to $40. What should be the price of a call on this stock with strike price $50 so that the expected value is zero?

3   If a film earns $10, $13, $15, and $16 million in the next 3 months, the price of the film company's stock will go to $20, $30, $35, and $40, respectively, at that time. If all outcomes are equally likely, what is the fair price of a call if the strike price is (a) $25? (b) $30?

4   If an invention is a failure, successful, or very successful, the price of the stock of a company holding the patent will be $10, $40, or $100, respectively, in two months. What is the expected value of a two-month call with strike price $30 if the probability of each of the three prospects is $\frac{1}{3}$?

5   The price of a stock at the end of the expiration period of an option is 60. What would the option be worth if the option were
(a)   a call with strike price   (i) 70?   (ii) 65?   (iii) 55?
(b)   a put with strike price   (i) 70?   (ii) 65?   (iii) 55?

6   You own a call on a stock with strike price 70 and a put on that stock with strike price 80, both with the same expiration time. What would these be worth if the stock price at expiration were
(a)   60?   (b)   70?
(c)   85?   (d)   100?

7   You have a call option with strike price 50 and two call options with strike price 60 for which you paid $1500. If all your options have the same expiration date, what must the price of the stock be when the option expires for you to break even?

8   Each month a stock rises 5 points (with probability .6) or falls 5 points (with probability .4), independently of other months. If the strike price is the same as the market price, show that the value of a 5-month call is $756.00 and the value of a 5-month put is $256.00.

9   The stock of an aircraft company is currently selling for $50. It recently bid on two contracts; there is a 10% chance of getting both contracts, and if they do the stock will go to 80 at the end of the period. There is a 20% chance that the company will get one contract; if it does, the price of the stock will be 60 at the end of the period. If the company gets no contract—which has a 70% probability of occurring—the price of the stock will be 40 at the end of the period.
(a)   What is the value of a call with a strike price of
      (i) 35?   (ii) 55?   (iii) 75?
(b)   What is the value of a put with strike price
      (i) 35?   (ii) 55?   (iii) 75?

10   Suppose that at the end of a 3-month period the probability that a stock will be selling at 40 is .15, at 45 is .15, at 50 is .25, at 55 is .20, at 60 is .15, and at 65 is .1. What is the expected value of
(a)   a call with strike price   (i) 40?   (ii) 50?   (iii) 60?
(b)   a put with strike price   (i) 40?   (ii) 50?   (iii) 60?
(c)   your profit if you buy 100 shares at 50?

A Word of Caution

An aspirin manufacturer advertises that two out of three people receive quicker relief from its product than from the products of any of its four competitors. This sounds persuasive, but before buying a year's supply let's explore what this claim might mean.

The manufacturer's statement seems to imply that it picked a random sample of people who suffer from headaches, had them try each of the five products, and published what they said in the local paper. In fact, the "two out of three people" mentioned in the advertisement might have been just that—two out of three people. If that were the case, there would be a fair chance (more than 10%) that at least two of the three people would favor the manufacturer's product even if all the aspirins were the same. But it might be worse yet; there might have been earlier studies that favored another brand, and these might have been discarded.

In statistics, as in everyday life, you have to draw conclusions on the basis of partial information. When you do, you risk being deceived, especially if someone (like the aspirin manufacturer) is *trying* to deceive you. In this concluding section we will mention a few of the pitfalls that should be avoided.

In 1936, in the midst of a depression, *The Literary Digest* took a poll to see how people would vote in the upcoming presidential election. After calling numbers selected at random from the telephone directory, they concluded the Republican candidate, Alfred

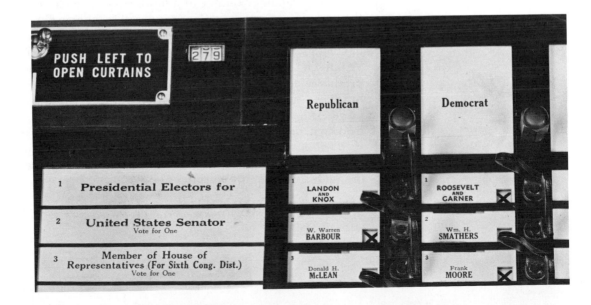

Landon, would win; he lost by a landslide. The trouble was that people with telephones were not typical of the population as a whole, and the difference was reflected in the vote. If some members of a population are more likely to be in your sample than others (in which case your sample is said to be *biased*), your conclusions may be distorted. You wouldn't judge the knowledge of an average student in a school by questioning students in an honors class, nor would you gauge the average student's height by observing students on a basketball court.

When you use a sample to draw conclusions about a larger population, it is important that you measure what you think you are measuring. If you count the number of traffic tickets issued during a recent two-week period and compare that with the number of tickets issued a year ago, you might conclude drivers are more reckless today, but in fact you might just be observing a job action by the police. Again, the police may issue more summonses at one time to prove their efficiency and fewer at other times so that the Chamber of Commerce can claim it's safe to live in their town. In either case the number of tickets issued would tell you little about motorists' driving habits. An apparent increase in the occurrence of a certain disease may only reflect better diagnostic techniques. Voters may mislead their questioners if they resent having their privacy invaded. And even if you know what you're measuring, you may be misled by purely mechanical errors. In the early 1960s an amusing paper was written about an apparent excess of teenage widows that appeared in a recent census. It seems that a fair number of keypunch operators punched a hole in the computer cards a column away from where it should have been when tabulating the results.

A spectacular example of jumping to the wrong conclusion on the basis of a few observations occurred in what is now a classical experiment. The experimenters wanted to know what effect changing a worker's environment would have on his productivity. When the lighting in the factory was improved, the workers produced more; when rest periods were introduced, the workers produced more; with every new change, the workers produced more. And when the original conditions in the factory were reintroduced, once again the workers produced more! Apparently the very act of selecting the sample of workers and setting them apart from the rest made them special and no longer representative of the larger, indifferent working population.*

---

* See the Hawthorne Study, Hawthorne works of the General Electric Co., 1927–1932 by Elton Mayo.

Another common source of error which we haven't discussed before but which is very important is the relationship between cause and effect.

In his book *The Golden Bough* James G. Frazer described the rituals which primitive man performed to modify his environment. You would think that after a while people would have noticed that rain dances didn't really bring rain and yet somehow the custom of rain dances persisted. Frazer's explanation was that sooner or later rain *did* come. It is not always easy to see that two kinds of events are related or unrelated, and if they are related, it may not be clear which is the cause of the other or whether they are both the effects of a third, independent cause. Bleeding was standard medical practice for centuries, although it probably killed more than it cured; and there are still those who believe that walking under ladders and crossing black cats' paths bring bad luck. You only have to compare the way the Surgeon General and the tobacco industry interpret the same set of facts to see that drawing inferences from samples is a hazardous business. It's difficult enough when the interpreters have the best of intentions, and they often don't.

1   In each of the following examples why might the sample selected not be representative of the population from which it was chosen?

(a)   You want to know the average income of families living in a certain building. You list all of the families, select twenty of them at random, and then try to visit each of them in their apartment some evening. When you come to an apartment in which there is no one home, you visit a neighbor instead. (Since one family was as likely to be picked as another, you reason, the substitution should make no difference.)

(b)   You read every story in your local newspaper and find that twelve different neighborhoods are mentioned. In ten of them a murder, a fire, or some such catastrophe has occurred. You conclude that your town is very unsafe to live in.

(c)   Zork, a topnotch investigative reporter from the planet Jupiter, visits Earth to find out what Earthlings are like. After interviewing a psychiatrist employed in a mental institution who tells him how disturbed people are, a lawyer who assures him dishonesty is rampant, and a plumber complaining about the poor quality of modern buildings, Zork decides the population of Earth consists mainly of deranged thieves with leaky sinks.

2 In each of the following cases the conclusions about a population may not be justified by the observations about the sample. Why not?

(a) You want to find out whether students prefer Shakespeare to musicals, whether they use narcotics, and whether they feel inadequate with members of the opposite sex, so you ask them.

(b) To find the average salary of the alumni from your college, you carefully select a sample and send out a questionaire.

(c) During World War II, submarine commanders using radar for the first time feared that enemy planes had radar-sensing devices. The source of this idea was the fact that enemy aircraft were observed (by radar) much more frequently after radar was introduced than they were observed (by eye) before.*

(d) Yvonne arrives at random times at the train station near her home and observes that 65% of the time she sees a train outbound from the city before she boards her own train going into the city. She concludes that trains leave the city more often than they arrive.

(e) During a war the death rate in the army is lower than in an old people's home, so you conclude that the home is more dangerous.

(f) Your stock goes up 10% in a trial period of one month. You hold on to it because at this rate a dollar invested for ten years will grow to more than $90,000.

3 What might be wrong with the following reasoning?

(a) Since movie stars are often seen surrounded by crowds, movie stars must be more sociable than most people.

(b) Restrooms are used more as beer sales go up. So installing more restrooms should increase the sale of beer.

(c) Since the percentage of smokers who die from cancer is greater than the percentage of nonsmokers who do, it seems natural to conclude that smoking causes cancer. But what do you think of the following argument? Since everyone dies eventually, it follows that a higher percentage of nonsmokers die from causes *other than* cancer. Mightn't you conclude with equal justification that a failure to smoke causes diseases other than cancer? Could you conclude that cancer-prone people tend to smoke more? (Perhaps nervous people tend to get cancer and these same nervous people tend to smoke as well.)

(d) Zork, the reporter from Jupiter, comes back to Earth for another story; this time his subject is to be "Footwear on Earth." He notices that, in a certain neighborhood, whenever the mice do not come out of their holes at night the residents seem to go barefoot the next morning. His conclusion is that the failure of mice to come out of their holes causes Earthlings to go barefoot. (Actually, on nights when the local cat serenades the neighborhood, the mice stay under cover and people throw their shoes at the cat.)

---

*Based on Phillip M. Morse and George E. Kimball, *Methods of Operations Research* (Cambridge, Mass.: M.I.T. Press, 1950), pp. 95–98.

(e)   By comparing salaries of college graduates with those of the general population you conclude it pays to go to college. (Perhaps wealthy parents send their children to college *and* then get them well-paid jobs.)

4   The police find there are more people mugged at 7:00 P.M. than at 2:00 A.M. Should you conclude it's safer to take a walk at 2:00 A.M.?

5   You conclude trains are safer than planes when you find that for every million man-hours spent traveling there is one death on a train but two deaths on an airplane. Can you use the same figures (but different logic) to reach the opposite conclusion?

6   A well-known stock market commentator asserts that when the price of a stock has a certain pattern over time, a large price increase is imminent. If this conclusion is not justified on the basis of past observations, might it still turn out to be a valid prediction for the future? Why?

Miscellaneous Problems

So far we have talked about the basics—those topics which are usually found in most elementary books because they are commonly used in applications or because they serve as an introduction to more advanced topics. Now we will look at a variety of different problems which are interesting in themselves because they lead to paradoxical solutions and which, for that reason, point up the need for accurate analysis as well as intuition.

The first three problems are based on an article by Martin Gardner.*

1.   Mr. Jones teaches the two first-grade classes, 1a and 1b, in a certain school, and Ms. Smith teaches the two second-grade classes, 2a and 2b. Class 1a has a higher percentage of boys than 2a, and class 1b has a higher percentage of boys than 2b. Mr. Jones's two first-grade classes are combined to form a single first-grade class, and Ms. Smith's two second-grade classes are combined to form a single second-grade class. Is

_____

* Martin Gardner, "Mathematical Games," *Scientific American*, March 1976.

it possible that Ms. Smith will have a higher percentage of boys in her new combined class than Mr. Jones has?

2. In a certain high school a student picked at random from the freshman class is more likely to be on the football team than a student picked at random from the senior class, and a student picked at random from the freshman class is more likely to be an honor student than one from the senior class. Is it possible that a student from the senior class is more likely to be both an athlete and an honor student than one from the freshman class?

3. A, B, and C are three horses about to race. You can't predict the exact speed of the horses, but you do know the probability that each horse will run at any given speed. If your horse comes in first, you get $100. If you know (a) A beats B 35% of the time, (b) B beats C 50% of the time, and (c) A beats C 35% of the time, is it conceivable you would want to bet on A?

4. You drop in on your local bridge club for a friendly game, and on the first hand you are dealt 13 spades. Should you immediately assume you were the butt of a practical joke?

5. If you want to be sure of winning a modest amount of money in a gambling game, keep doubling your bets. By betting $X the first time, $2X the next time, and so on, you must inevitably win $X, since the probability of N losses in succession gets closer and closer to 0 as N gets larger and larger, and when you finally win you're $X ahead. Why don't more people try this system?

6. In a certain casino you can play the following game. You first pay $X and then toss a fair coin repeatedly until you get a tail. If you toss N tails before getting your first head, you win $$2^N$. If the house wants to make an average profit of $1 on each bet, what should it charge?

7. Two duelists, A and B, are to fire alternately at one another. A's chance of hitting B is $\frac{1}{3}$, and B's chance of hitting A is $\frac{4}{7}$. If A shoots first, what is his chance of winning the duel?

8. Find the solution to problem 7 if A's chance of hitting B is a and B's chance of hitting A is b.

9. Suppose there are three duelists, A, B, and C, in a three-way duel. Let a, b, and c be the probability that A, B, and C hit what they aim at and assume they fire in that order: $A,B,C,A,B, \ldots$. Can you predict what will happen? Does it ever pay not to fire at all?

Summary

When the major networks plan next year's television lineup, when an advertiser introduces a new product, or when a politician tries to determine what the voters are thinking, they often use a statistical model. A statistical model transforms an overwhelming mass of numbers into a few manageable, key statistics. Such a model may be, at the same time, more and less effective than intuition would suggest. If, in a population of 10 million voters, a truly random sample of 10,000 voters is selected and 5200 of these favor candidate X and 4800 favor candidate Y, then the probability that Y will win the election is much less than most people would suspect. On the other hand, statistical models must be used with caution if erroneous conclusions are not to be reached, as we observed in the last section.

GAMES ADULTS PLAY

GAME THEORY

1 Introduction

Almost every topic taught in elementary mathematics courses today has been known for hundreds of years; not so game theory. About thirty years ago the mathematician John von Neumann and the economist Oskar Morgenstern constructed a type of mathematics that was tailored for solving economic problems. They felt that the mathematics which has been devised to solve physical problems could not be simply warmed over and applied to economics. What was called for was a new kind of mathematics which recognized that there was a difference between the behavior of an impartial, indifferent universe and that of highly motivated competitors. One discovers that competing human beings are very similar to players in a parlor game, and, not surprisingly, this area of mathematics is called *the theory of games*. Since its birth, game theory has been appled to political science, sociology, psychology, marketing, finance, warfare, mathematical logic, and statistics as well as economics.

Some General Comments About Games

Before we say just what a "game" is and how you go about analyzing it, let's look at an example of one:

A besieged island is scheduled to be reinforced tomorrow, and so an attacker decides to invade it today. There are four possible places for the attack: ports *A*, *B*, *C*, and *D*. The defender has a large mobile cannon that can be placed in any of the four ports, but it must be committed to one of them before the invasion; after the invasion begins, it can't be moved. Since the gun will be hidden, the attackers won't know its location when they invade, and once the invasion starts they can't switch to a different invasion site.

Because of differences in the terrain, the effectiveness of the gun varies from port to port. Shown in the following chart are the probabilities of *repelling an invasion* at each of the four ports for all possible positions of the cannon.

Invaded Port

		A	B	C	D
	A	.4	.2	.2	.2
Position of the Defending Cannon	B	.2	.6	.2	.2
	C	.2	.2	.8	.2
	D	.2	.2	.2	1.0

Probability of Repelling an Attack

In essence, the invader must choose a column from this chart and at the same time the defender must choose a row, without either knowing of the other's decision. There are four alternatives or strategies (**pure strategies**) for the invader, corresponding to the four ports where he may invade (a column represents each of them), and there are four alternatives for the defender, corresponding to the potential cannon sites (these are represented by the rows). The number in the chart that is common to the row and column chosen represents the probability that the attack will be repelled if these strategies are adopted. Thus if port *B* is invaded, a cannon placed at *B* will repel the invasion 60% of the time and a cannon placed anywhere else will repel the invasion 20% of the time.

In everyday language the word "game" refers to a sport or parlor game, but we will use the term in a broader sense here. A **game** can be a business competition, an election campaign, the process of selecting a jury, a bargaining session, or a battle such as the one we just described.

Involved in our games, as in ordinary parlor games, are "players." In the game we just described there were two players: an attacker and a defender. In general, a **player** can be a single human being or a group of human beings provided each member of the group is assumed to have the same attitude to the outcome of the game. So the players not only can be actual people but also can be IBM, Switzerland, a German submarine, or the New York Mets.

During the course of a game the players make certain decisions. An overall plan of action that describes how a player will behave in any situation that may arise is called a **pure strategy**.* In the game we considered the defender had four strategies—to place the cannon at *A*, *B*, *C*, or *D*; the attacker also had four strategies—to invade any one of the four ports. In most parlor games there are considerably more than four strategies, and a game such as chess has an astronomical number of them.

A chart such as the one in our example is called the **payoff matrix**. After the game is over a **payoff** is made to each of the players. The payoff is determined by the strategies the players picked and possibly by the actions of chance as well. This payoff may be some number of dollars or points, it may be the jury that is finally selected, the profit of a business, or an elected office. In our example, the payoff was success or failure of the invasion. The players' choice of strategies only determined the *probability* that the invasion would be successful; chance had the last word on whether it was actually a failure or a success.

In general, there are two things you want to know about a game:

1. Which strategies would "rational" players use?
2. What will the outcome of the game be if the players act rationally?

The answer to the first question is the key to finding the answer to the second, but the first answer is often not easy to find. Decisions made in a game are usually different from most other kinds of decisions. Ordinarily your decision determines what happens: If you pull the right plunger, you get peppermint gum; if you order hard-boiled eggs, they are served for breakfast; and if you say you don't want mustard on your sandwich, you don't get it. Even if chance is involved, you can make a rational decision; you can estimate the chance of having an accident and then decide whether avoiding the risk of having to pay for one is worth the insurance premium. But in a game in which a "player" is a nation at war, a member of the Senate, or a corporation negotiating a contract, there are other players with preferences that may be very different from yours. So how do you make an intelligent decision when the outcome of a game is affected by the decisions of others,

---

* Pure strategies must be distinguished from mixed strategies, which we will define and discuss later. For the moment, when we refer to a "strategy," it will be understood to be a pure strategy.

decisions you don't control or even know? There is no universal answer, of course. The answer, if it exists at all, will vary from game to game. When the answers to the two questions listed above are found, they are called a **solution**.

There are many possible situations that can be viewed as games. The number of players, the interests of the players (are they cooperating, competing, or something in between), the information available to them, and their ability to communcate vary from one game to another and determine the nature of the game. In the following sections we will take a closer look to see how.

exercises

Each situation described below may be viewed as a "game." In each case
(a) list the "players" and their possible strategies.
(b) indicate the payoffs associated with the strategies.
(c) state whether the players have identical or opposed interests or a mixture of the two.

1 A destroyer has a depth charge with 2000-foot and 5000-foot settings. If the charge is set for 2000 feet, the submarine will be destroyed with probability $\frac{1}{4}$ if it dives and $\frac{3}{4}$ if it fails to dive. If the setting is 5000 feet, the respective probabilities in these two cases are $\frac{2}{3}$ and $\frac{1}{3}$.

2 A retailer can sell a car for $4000, and a wholesaler can buy it for $3000. They negotiate to set the price at which the retailer will buy the car from the wholesaler. In effect, this price will apportion the potential $1000 profit between them.

3 Of three companies (*A*, *B*, and *C*) two will jointly obtain a government contract and the third will be left out in the cold. The profit to be made varies with the companies; if the two companies are *A* and *B*, *A* and *C*, and *B* and *C*, their profits will be $3, $4, and $5 million, respectively. As soon as any two of the three companies agree on a profit split, they inform the government and the contract is theirs.

4 Two competing gas stations independently set a price for gas at the start of the day. They charge 70¢, 75¢, or 80¢ for a gallon, and once they fix the price they don't change it that day. Together, they sell 10,000 gallons whatever price they set. If they both set the same price, they each sell 5000 gallons; if one station offers a lower price than the other, it will sell 8000 gallons and its competitor will only sell 2000 gallons.

5 The prosecutor offers the defendant a chance to plead guilty to second-degree murder (sentence 10 years) rather than standing trial and having a 60% chance of being found guilty of first-degree murder (sentence 30 years)

6 On a TV panel show, a husband must state whether he prefers to read, bowl, or go to the movies; his wife must guess his choice. If she and her husband agree, they win $500 each; otherwise, they get nothing.

Two-Person, Zero-Sum Games with Equilibrium Points

A community is served by two television channels, Channel I and Channel II. Both must select a program for the noon hour, and both are considering three kinds of shows: A comedy (C), a soap opera (S), and a talk show (T). Each channel must decide before learning of its competitor's decision, and the decisions can't be changed once they are made. The size of the total audience is fixed, and the percentage of the audience that each channel attracts depends on which programs are selected. The following chart indicates the percentage of the audience that *Channel I* obtains for each pair of strategies (programs) selected:

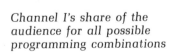

Channel II

		C	S	T
	C	35%	45%	40%
Channel I	S	45%	50%	60%
	T	30%	40%	50%

Channel I's share of the audience for all possible programming combinations

Channel II's share of the market is easily calculated by subtracting Channel I's share from 100%. For instance, if I selects T and II selects S, II will get (100% — 40%) or 60% of the audience.

What type of program should each channel select to get the largest possible share of the audience?

This is an example of a two-person, zero-sum game. It is **two-person** because there are two players, the two channels. The term **zero-sum** comes from parlor games such as poker in which wealth is neither created nor destroyed; after an evening's play the total of all the winnings and losings is always zero. Since the only way to win anything is to take it from someone else, such games are completely competitive. The game we are considering could really be called constant-sum rather than zero-sum, since the sum of the two audiences always comes to 100%; but it too is strictly competitive, in the sense that one channel can increase its audience only by taking part of its competitor's audience.

The Analysis

Earlier we asked how you can choose a strategy intelligently when the outcome of a game is affected by the decisions of others. Although there is no general answer to this question, we can find an answer for this example: Each player should play what is called his *undominated strategy*.

Suppose that no matter how any other player or players act, a player gets at least as good a payoff using strategy A as he would get using strategy B and at least some of the time he actually does better; in such a case we say that strategy A **dominates** strategy B, or strategy B is dominated by strategy A.

A player should never select a dominated strategy, since he does at least as well with the strategy that dominates it. By the same token a player should assume that his opponent won't pick a dominated strategy; if he's wrong it can only be a pleasant surprise. In general, you won't go wrong if you assume that dominated strategies—both yours and your opponent's—are never used.

Domination may be defined in a more formal way in terms of inequalities.

Suppose player II has N strategies: $Q_1, Q_2, \ldots, Q_N$. Suppose further that if player I uses strategy A and player II uses strategy Q_i, then player I receives a_i from player II, and if player II uses strategy Q_i and player I uses strategy B, then player I receives b_i

from player II. The payoff matrix (or at least part of it) would look something like this:

Player II

	Q_1	Q_2	Q_3	$\cdots$	Q_N
$\vdots$					
A	a_1	a_2	a_3	$\cdots$	a_N
B	h_1	b_2	b_3	$\cdots$	b_N
$\vdots$					

Player I (label at left)

If $a_i \geq b_i$ for every i and if $a_j > b_j$ for at least one j, then strategy A is said to **dominate** strategy B.

In this example, Channel I does better by choosing S than it would by choosing either C or T, *whatever Channel II does*, so S dominates both C and T. For the same reason Channel II's strategy C dominates both S and T. If this pair of strategies is chosen, Channel I will get 45% of the audience and Channel II will get the remaining 55%. These strategies are called an equilibrium pair of strategies, and the payoff corresponding to them is called an equilibrium point.

In general, if two strategies form an **equilibrium strategy pair** and one player changes his strategy while the other does not, the player changing his strategy can never gain. If for some reason Channels I and II agreed in advance to choose S and C, respectively, neither would be tempted to switch. And in fact no such agreement is necessary; knowledgeable players should hit on the equilibrium strategies without prior consultation. The payoff corresponding to an equilibrium pair of strategies is called an **equilibrium point.**

When a game of this type has equilibrium strategies, it is generally accepted that rational players will choose them and the outcome of the game will be an equilibrium point. The equilibrium point is often called the **value** of the game; in this example the value of the game is 45% of the audience for Channel I and the remaining 55% for Channel II.

When an equilibrium strategy pair exists, it, along with the value of the game, is considered a **solution**. The reason, briefly stated, is this:

Each player can obtain the value of the game by his own efforts; his opponent can stop him from doing any better.

One can sometimes analyze a game even though a player may have no single dominating strategy. In the game shown below, player I would want to pick strategies (iii), (i), or (ii), respectively, if he knew that II intended to play *A*, *B*, or *C*. (*We adopt the usual convention that the matrix entries indicate what player II pays player I.*)

Player II

		A	*B*	*C*
	(i)	10	6	1
Player I	(ii)	6	5	2
	(iii)	12	4	−1

Player II has an easy choice, however. Strategy *C* dominates both *A* and *B*, so *both* players should behave as though *A* and *B* won't be used. Player I's obvious choice then becomes (ii). Strategies (ii) and *C* form an equilibrium pair, and II should wind up paying 2 to player I. The value of the game is therefore 2.*

Sometimes the situation is even more complicated. In the following game neither player has a single dominating strategy:

Player II

		A	*B*	*C*
	(i)	2	−3	−2
Player I	(ii)	0	0	−1
	(iii)	3	−1	−4

---

*By the value of a game we mean its value to player I. So if the value of a game is 5, player II pays 5 to player I, and if the value of the game is −3, player I pays 3 to player II.

But C dominates A, so we can eliminate A. After A is eliminated, (ii) dominates (i) and (iii); and finally, after (i) and (iii) are eliminated, C dominates B. So the final strategies should be (ii) and C—an equilibrium strategy pair—and I should pay 1 to II. The value of the game is -1.

1 In each of the following games identify one strategy for each player that dominates all his other strategies. Find the equilibrium point, that is, the payoff that is equal to the value of the game.

(a)   Player II

	A	B	C
(i)	−2	−3	0
(ii)	−1	−2	5
(iii)	−3	−4	3

Player I (for rows (i), (ii), (iii))

(b)   Player II

	A	B	C	D
(i)	2	−6	−4	−1
(ii)	3	−5	−2	1
(iii)	4	−3	−1	2

Player I (for rows (i), (ii), (iii))

(c)   Player II

	A	B	C	D
(i)	1	−1	0	2
(ii)	0	−2	−1	1

Player I (for rows (i), (ii))

(d)   Player II

	A	B	C
(i)	4	−2	−1
(ii)	3	−3	−2

Player I (for rows (i), (ii))

In Exercises 2–4 the best strategy for each player is derived by successively eliminating dominated strategies. In each case:

(a) Find the remaining dominated strategies and eliminate them.
(b) Verify that the undominated strategy pair is in equilibrium.
(c) Find the value of the game.

2

Player II

		A	B	C
	(i)	4	−3	0
Player I	(ii)	5	3	1
	(iii)	−1	6	−2

3

Player II

		A	B	C	D
	(i)	−1	2	1	2
Player I	(ii)	−2	1	−4	3
	(iii)	−3	−5	−3	5

4

Player II

		A	B	C	D
	(i)	−2	8	−3	10
	(ii)	3	4	2	3
Player I	(iii)	1	−4	0	6
	(iv)	0	−10	1	5

In each of Exercises 5–8 derive the equilibrium strategy pair by successively eliminating dominated strategies. Show that the strategy pair obtained is in equilibrium. Find the value of each game.

5

Player II

	A	B	C	D
(i)	8	3	4	3
(ii)	6	3	7	3
(iii)	−12	2	−10	0
(iv)	5	−1	−14	2

Player I

6

Player II

	A	B	C	D
(i)	1	6	5	0
(ii)	−4	3	10	−2
(iii)	3	3	2	1

Player I

7

Player II

	A	B	C
(i)	1	3	0
(ii)	−5	4	−3
(iii)	4	6	−4
(iv)	0	6	−1

Player I

8

Player II

	A	B	C
(i)	−6	12	−2
(ii)	6	10	4
(iii)	12	16	−4

Player I

1 If a game has two different equilibrium points, show that the payoffs associated with them must be equal.

2 For what values of the constant *a* does player I have a dominating strategy? What is it? For what value of *a* does II have a dominating strategy? What is it? For what values of *a* does neither player have a dominating strategy?

(a)   *Player II*

		C	D
	A	10	a
Player I	B	3	6

(b)   *Player II*

		C	D
	A	4	2
Player I	B	a	6

3 In the matrix below for what values of *a* and *b* does
  (a) strategy A dominate strategy B for player I?
  (b) strategy B dominate strategy A?
  (c) player I have no dominating strategy?
  (d) strategy C dominate strategy D for player II?
  (e) strategy D dominate strategy C?
  (f) player II have no dominating strategy?
  (g) neither player have a dominating strategy?

                              Player II

		C	D
	A	3	b
Player I	B	a	8

4   Suppose that in the following game (B, (ii)) is an equilibrium strategy pair:

Player II

	A	B	C
(i)	p	q	r
(ii)	s	5	t
(iii)	u	v	w

Player I labels rows (i), (ii), (iii).

(a)   What can you say about the values of q, s, t, and v?
(b)   Can you say anything at all about the values of p, r, u, and w?

5   (a)   In the following matrix show that there is always an equilibrium pair, whatever the value of x:

Player II

	A	B
(i)	7	x
(ii)	6	5

Player I labels rows (i), (ii).

Give the equilibrium strategy pair if (i) x < 5, (ii) 5 < x < 7, and (iii) x > 7.

(b)   In the following matrix calculate those values of x for which there is an equilibrium strategy pair and state what that equilibrium strategy pair is. (There are two cases.)

Player II

	A	B
(i)	7	x
(ii)	8	5

Player I labels rows (i), (ii).

(c)   In the following matrix, for what values of x will there be an equilibrium pair of strategies? What will the corresponding payoffs be? (There are two cases.)

Player II

		A	B
Player I	(i)	7	x
	(ii)	4	5

Two-Person, Zero-Sum Games Without Equilibrium Points

Presidential candidates Smith and Jones are about to complete their campaigns and have only time for one more speech. Each of them has narrowed his choice to New York or California and must pick one or the other before learning of his rival's decision. On the basis of past experience it has been calculated that the number of electoral votes Smith will gain is roughly determined by where these last speeches are given, and this dependence is reflected in the following matrix:

Site of Jones's Speech

	New York	California
New York	30	21
California	22	25

Site of Smith's Speech (label to the left of the table rows)

Electoral Votes Obtained by Smith

From the matrix you can see that there are no dominated strategies for either player, nor is there an equilibrium strategy pair; if the two players tentatively agreed on any pair of strategies,

one or the other could always do better by switching. So we must reconsider the question we asked earlier: How do you decide which strategy to choose when the final outcome depends not only on your choice but on an unpredictable opponent as well?

Since neither candidate has a clearcut strategy, it might be best to play conservatively. And since speaking in California ensures that Smith will get at least 22 votes (he might get only 21 if he speaks in New York), that seems to be the place for him to go. It can also be argued that Jones should make the same choice for much the same reason. But this pair of decisions, in which both candidates speak in California, can easily be shown to be unstable (in the sense that either player might benefit by changing his strategy). Jones may foresee Smith's conservative strategy and restrict him to 22 votes by switching to New York, and Smith may guess that Jones will do that and obtain 30 votes by shifting to New York as well.

This kind of second-guessing can go on indefinitely and is useless. What is really needed is a general theory that spells out a sensible way of making this kind of decision. But the very existence of such a theory leads to a paradox. Suppose Smith knew of such a theory and used it to make his choice. A theory derived by one person can be worked out by another, so Jones might deduce Smith's theory and guess what he will do. But if Smith's strategy is revealed to Jones, Smith will be at a substantial disadvantage; it would seem to be better for Smith to ignore the theory and keep his own counsel after all. In short, constructing a theory for this kind of game seems to be hopeless in principle. In fact, such a theory can be constructed, but you must first look at the problem in the right way.

Ultimately, Smith must decide to speak either in New York or in California—there's no changing that. What Smith can do, however, is determine *how* he makes his decision. By using some random device such as a roulette wheel, a coin, or dice, he can arrange it so that he speaks in New York with probability p and in California with probability $(1 - p)$ for any p of his choice. Such a strategy, in which a player uses each of his possible choices a certain percentage of the time, is called a **mixed strategy**.

However, the difficulty we mentioned earlier still has all of its force. If Smith is clever enough to pick out an advantageous value of p, then he must assume that Jones may be clever enough to figure out what p is. Of course, once the number p is determined, Jones has no way of knowing what the random device will *actually* do; that depends on the vagaries of chance.

So let's put ourselves in Smith's shoes and reconsider the problem. We make the following assumptions:

1. *Both players want to obtain the highest possible average electoral vote.* In the earlier examples, in which each player used a single pure strategy, the outcome was a simple payoff for each player. If either or both players adopt mixed strategies—strategies based on the use of a random device—the payoff will not be a single, definite outcome but a set of different possible outcomes, each with its own probability of occurring. In such a case we assume that the players want to maximize their *average* return.

2. *Smith will speak in New York with probability p and in California with probability (1 − p).* We will talk about what *p* should be presently.

3. *Jones is looking over Smith's shoulder.* This means that, whatever value Smith assigns to *p*, Jones will be clever enough to guess it and use the information to his own best advantage.

To simplify our explanation we will begin by stating the "right" strategies for the players and the "rational" outcome and then justify them. We will postpone for the moment discussion of how we hit on these strategies in the first place.

Suppose, then, that Smith speaks in New York and California with respective probabilities of $\frac{1}{4}$ and $\frac{3}{4}$. Since Jones is assumed to know this, he will calculate the outcome if he (Jones) speaks in New York:

$$\left(\frac{1}{4}\right)(30) + \left(\frac{3}{4}\right)(22) = \text{an average of 24 electoral votes for Smith}$$

and the outcome if he speaks in California:

$$\left(\frac{1}{4}\right)(21) + \left(\frac{3}{4}\right)(25) = \text{an average of 24 electoral votes for Smith}$$

and find that they are the same.

So it turns out that Jones's choice is no choice at all! By choosing this particular mixed strategy, Smith will gain an average of 24 electoral votes *whatever Jones does.* Thus Smith can convert a game in which the outcome seems to depend on the actions of both players to one in which he averages 24 votes whatever Jones does. But can he do even better?

The answer can be found by looking at the game through Jones's eyes. If Jones speaks in New York and California with probabilities of $\frac{1}{3}$ and $\frac{2}{3}$, respectively, the outcome will be the same *whatever Smith does:* an average gain of 24 electoral votes for Smith. So either player, by his own efforts alone, can ensure that the result will be no worse to himself than that Smith will get an average of 24 electoral votes. This is considered to be the value of the game.

Our solution to the presidential campaign game is a particular application of a more general, and very important, theorem known as the **Minimax Theorem**. The proof of the theorem is fairly involved, but its meaning and importance are easy to understand.

> **The Minimax Theorem**: In any two-person, zero-sum game whose payoff is described in matrix form, there is a number V called the *value* of the game with the following two properties:
>
> 1. Player I has a strategy (generally a mixed strategy) that guarantees him an average payoff of V whatever player II does and whether or not player II knows player I's strategy.
>
> 2. Player II has a strategy that prevents player I from obtaining an average payoff greater than V whatever player I does and whether or not player I knows player II's strategy.

To see the importance of this theorem, consider two variations of this game:

Variation 1: Player I picks his mixed strategy first and tells it to II, and then II picks his strategy. Whichever strategy I picks, II will counter with a strategy that minimizes I's payoff. So it is up to I to find a strategy that maximizes this minimum payoff. Such a strategy is called the **maximin strategy**.

Variation 2: This is the same as variation 1 except that the players reverse roles. Player II reveals his strategy first, and I counters with a strategy that maximizes his own payoff. So II must pick an initial strategy that minimizes this maximum payoff to I, a **minimax strategy**.

Hence, maximin is the very *least* player I can expect from the original game, since it assumes that I tells II his strategy in advance. And for similar reasons the minimax is the *most* I can expect. The Minimax Theorem states that the maximin and the minimax are equal.

We said before that if there is an equilibrium pair of strategies, then the equilibrium point is considered a solution, since each player can obtain the value of the game by his own efforts and his opponent can stop him from doing any better. The effect of the Minimax Theorem is to extend this concept to games that do *not* have equilibrium points. The only difference is that the equilibrium strategies are mixed rather than pure and a player can no longer be sure of receiving at least the value of the game; he can only be sure of receiving it *on the average.*

To see the implications of the Minimax Theorem a little better, let us imagine that two countries are engaged in what amounts to a two-person, zero-sum game of the type we have been discussing. Each has built up a spy network inside the other's borders, and the highest councils on both sides are thoroughly infiltrated. The Minimax Theorem asserts that in such a case the spies might just as well pack up their trenchcoats and miniature cameras and go home. If both countries are up on their game theory and act appropriately, the average payoffs will be the same no matter what information the spies gather about their opponent's plans.

We have said that a dominated pure strategy—yours or your opponent's—should be ignored. The same principle can be extended further. A pure strategy may not be dominated by any other pure stategy and yet may be dominated by a combination of them (that is, by a mixed strategy). In such a case it should also be discarded. Suppose that Gina and Lou are playing a game with the following payoff matrix:

Lou

		A	B	C
	(i)	12	3	−3
Gina	(ii)	4	50	1
	(iii)	0	4	5

Observe that B is dominated by neither A nor C. If Gina plays (i), then B is better for Lou than A, and if Gina plays (iii), then B is better than C. But if Lou plays A $\frac{1}{3}$ of the time and C $\frac{2}{3}$ of the time, Lou will lose 2, 2, and $\frac{10}{3}$ if Gina plays (i), (ii), and (iii), respectively, and this is a better outcome for Lou in all cases than if he had played B. Since B is not a viable strategy for Lou, we can reduce the payoff matrix to the following one:

Lou

	A	C
(i)	12	−3
(ii)	4	1
(iii)	0	5

Gina

This new matrix in turn can be reduced by observing that Gina's strategy (ii) is dominated by playing (i) $\frac{2}{5}$ of the time and (iii) $\frac{3}{5}$ of the time. We leave the verification of this for you as an exercise.

Lou

	A	C
(i)	12	−3
(iii)	0	5

Gina

It is simple to check whether a pure strategy dominates another pure strategy, but it is more difficult to see that a mixed strategy dominates a pure strategy. The problems we consider will be simple enough so that this kind of domination can be determined by trial and error.

We must wait until the next section before solving the reduced game between Gina and Lou, but you should confirm these facts:

1. If Gina plays (i) $\frac{1}{4}$ of the time and (iii) $\frac{3}{4}$ of the time, Lou will gain 3 no matter what Gina does.
2. If Lou plays A $\frac{2}{5}$ of the time and C $\frac{3}{5}$ of the time, Gina will gain 3 no matter what she does.

You should also verify that these strategies are effective in the original game as well. Observe that if Gina plays the strategy described in (1), Lou loses 3 unless he plays his dominated strategy, B, in which case he loses $\frac{15}{4}$. And if Lou plays the strategy decribed in (2), Gina gains 3 unless she plays her dominated strategy, (ii), in which case she gains only $\frac{11}{5}$. So the value of the game is 3.

We mentioned earlier that the Minimax Theorem states, in effect, that knowledgeable players lose nothing by telling their opponents their strategies in advance. Let's confirm this at least for one particular case.

(**EXAMPLE 1**)

Players A and B are engaged in a two-person, zero-sum game with the following payoff matrix:

Player B

		P	Q
Player A	R	5	-9
	S	-2	5

(a)   Suppose that A plays first and B plays after he sees what A has done. Show that if A must choose a pure strategy, the best he can do is limit his loss to 2.

(b)   Suppose that A is allowed to announce a mixed strategy and says that he will play R with probability p and play S with probability $(1 - p)$. Show that B should play P if p is less than $\frac{1}{3}$ and Q otherwise. Deduce that the maximin for A is obtained by setting $p = \frac{1}{3}$, and show that in this case A will receive an average payoff of $\frac{1}{3}$ whatever B does.

(c)   Show that if the roles of the players are reversed, so that B must pick a pure strategy first and tell it to A, the outcome should be that A will win 5.

(d)   Show that if B plays P with probability q and Q with probability $(1 - q)$, then A should play R if q is more than $\frac{2}{3}$ and play S otherwise. Deduce that the minimax is obtained for B when $q = \frac{2}{3}$ and that the payoff in this case will be $-\frac{1}{3}$ for B (and $\frac{1}{3}$ for A), on the average.

Solution

(a)   If A plays R, then B will play Q and A will lose 9. By playing S, A can lose only 2 (when B plays P).

(b) Suppose A plays R with probability w and S with probability $(1 - w)$. If B plays P, A's expected payoff will be $5w + (-2)(1 - w) = 7w - 2$. If B plays Q, A's expected payoff will be $(-9)(w) + 5(1 - w) = -14w + 5$. Verify that if w is set equal to $\frac{1}{3}$ by A, A will get $\frac{1}{3}$ on the average whatever player B does. If A sets w greater than $\frac{1}{3}$, B can lower A's expected gain by playing Q; and if A sets w less than $\frac{1}{3}$, B can lower A's expected winnings by playing P. (This is most easily seen if you graph the two equations that express A's expected gain in terms of w.) If A assumes B can second-guess him, or if A must tell his strategy in advance, he does best to set $w = \frac{1}{3}$ and win $\frac{1}{3}$.

(c) This is easily verified by inspection.

(d) In this case if B plays P with probability t, then A's payoff is $5t + (-9)(1 - t) = 14t - 9$ if A plays R and $(-2t) + 5(1 - t) = -7t + 5$ if A plays S. By setting $t = \frac{2}{3}$, B can be sure that A will get no more (and no less) than $\frac{1}{3}$, whatever he does. If B plays P more than $\frac{2}{3}$ of the time, A will get more than $\frac{1}{3}$ on the average by playing R; and if B plays P less than $\frac{2}{3}$ of the time, A will get more by playing S.

1 Two competing automobile supply chains, A and B, control the entire market. Each plans to open one store either in town X or town Y, and each must decide which town to choose without knowing what its competitor will do. There is a fixed market, so the only way to gain customers is to get them from your competitor. The entries in the following matrix are the customers gained (or lost) by A from B:

Player B

		X	Y
	X	2000	-400
Player A	Y	-1000	-200

Show that A should open a store in X $\frac{1}{4}$ of the time and B should open a store in X $\frac{1}{16}$ of the time. What should the outcome be (on the average)?

2   A fugitive has two escape routes. He can go either north or south, and the pursuing police can cover only one of the two routes. If the northern route is taken by both, the fugitive will certainly be caught, and if both take the southern route, he will be caught with probability Q. Otherwise the fugitive will get away:

Fugitive

		North	South
		North	South
Police	North	1	0
	South	0	Q

Probability of Detection

Show that both should go north with probabilities $\frac{1}{11}$, $\frac{1}{3}$, $\frac{9}{19}$, and $\frac{1}{2}$ if Q is .1, .5, .9, and 1, respectively.

3   In the matrix

II

		A	B
I	(i)	$-100 + x$	$5 + x$
	(ii)	5	-10

show that if $-15 < x < 105$, then player I should choose (i) $\frac{1}{8}$ of the time and II should choose A $(15 + x)/120$ of the time. Show also that the value of the game should be $(x - 65)/8$. What value of x makes this a fair game (that is, a game in which each player can expect to break even, on the average)? What can you say if x is less than -15 or more than 105?

In Exercises 4–6 it is assumed that player I plays strategy (i) with probability p and (ii) with probability $(1-p)$ while player II plays A with probability q and B with probability $(1-q)$. Verify that the minimax strategies and the value of the game are as stated.

4

Player II

	A	B
(i)	3	1
(ii)	2	4

Player I

$p = \frac{1}{2}; q = \frac{3}{4}$
Value: $2\frac{1}{2}$

5

Player II

	A	B
(i)	0	8
(ii)	2	0

Player I

$p = \frac{1}{5}; q = \frac{4}{5}$
Value: $\frac{8}{5}$

6

Player II

	A	B
(i)	10	5
(ii)	4	8

Player I

$p = \frac{4}{9}, q = \frac{1}{3}$
Value: $\frac{20}{3}$

7 Consider the game defined by the following matrix:

Player II

	A	B
(i)	5	4
(ii)	1	5

Player I

Since the entries represent what II must pay I, would you guess that A or B is a more appealing strategy for II? Verify that the value of the game is $\frac{21}{5}$, which can be enforced either by I [playing (i) with probability $\frac{4}{5}$] or by II [playing A with probability $\frac{1}{5}$]. Notice that II plays his apparently worst pure strategy 80% of the time.

In Exercises 8 and 9, player I plays (i), (ii), and (iii) with probabilities p, q, and $1 - p - q$, respectively, and player II plays strategies A, B, and C with probabilities r, s, and $1 - r - s$, respectively. Confirm that the minimax strategies and the values of the game are as stated.

8

Player II

		A	B	C
	(i)	-4	-4	4
Player I	(ii)	4	12	-6
	(iii)	8	-8	-4

$p = \frac{4}{7}$, $q = \frac{2}{7}$
$r = \frac{3}{8}$, $s = \frac{1}{8}$
Value: 0

9

Player II

		A	B	C
	(i)	2	3	6
Player I	(ii)	12	-15	12
	(iii)	6	3	-6

$p = \frac{3}{4}$, $q = 0$
$r = \frac{1}{2}$, $s = \frac{1}{3}$
Value: 3

10 In the following payoff matrix one of the pure strategies of each player is dominated by one of his mixed strategies:

Player II

		A	B	C
	(i)	9	8	3
Player I	(ii)	0	3	6
	(iii)	2	4	5

Indicate the pure mixed strategy for each player. Show that the optimal strategy for player I is to play (i) and (ii) each half of the time and the optimal strategy for II is to play A $\frac{1}{4}$ of the time and C $\frac{3}{4}$ of the time. Find the value of the game.

11 In the following payoff matrix one of the pure strategies of each player is dominated by one of his mixed strategies:

Player II

		A	B	C
	(i)	6	−5	4
Player I	(ii)	−12	0	−6
	(iii)	−8	−2	−4

Indicate the pure and mixed strategy for each player. Show that the optimal strategy for player I is to play (i) $\frac{12}{23}$ of the time and (ii) $\frac{11}{23}$ of the time and the optimal strategy for II is to play A $\frac{5}{23}$ of the time and B $\frac{18}{23}$ of the time. Find the value of the game.

Calculating Optimal Mixed Strategies

It is always possible to calculate optimal strategies for two-person, zero-sum matrix games, but it can be difficult.* There is also a much simpler, but not completely reliable, way of calculating these strategies. Still, the simpler approach is worth trying, since it's easy to check whether the strategies obtained are really optimal; if they are not, there's little lost. The basis for the technique is this:

A mixed strategy that yields the same payoff whatever your opponent does will often be optimal.

---

*The calculation involves a computational technique called *linear programming*, which we will not discuss here.

To see how this principle works let's apply it to finding an optimal strategy for the following matrix game:

Gould

		A	B
Curry	(i)	17	−4
	(ii)	−1	5

Assume that Curry plays (i) with probability p. Then if Gould plays A, Curry's payoff will be $17p + (-1)(1 - p)$, and if Gould plays B, Curry's payoff will be $-4p + 5(1 - p)$. We want to find a strategy (that is, a value of p) for which Curry's payoff is the same whatever Gould does. So we set the two expressions for Curry's payoffs equal:

$$17p + (-1)(1 - p) = -4p + 5(1 - p)$$

We find by solving this equation that p must be $\frac{2}{9}$, and Curry's payoff is 3. Gould's optimal strategy can be calculated in the same way: If Gould plays A with probability q, Curry will get $17q + (-4)(1 - q)$ using (i) and $-q + 5(1 - q)$ using (ii). Setting the two equal, we find $q = \frac{1}{3}$, and the payoff to Curry is 3 whatever he does. Since Curry can be sure of averaging 3 and Gould can stop him from averaging any more, we see that these are really the optimal strategies and the value of the game.

This method can fail in either of two ways. Sometimes the equations cannot be solved at all. (For example, in the game

Player II

		A	B
Player I	(i)	6	3
	(ii)	3	0

player I gets 3 more by playing (i) than by playing (ii) whatever II does.) And sometimes some of the probabilities turn out to be greater than 1 or less than 0.

EXAMPLE 1

A band of smugglers intends to bring some contraband to shore and can use either of two inlets; one is generally foggy, and the other is clear. The police have one patrol boat and must guess in advance which inlet the smugglers will use because if they go to the wrong one first they will not have time to go to the other. Even if the police hit on the right inlet, they may still be evaded; if both police and smugglers are at the foggy inlet the chance of detection is a, and at the clear inlet it is b. If the smugglers and police go to different inlets, there is no chance of detection at all. Calculate the appropriate strategies for both players and the value of the game.

Solution

The payoff matrix is

Smugglers

		Foggy	*Clear*
Police	*Foggy*	a	0
	Clear	0	b

We assume, as we did earlier, that the proper strategy for the police is the one that yields the same probability of detection whatever the smugglers do. If the police go to the foggy inlet with probability p, we have $ap = b(1 - p)$, or $p = b/(a + b)$. So the police will go to the clear inlet with probability $a/(a + b)$. By the identical calculation we find that the smugglers should do the same. The expected value of the probability of detection if either party follows the suggested strategy is $ab/(a + b)$.

It is interesting that the police tend to go to the environment that favors them least; and the less favorable it is, the more likely they are to go. If the chances of detection at the foggy and clear inlets are $\frac{1}{10}$ and $\frac{9}{10}$, respectively, the police will be found in the fog 90% of the time. The smugglers, not surprisingly, will tend to go to the foggy inlet as well.

EXAMPLE 2

Two planes are scheduled to land at Kennedy and Dulles airports at the same time. Two customs guards are to be assigned to these airports, and two smugglers will enter the country using one of them or the other. If there are as many guards as smugglers at an airport, nothing gets through; if there is one more smuggler, 10 pounds of narcotics gets through; and if there are two more smugglers, 15 pounds of narcotics gets through. Assume that the smugglers want to maximize the amount of narcotics smuggled and the guards want to minimize it. Find the proper strategies for smugglers and guards and the average amount of narcotics smuggled if the players adopt these strategies.

Solution

We use *KK* to represent each of the strategies "both guards go to Kennedy" and "both smugglers go to Kennedy." *KD* and *DD* are defined similarly.

Guards

		KK	*KD*	*DD*
	KK	0	10	15
Smugglers	*KD*	10	0	10
	DD	15	10	0

As the matrix shows, there is no practical difference between the two airports; only the presence or absence of smugglers and guards affects the amount of narcotics smuggled. So it is clear that the guards should play *DD* and *KK* with the same probability, and the same holds for the smugglers. If the guards play *DD* and *KK* each with probability p and *KD* with probability $(1 - 2p)$, then an average of $20p$ will get through if the smugglers play *KD* and $10(1 - 2p) + 15p$ will get through if the smugglers play either *KK* or *DD*. Setting these equal, we obtain $p = \frac{2}{5}$. A similar calculation shows that the smugglers and guards alike should split $\frac{1}{5}$ of the time and should send both guards (smugglers) to one airport $\frac{2}{5}$ of the time and to the other airport the remaining $\frac{2}{5}$ of the time. On the average, 8 pounds of narcotics should get through.

1   In each of the following games, calculate the optimal mixed strategies for both players.

(a)

Player II

		C	D
Player I	A	4	3
	B	1	6

(b)

Player II

		C	D
Player I	A	4	8
	B	10	0

(c)

Player II

		D	E	F
Player I	A	18	15	9
	B	10	16	8
	C	12	0	36

(d)

Player II

		C	D
Player I	A	1	a
	B	b	1

(a and b both greater than 1)

2   Find the optimal strategies and the value of each of the following games:

(a)

Player II

		A	B
Player I	(i)	4	10
	(ii)	8	6

(b)

Player II

		A	B
Player I	(i)	5	−3
	(ii)	−2	4

(c)

Player II

		A	B
Player I	(i)	3	2
	(ii)	1	4

3 For each of the following games find a value of x that makes the game fair, that is, a value of x that makes the value of the game zero. If no such value exists, indicate why.

(a)   *Player II*

Player I

	A	B
(i)	7	−2
(ii)	x	0

(b)   *Player II*

Player I

	A	B
(i)	15	−2
(ii)	x	−1

(c)   *Player II*

Player I

A	B
−5	x
4	−1

(d)   *Player II*

Player I

A	B
−2	6
x	−4

4 A hawker has two locations, X and Y, at which he sells his wares. If the police officer on the beat turns up at the location he chooses, he makes nothing that day. If the officer goes to a different location he makes $20 at X and $40 at Y. The payoff matrix for this game is

Officer's Location

Hawker's Location

	X	Y
X	0	20
Y	40	0

(a) Find the optimal strategy for the officer (who wants to minimize the hawker's profits) and the hawker. Find the value of the game.

(b) Calculate the change in optimal strategy of both players and the change in the value of the game under each of the following conditions:

  (i)   The $20 profit is increased to $30 (and all other entries remain the same).

  (ii) The $40 profit is increased to $50.

(iii)   Some sales are possible before being spotted by the police officer, so that the payoff matrix is either

Officer's Location

	X	Y
X	10	20
Y	40	0

Hawker's Location

or

Officer's Location

	X	Y
X	0	20
Y	40	10

Hawker's Location

(c)   Which increase in part (b) is most beneficial to the hawker?

5   In the earlier parts of this chapter we described a number of games and mentioned their optimal strategies without calculating them. Calculate the optimal strategies for

(a)   the invasion problem on pages 155–56.
(b)   the Presidential race on page 168.
(c)   the game between *A* and *B* on page 174.
(d)   Exercises 1–11 on pages 175–79.

6   Verify that choosing a strategy that gives your opponent the same payoff whatever he does will *not* lead to an optimal strategy in the game between Gina and Lou described on page 172. The reason it breaks down is that some pure strategies are dominated and should never be used. In this problem, Gina's strategy (ii) and Lou's strategy *B* are dominated, and it is to be expected that their use would lead to an inferior outcome. If you eliminate the dominated strategies, you obtain the reduced matrix shown on page 173 at the bottom; confirm that the method works once again.

7   In the following matrix calculate the optimal strategies for both players and find the value of the game:

Mendez

		A	*B*	*C*
	(i)	10	2	− 5
Engel	(ii)	− 2	12	3
	(iii)	4	− 4	8

8   Show that in the following game you cannot use the shortcut method for deriving the optimal strategies:

Lamont

	A	B	C
(i)	−2	4	8
(ii)	6	5	0

Wilde

Find a dominated strategy of Lamont and give a strategy that dominates it. Reduce the matrix by eliminating the dominated strategy and then find the optimal strategies for the reduced matrix. Show that these strategies are optimal for the original matrix as well. What is the value of the game?

9   In parts (a) and (b) find the optimal strategies for both players. (You may assume that the optimal mixed strategy for a player is the one that will yield the same return whatever his opponent does.) Verify that these mixed strategies are a solution and give the value of the game.

(a)   Player II

	P	Q	R
S	6	−4	0
T	−5	8	−7
U	−8	4	2

Player I

(b)   Player II

	O	P	Q	R
S	0	−2	0	14
T	11	0	−5	0
U	−4	4	0	0
V	0	0	5	−1

Player I

1 Suppose that Smith formally calculates a mixed strategy that gives Jones the same expected payoff whatever he does (in the way we described earlier).

Jones

		A	B
	(i)	10	x
Smith			
	(ii)	4	12

(a) Show that the probability of playing (i) should be $8/(18 - x)$.
(b) Show that $8/(18 - x)$ is between 0 and 1 if and only if $x \leq 10$.
(c) Show that A dominates B for Jones if $x > 10$.

2 Suppose Smith and Jones formally calculate mixed strategies as described earlier and that p and q are the respective probabilities of using (i) and A.

Jones

		A	B
	(i)	4	y
Smith			
	(ii)	5	6

(a) Show that $p = 1/(5 - y)$ and $q = (y - 6)/(y - 5) = 1 + p$.
(b) Show that $p \neq 0$ and if $p > 0$, then $q > 1$. Conclude that a mixed strategy is *never* appropriate in this game.
(c) Prove (b) in another way by showing that A dominates B if $y \geq 4$ and (ii) dominates (i) if $y \leq 6$.

3 Suppose Jones decides to calculate his optimal mixed strategy by playing
A, B, and C with probability p, q, and $(1 - p - q)$, respectively, and fixes
p and q so that he gets the same expected return.

Jones

		A	B	C
	(i)	2	4	2
Smith	(ii)	4	8	7
	(iii)	14	10	5

(a) Show that A and B will yield the same payoff only if $4 = 6p + 8q$.

(b) Show that B and C will yield the same payoff only if $5 = 3p + 4q$.

(c) Show that it is impossible to satisfy the equations in (a) and (b)
simultaneously.

(d) Show that mixed strategies are inappropriate, since C dominates B
and (i) is dominated by both (ii) and (iii).

4 In the final round of a bridge tournament you're about to play the last
hand. You and your opponents (in another room) both know you're
ahead. Each team must decide whether to bid a grand slam or small slam;
if both make the same decision, your team will win the tournament. If
you take different actions, the team that bids the small slam has a 60%
chance of winning the tournament. All this is summarized in the following payoff matrix. (The entries are *your* probabilities of winning the
tournament.)

		Your opponent's action	
		Bid small slam	Bid grand slam
Your action	Bid small slam	1	.6
	Bid grand slam	.4	1

What is your (and your opponent's) best action? How likely is it that
you'll win the tournament?

Two-Person, Non-Zero-Sum Games

Two burglars, *A* and *B*, are arrested by the police while breaking into a home and taken to two separate rooms for interrogation. The burglars have two choices: to confess or remain silent. When they make their respective decisions, they do not know what their associate is doing, but they are aware of the consequences of their joint action. If both burglars confess, each will be sent to prison for five years. If both remain silent, each will be charged with carrying a concealed weapon and sent to prison for a year. And if one confesses and the other does not, the one who confesses will gain his freedom by turning state's evidence while his partner will be sent away for twenty years. The payoff matrix for this game is

Burglar B

		Confess	Remain silent
	Confess	(5, 5)	(0, 20)
Burglar A	Remain silent	(20, 0)	(1,1)

This is an example of a **non-zero-sum game.** There is an obvious difference between this payoff matrix and those we used for zero-sum games: In non-zero-sum games there are two payoffs associated with each pair of strategies instead of one. This is because you can't deduce the payoff of one player from the payoff of the other as you could in zero-sum games. By convention the first number of the pair represents the payoff to the player who chooses the row (in this case burglar *A*) and the second number is the payoff to the player who chooses the column (*B*).

Let's assume that each burglar is interested solely in his own welfare—he doesn't want to hurt his partner and has no reason to help him. If each burglar has to make his own decision before he learns what his partner does, what should an intelligent burglar do and what should the ultimate outcome be?

This problem is a version of the celebrated *prisoners' dilemma,* a well-known example of a non-zero-sum game originally formulated by Professor A. W. Tucker of Princeton University. The essential distinction between this game and the zero-sum games we discussed earlier is that here the players have common interests as well as opposing ones (or, to put it another way, an improvement in one player's situation is not necessarily at the expense of the other player). The two burglars *both* do better if they remain silent together instead of confessing together. In a two-person, zero-sum game it can never happen that both players simultaneously do better by changing from one strategy pair to another.

Zero-sum games are purely competitive; a player trying to maximize his own gain is, by definition of "zero-sum," trying to minimize his opponent's. Non-zero-sum games may have a competitive aspect too, but they may also have a cooperative element.

Non-zero-sum games are generally a more accurate mirror of reality than are zero-sum games, since few situations in life are purely competitive. In the competitive business world, companies make "gentleman's agreements," rival political parties routinely agree to pair off the opposing votes of absent members, and even in wartime, agreements about the use of weapons and the treatment of civilians and prisoners of war are honored.

Non-zero-sum games are usually much more interesting than zero-sum games, and they offer insights into a surprisingly wide variety of problems. They are certainly more complex in that it is much harder to prescribe what a player should do or to predict what will happen. It is to be expected that the theory of non-zero-sum games will be less compelling and the concept of a "solution" less convincing; we certainly can't hope for the neat, plausible theory we developed for zero-sum games.

In our discussion of two-person, non-zero-sum games we will not be primarily concerned with solutions. What we will do instead is consider some of the ways these games come up in actual practice and look at some of the conceptual problems. As a start, we will go back to the prisoners' dilemma and view the game as one of the burglars might.

As often happens, the outcome of this game depends on the actions of both players, so it seems that a burglar's first step is to try to figure out what his partner will do. But in this particular game, he needn't bother; he has a clearcut action whatever his partner does.

Suppose you were *A* trying to work out some course of action. Your partner, *B,* must take one of two actions, and you try to explore them both. If *B* confesses, your best course is to confess along with him, since it is preferable to go to jail for five years rather than twenty. If *B* remains silent, you still do better to confess; confessing brings you freedom, while silence leads to a year in jail. So it turns out your best course is to confess, whatever your partner does.

The analysis seems to be reasonable, and you might wonder why we brought the whole thing up were it not for one disturbing afterthought: Two "clever" burglars acting in what is clearly their own self-interest wind up in jail for five years, while two "fools" can be out of jail in only one year by remaining silent! And that is the paradox.

There are many factors that determine the outcome of non-zero-sum games, and the effects of these factors are not always what you would expect. The exercises below are a bit different from the others you have seen, in that they are not meant to check if routine material has been mastered; instead these exercises raise questions (which need not have simple answers) to stimulate thought.

Before even looking at the Exercises, try to answer these questions. After you have finished working the exercises, answer these questions again and see if you have changed your mind.

1 Suppose a pair of players using one set of strategies receives a certain payoff. By jointly switching to a new set of strategies they can obtain a higher payoff for each of them. Is it plausible that reasonable people might persist in the lower payoff instead of moving to the higher one?

2 In some games players can improve their payoffs if they are allowed to communicate with one another. Is it conceivable that a player's payoff would be reduced when he is given the opportunity to communicate with other players? (Even if the facilities for communication are available, can't one act as if they weren't and refuse to talk?)

3 It is not hard to think of a game in which player A would want player B misinformed about his own payoff. (You wouldn't want the potential seller of a house to know how much you want it.) But can you imagine circumstances in which B is misinformed about A's payoff and it is to A's advantage to set B right?

4 Suppose that in a game the rules are changed so that at a certain point you lose some of your options. Can this restriction ever work to your advantage?

5 Suppose that the rules of a game are changed; originally both players made their decisions at the same time, but now A must decide and act first and B decides only after he sees what A has done. Can the change ever be to A's advantage and B's disadvantage?

1 Two competing supermarket chains, ShoppAll and Buy-It-Here, customarily set their prices a week in advance at company headquarters and send them out to their retail outlets. A quart of milk, which costs the chains 50¢, will be sold for 55¢ or 56¢, and each company sets its price for the following week without knowing what its competitor will do.

If both companies set the same price, they will each sell 10,000 quarts; if one undersells the other, it will sell 14,000 quarts to the other's 6000 quarts. The payoff matrix for this game is

*Buy-It-Here's
selling price*

		55	*56*
ShoppAll's selling price	*55*	(500, 500)	(700, 360)
	56	(360, 700)	(600, 600)

(The payoffs are given
in dollars of profit.)

(a)   Suppose both companies are curently charging 55. Does ShoppAll have any reason to raise its price to 56? If it does raise its price, do you think it more likely that Buy-It-Here will also raise its price or that ShoppAll will lower its price to 55 again? Is (55, 55) a stable pair of strategies?

(b)   Suppose the companies are currently charging 56. Would a company be tempted to lower its price? If it did lower its price, what do you think would happen? Specifically, would its competitor be likely to follow suit or would the price cutter tend to raise its price? Does (56, 56) seem a stable strategy pair?

(c)   Of the four strategy pairs (55, 55), (56, 55), (55, 56), and (56, 56), which seems most stable? Which would you consider most desirable from the point of view of maximizing the profits of both companies? Is one strategy pair more attractive to *both* companies than another? Comment on the relationship between the desirability of a strategy pair and its stability.

(d)   Suppose ShoppAll knows Buy-It-Here is going out of business next week, so the game will be played only once. Show that ShoppAll gets more profit playing 55 than 56 whatever Buy-It-Here does. Do you have reservations about playing 55 nonetheless?

Review your answer to question 1 on page 194 in the light of your answers to these questions.

Note: This last problem is another example of the prisoners' dilemma. Situations like this arise frequently in everyday life. In the following examples, each member seems to be acting in what is clearly his own self-interest, but each individual winds up worse off than if he had acted unselfishly.

(a)   A person walks on the grass or litters because the immediate convenience is more important than the negligible harm to the environment. But the effect on the quality of the individual's life if everyone acts the same way is another matter.

(b)   The cost of antipollution devices is more important to a company than the marginal effect of its own pollution. But the combined effect of all polluters may be so bad that each would prefer to pay its share for clean air. The trouble is, the companies may not have this choice. They can pay for their own devices, but they can't make others pay for theirs. So a company has two real options: buy equipment and have an almost imperceptible decrease in pollution, or save money with no noticeable change for the worse. Of course, the government may force all companies to comply, and ironically this may be to everyone's advantage. In fact, it has been stated that the role of government is to force cooperative solutions to the prisoners' dilemma.

(c)   When there is a shortage of power (or fuel or water), each person may still use his or her air conditioner. The individual's gain in comfort is clear, and the difference in total consumption is negligible.

It is easy to see what happens when everybody follows this "rational" strategy, however.

2   A building syndicate plans to build a sports arena and has options to buy all the houses in the area except one, the Finch family home. The market value of their house is $50,000, but it is worth $500,000 to the syndicate. The options expire in a week, so an agreement must be reached soon.

(a) Suppose the Finches leave town, tell their attorney to accept only $300,000 or more, and remain inaccessible for a week. Do they necessarily lose by tying their hands in advance? Are they hurt by the lack of communications? Compare your answers to those you gave to questions 2 and 4 on page 195.

(b) Suppose the syndicate's potential profit were $75,000 rather than $500,000 but the Finches thought it was $500,000. Would the syndicate (and the homeowners) gain if they were enlightened? See question 3 on page 195.

3   Tina and Janet are traveling through Europe together. Tina would prefer to see Paris, and Janet would prefer to see London. Each would prefer to be with her companion in her least preferred city than to be alone in her most preferred one. Subsequently, Janet learns that she cannot return to Paris because she failed to pay a parking ticket there two years earlier. In effect, Janet has lost one of her alternatives. Is Janet better off? Is Tina better off? Compare your answers to those you gave to question 5 on page 195.

4   The July 1976 issue of *Consumer Reports* quotes the following two excerpts from *Pest Control* and *The Wall Street Journal*, respectively:

(a) "Raise your prices if you haven't raised them in the past few years. There is absolutely no reason to fear a damaging loss in business if you properly notify your customers . . . . People expect to pay $15 just to have a serviceman of any trade come to their home to make repairs! . . . No one can legally set a price for this industry to charge. But it sure would be good for all of us if we agreed to cut out the undercutting competition and set a minimum charge in the neighborhood of $15."

(b) "A marketing official of a large maker of electrical gear . . . says: 'In the current competitive situation, we feel we aren't forced to move prices back, so why should we?' Another company in a usually competitive field says it has raised a number of prices recently and publicized them, 'just as much to let our competitors know as our customers.' He adds, 'Our competitors generally follow pretty quickly, and we follow them pretty quickly when they kick off an increase."

These two realistic examples are very similar to the prisoner's dilemma. How is the prisoners' dilemma being resolved? Can players tacitly communicate [see (b)]? Are there any legal restrictions on overt communication?

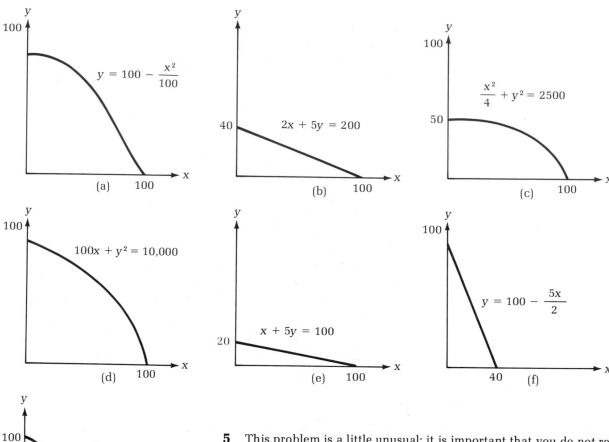

5   This problem is a little unusual; it is important that you do *not* read the "Discussion" until you have finished all eight parts.

In each of the eight diagrams in Figure 1, a graph is shown along with the algebraic equation it represents. In each case you pick x and some other player, with whom you cannot communicate, picks y. If the point (x, y) lies above the graph, neither player gets anything; if the point lies below (or on) the graph, you get x dollars and your partner gets y dollars. For example, in case (c), where the equation is $x^2/4 + y^2 = 2500$, if you pick x = 80 and your partner picks y = 20, you get $80 and your partner gets $20 because the point (80, 20) is below the graph. If you pick x = 80 and your partner picks y = 40, you both receive nothing, since (80, 40) is above the graph.

Since you would like to receive as much money as possible, you might be tempted to pick a large value of x (because you get x dollars when the point (x, y) is below the graph). On the other hand, if you make x too large (or your partner makes y too large), the point (x, y) will be *above* the graph, and both you and your partner will get nothing. Your job, then, is to find a reasonable middle ground (it will probably be different for each graph).

Figure 1

Discussion: This question is more like a psychological test than a routine question with a right and wrong answer. If you want to pick x intelligently, two obviously conflicting considerations must be balanced. You want x large because that is the amount you will get if the point obtained is below the graph; you want x small to increase the probability that point will be below the graph. How, then, do you pick the "right" x? And if you find it, how do you know a greedy partner won't wreck your chances by bidding too high?

This might seem to be a difficult problem, but in this case the situation is clear; you were really *playing against yourself.* To see why, look at graphs (b) and (f). The person choosing x in (f) is in exactly the same position as the person choosing y in (b); so if you view (b) and (f) as one game in which your choice of x in (f) is your "partner's" choice of y in (b), you are playing with yourself as partner. In the same way (a) and (d), (c) and (g), and (e) and (h) may be combined.

It is not entirely clear what the "right" strategy is here. Some suggest* that you pick x (and assume your partner will pick y) so that the product xy is a maximum. A reasonable measure of success might also be the sum x + y of your actual payoffs (although to maximize x + y you might have to make x or y zero, which would be absurd for the player receiving that amount). But whatever strategy you chose, the outcome will be a clue to your own personality; if you find the point (x, y) consistently above the graph you are probably too greedy and if you find the point too far below the graph you may be too modest.

Summary

Game theory, originally devised to deal with economic problems, led to the development of models applicable to many other types of problems as well. The confrontation between an invading force and the defenders of an island, the competition between two television networks or the competition between two political candidates can all be reduced to choosing a row or column in the appropriate payoff matrix. A game-theoretic model will indicate which strategies will leave you least vulnerable; these strategies may not have been recognized as "best" even by sophisticated players.

A game-theoretic model can show the advantage of inserting randomness, or "systematic ignorance," into one's strategy (superficially this seems irrational); as we saw in the example of the smugglers, it may even find an optimal strategy that *appeared* to be the least advantageous one. In both these cases the model takes us beyond conclusions reached by unaided intuition.

---

*Nash, John F., "The Bargaining Problem," *Econometrica* 18 (1950), 155–62 (assuming that money and utility value are identical).

TO PICK AND CHOOSE

AN APPLICATION TO VOTING

Introduction

Every Election Day you and other citizens are faced with a choice. Your joint decision will determine who is to represent you in a variety of political bodies. If your goals and the goals of the other voters differ (as they almost certainly do), how do you make a rational choice? For instance, should you vote for your most preferred candidate even if it is generally conceded that he has no chance of winning, or should you try to make your vote have as much influence on the outcome as possible?

This question may sound familiar, and it should; it is essentially the same question we discussed in Chapter 5, on game theory. Now we will discuss it further, but in a much more limited way, by restricting ourselves to the confines of a (figurative) voting booth.

Aside from this central problem—the one facing the voter—we will also examine another, related problem: the one facing the social planner.

A *society* (which we may take to mean an assortment of individuals grouped together) often must make a collective decision. For instance, a country may need to increase its revenue, and its possible methods for doing this may include adopting an income tax, adopting a sales tax, floating a bond, and so on. Each individual in that society has his or her own ideas about which alternative to adopt: Perhaps a given person would prefer an income tax, would consider a bond or sales tax the next best choice (possibly being indifferent between the two), and would consider a tax on cigarettes less desirable than any of the other options. Such an arrangement of options, in which the most preferred are put first and the least preferred are placed last (whether by a society or by an individual), is called a **preference ordering**.

The job of the social planner is to translate the individual preference orderings into an overall preference ordering for society as the basis of society's decisions. The planner's job is not to determine which option is best but to determine which preference ordering for the society as a whole most closely conforms to the preference orderings of the individuals in the society, and this must be done in the "fairest" possible way.

The same problem can come up in different ways in everyday life. The "society" doesn't have to be political; it might consist of the members of a family, stockholders in a company, or the members of a trade union, as well as U.S. citizens. Sometimes the individuals in a society can measure precisely how much they prefer one alternative to another (a senator may calculate exactly how much each of several federal facilities is worth to his home state), but as a rule all one can say is that Jones likes option A better than option B and that better than option C. (Mary may prefer to have her house painted yellow rather than pink, but she may not be able to say by how much.)

The kinds of decisions that society makes aren't always the same, either. Society may be called on to form a preference ordering based on the preferences of its members (a legislature must set up spending priorities among crime control, medical services, education, and so on), or it may simply have to pick out a single option (a political convention names only a single candidate for president from all of the aspirants).

A common way of bridging the gap between what individuals want and what society chooses is to take a vote in which each person indicates his most preferred alternative and ignores the rest. Often each person has a single vote, but this is not the only possible arrangement. The number of votes a stockholder controls depends on the number of shares he owns, and the representation of a state in the House of Representatives or the Electoral College reflects the size of its population. The rules that determine whether a given measure is accepted or rejected will also differ from one forum to another; as a matter of fact, one of the questions we will examine is which rules are most equitable. For some decisions—conviction by a jury or admission to a private club, for example—a unanimous vote may be required. Sometimes, and under certain conditions, $\frac{2}{3}$ of a body is required for approval; but generally, a majority consensus is sufficient.

Once the mechanism for voting is established, there is still a wide gamut of strategies the voters can adopt. They can make logrolling agreements ("you scratch my back and I'll scratch yours"), in which they vote against their own interests on one issue in return for someone else's vote on an issue they consider more important. They may even follow a strategy that seems absurd on the face of it but in fact may prove very useful: voting against what appears to be their own best interest.

Another concept we will be concerned with is that of "power"; if the voters have unequal numbers of votes, or if they have the same number of votes but some part of them agrees to form a bloc or coalition, how strong is that bloc? Is a bloc of 1 million votes at a stockholders' meeting exactly ten times as strong as a bloc of 100,000 votes? The question is meaningless, of course, without a plausible definition of "power". But how *do* you define power in a sensible and persuasive way?

At this point one of the most perplexing questions may be why there should be any problem at all. The obvious course seems to be to take a vote in which everyone indicates what he or she desires and then to pick out the most popular of the alternatives. If you want to know the power of a bloc of votes, just count them. This seems to be a straightforward, no-nonsense approach to the problem. As a matter of fact, there are a number of things you can say about voting procedures that seem plausible and can be derived using only common sense; we have listed a number of them below. Read these statements now, and reread them after you have studied the rest of the chapter to see how well they hold up under critical examination.

Some "Facts" About Voting Procedures

1. In any election, *always* vote for the candidate you like best. There is never any reason to vote as if you preferred Jones to Smith when you really prefer Smith to Jones.

2. Suppose you are a member of a legislature about to consider a number of different issues. If you can trade votes with someone (vote his way on an issue you consider unimportant so he'll vote your way on another, more pressing issue), do it! The trade must be a good thing, since both parties gain; they wouldn't agree to trade if they didn't. And the more people who make trades, the more everyone gains.

3. Alice, Bob, and Cora are appointed to determine the site of an annual meeting; the cities of Omaha, Boston, and Chicago are being considered. If a majority vote were required to reach a decision, the outcome might well be a deadlock, so Alice is made chairperson with this understanding: If any city gets two of the three votes, that will be the site; but if each city gets one vote, the chairperson's vote will decide the issue. Though we haven't said what we mean by "power", under any reasonable definition the chairperson will be regarded as the most powerful of the three.

4. Suppose a group has to take one of several different actions and the individuals in the group have different opinions

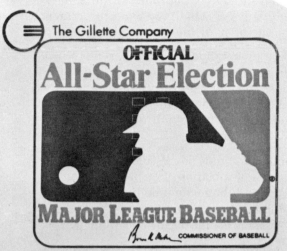

The Gillette Company

OFFICIAL
All-Star Election
MAJOR LEAGUE BASEBALL

COMMISSIONER OF BASEBALL

HOW TO VOTE

1. Vote for **three** outfielders and **one** player from each of the other positions in each league. Punch out the box next to the name of each player you have selected. To vote for a player other than a nominee, punch out the "write-in" box, then print the player's name.

2. Deposit or mail your ballot on or before July 4, 1979. 3. Official ballots will be tabulated by Action Marketing Ltd., an independent judging organization. 4. The decision of the Baseball Commissioner will be final on all matters concerning the balloting.

50th ALL STAR GAME

Deposit ballots in Gillette displays at participating retailers or in All-Star Ballot Boxes at ball parks or place in an envelope and mail to:

**THE COMMISSIONER OF BASEBALL
ALL-STAR ELECTION HEADQUARTERS
P.O. BOX 2000
PARAMUS, N.J. 07652**

Gillette
1979 All-Star Game Official Ballot

Vote For ONE player for each position in each league

Vote For THREE players in each league

NATIONAL LEAGUE		AMERICAN LEAGUE	
1st Base		**1st Base**	
Buckner	T. Perez	Carew	Mayberry
Driessen	Rose	Chambliss	E. Murray
Garvey	Stargell	Cooper	G. Scott
K. Hernandez	Watson	Ron Jackson	Thompson
Montanez	Madlock	Grich	Thornton
2nd Base		**2nd Base**	
Cash	Morgan	Kuiper	Remy
Flynn	Stennett	Molitor	Whitaker
Howe	Trillo	Orta	F. White
Lopes	Tyson	Randolph	Wills
Shortstop		**Shortstop**	
Bowa	O. Smith	Belanger	Patek
Concepcion	Speier	Burleson	Smalley
DeJesus	Taveras	Dent	Trammell
Royster	Templeton	Guerrero	Veryzer
Russell	Hebner	Bando	Yount
3rd Base		**3rd Base**	
Cabell	Horner	B. Bell	Hobson
Cey	Larry Parrish	G. Brett	Lansford
Darr. Evans	Reitz	DeCinces	Money
Garner	Schmidt	Harrah	Nettles
Catcher		**Catcher**	
Bench	Simmons	Alexander	Munson
Boone	Stearns	Dempsey	Lan. Parrish
Carter	Tenace	Downing	Porter
Murphy	Yeager	Fisk	Sundberg
Ott	G. Maddox	Bailor	Wynegar
Outfield		**Outfield**	
D. Baker	Matthews	Baylor	Manning
Burroughs	Mazzilli	Bonds	McRae
Cedeno	McBride	Cowens	A. Oliver
Clark	Monday	Dw. Evans	Otis
Cromartie	Murcer	Ford	Page
Jose Cruz	Parker	Hisle	Piniella
Dawson	Puhl	Reg. Jackson	Rice
Foster	Richards	Rup. Jones	Rivers
Griffey	Bill Robinson	Kemp	L. Roberts
S. Henderson	R. Smith	LeFlore	Singleton
Hendrick	E. Valentine	Lemon	G. Thomas
Kingman	Whitfield	Lezcano	Yastrzemski
Luzinski	Winfield	Lynn	Zisk

PUNCH OUT ONLY IF YOU WRITE IN VOTE BELOW

Pos.	Player	Pos.	Player

PRINTED IN U.S.A.

In baseball, the members of the All-Star teams are chosen by plurality vote of the fans.

about what that action should be. The "fairest" thing to do is to have everybody vote for what they want to do most and then do what a majority of the people voted for.

5. There are a number of candidates for a certain office, among them Smith and Jones. If a majority of the voters prefer Smith to Jones, no fair voting system would ever allow Jones to be chosen. In general, the candidate chosen for an office should be able to outpoll any of the other candidates in a two-way race.

6. In a legislature a number of individual voters form a bloc; that is, they decide which position they will take on an issue by majority vote and then vote unanimously for that position. The combined power of these individuals is always greater after such a bloc is formed than before.

7. Assume that when a new state is added to the United States, the old states keep their former representation but the size of the House of Representatives is enlarged to accomodate the new members. Since the new members acquire a certain fraction of voting "power" in the house, the fraction of "power" of the old members may remain constant or decrease but can never become larger.

8. In order to make an intelligent choice a person's preferences must have a certain amount of structure. If you say you prefer salad to fish, fish to meat, and meat to salad, you are likely to find your preferences ignored altogether! Such preferences are called *intransitive,* and it is generally assumed that rational people have *transitive* preferences. (That is, if you like fish better than salad and salad better than meat, we assume you like fish better than meat.)

   Just as it is hard to make decisions for individuals with intransitive preferences, it is hard to make decisions for societies with intransitive preferences. However, *if society is composed of rational people with transitive preferences, society's preferences must also be transitive.*

9. You have just formed a club and are drafting a constitution. When issues arise in the future, they are to be decided by the members' vote. There are a number of ways of translating a vote into a decision, however; one may insist that there be a unanimous vote, 75% of the vote, a majority of the vote, etc., for a resolution to pass. All of these percentages might be "reasonable" under certain circumstances. One might also adopt "unreasonable" decision rules; for

example, if there was a rule which said a resolution may only be adopted if everyone votes against it, that would generally be considered unreasonable.

(a) Can you think of any general criteria which might be used to distinguish rules which you consider reasonable from those you consider unreasonable?

(b) Under what circumstances would you want a unanimous decision or a high percentage of the vote to reach a decision rather than a majority vote? Under what circumstances in everyday life is a unanimous decision required? Why do you suppose it is required?

(c) Would you ever allow a 40% vote to approve an issue? What implication would this have for the stability of your laws?

Analyzing the process of voting is useful even if you sometimes arrive at paradoxical conclusions. It is intriguing to mix the wisdom gained from the practical politician's experience with academic theory. The basic problems are generally easy to understand and never very far from our everyday experience.

The Voting Mechanism

The social planner who has the job of picking out one order of preferences for society from the varying and often contradictory wishes of its members is attacking a problem that has challenged many political philosophers in the past. In discussing this problem political philosophers have tended to use general terms, and as a result their analysis is often hard to apply to specific cases. Rousseau, for example, stated that "Each man, in giving his vote, states his opinion on that point; and the general will is formed by counting votes . . ." (*The Social Contract*). Jeremy Bentham asked, and then answered, "The interest of the community then is, what?—the sum of the interests of the several members who compose it" (*An Introduction to the Principles of Morals and Legislation*). And Locke asserted, "The end of government is the good of mankind. . ." (*Concerning Civil Government*).

The trouble with all of these general propositions becomes obvious when you put yourself in the place of the decision maker faced with violently different points of view. What precisely is

"WE'RE GOING TO A PICNIC..."

"NO! WE'RE GOING TO A BASEBALL GAME!"

If two members of a family want to go on a picnic and the other two prefer the ball game, how do you go about summing the interests of the several members?

"the general will," and what is "the good of mankind"? If two members of a family want to go on a picnic and the other two prefer the ball game, how do you go about ". . . sum [ming] . . . the interests of the several members . . ."?

In practice, the way you balance different individual interests to arrive at a group decision varies from case to case. Juries often need a unanimous vote to reach a guilty verdict. To amend the U.S. Constitution you do not need a unanimous vote, but you do need considerably more than a bare majority. In both cases the rules of voting reflect the importance of the issue being decided. In Congress the size of the vote required to pass legislation depends on whether the President is willing to sign the bill. And the votes need not be distributed uniformly; large stockholders have more votes than small ones, and large states have a greater representation in the Electoral College.

In any case, voting procedures vary—sometimes by historical accident, sometimes through design—and we ask again which voting procedure best reflects "the general will" and promotes "the good of mankind." As a step in translating these general ideas into specific ones, we will try out a number of different rules and see how well they work.

Principle 1: If society must pick one of several alternatives and the members of society have varying preferences, society should elect the alternative that a majority of the members prefer most.

Let's try to apply this principle to a specific problem. A company must decide whether to hold its annual convention in Acapulco, Honolulu, or San Francisco and appoints an executive committee of 48 to make the choice. There are three distinct viewpoints among the members of the committee. Sixteen of them prefer Honolulu most and prefer Acapulco to San Francisco. We abbreviate this by saying $N_{HAS} = 16$. Similarly, $N_{AHS} = 15$ and $N_{SAH} = 17$, which may be interpreted in an analogous way. Which meeting site should be selected?

If we try to apply it to the present problem, we immediately see at least one thing wrong with principle 1: There may not be an alternative that "a majority of the members prefer most." So we modify principle 1 and obtain

Principle 2: If society must pick one of several alternatives, every voter should select his or her first preference; the alternative that receives the most first preference votes (a *plurality*) should be selected for society.

If we apply principle 2 to this problem, we do get an answer. Since San Francisco is desired as a first choice by more people than any other place, that will be the site. But is it a satisfactory answer? We have already seen that principle 1 (which essentially was "fact" 4) was inapplicable at times. It turns out that principle 2 is in conflict with "fact" 5, which states that if A and B are two of several possible alternatives for society and a majority of the population would prefer A to B, then B should never be selected as society's choice. In the present case, if a vote were taken between San Francisco and *either* of the other cities, San Francisco would lose 31 to 17. The plurality vote procedure is not entirely convincing, and this suggests still another rule.

Principle 3: To pick the best of several alternatives, society should adopt the same procedure that it would if it were running a tennis tournament. Have head-to-head elimination matches (elections) between the alternatives—always two at a time—eliminating the loser and allowing the winner to compete further. The final winner, the alternative that never loses a vote, becomes society's ultimate choice.

How will this work in our convention site problem? In a two-way vote San Francisco would lose to either Acapulco or Honolulu (31–17), and Acapulco would outvote Honolulu (32–16).

So our investigation thus far leads us to this conclusion: If you have everyone vote their first preference and base your decision on this, the preference order of the sites will be San Francisco, Honolulu, and Acapulco. If you match the cities two at a time and eliminate the losers, society's preference ordering will be the same, but *in reverse*.

When principles 1 and 2 were applied to the convention site problem, the results were not completely satisfactory. Principle 1 couldn't be used at all, and principle 2 (the plurality choice) would have chosen a site that would lose to either of the two alternatives in a two-way vote. Our next suggestion, principle 3 (the "tournament method") seemed to lead to a more satisfactory conclusion. Let's analyze this method further by continuing with the problem of picking a convention site but changing the numbers a bit. We will assume that the decision is to be made by the majority vote of a committee of three (Klein, Gomez, and Green) with all votes weighted equally. We will also assume that the orders of preference of the three committee members are as shown in Table 1.

Table 1

Commitee member	Preference order		
	First	Second	Third
Klein	San Francisco	Acapulco	Honolulu
Gomez	Honolulu	San Francisco	Acapulco
Green	Acapulco	Honolulu	San Francisco

If we apply principle 3, we must match two of the sites in a two-way vote—say San Francisco and Honolulu—and we find that Honolulu is more popular; only Klein prefers San Francisco to Honolulu. Honolulu is then matched against Acapulco, and Acapulco wins by a two-to-one vote and is the ultimate choice of the committee.

This seems a satisfactory solution to the problem until you look a little closer. If you study the preferences of the committee, you will find they are completely symmetrical: Each city is a first, second, and third choice of exactly one of the members, so it would seem that the outcome should be a standoff. How, then, did Acapulco run off with first prize?

It turns out that the order in which the potential convention sites enter the "tournament" is crucial. If San Francisco and Acapulco competed first, then San Francisco would be the initial winner, only to be beaten by Honolulu in the second round; and if Acapulco and Honolulu competed first, then San Francisco would be the ultimate winner. In short, the ultimate site will be the city that does *not* compete on the first round. So principle 3 is not very convincing either, since the candidate it suggests may just be an accident of the way the vote was taken.

There's another interesting point that should be noted: When two cities compete, the preferred outcome is clear; it is the city preferred by a simple majority. In this case, when Acapulco and San Francisco compete, the committee prefers San Francisco; and when Honolulu and San Francisco compete, the committee prefers Honolulu. According to "fact" 8, it seems that society's preferences should be transitive, since all the individual committee members' preferences are transitive; and therefore Honolulu should be preferred to Acapulco. But it takes only a moment to confirm that Acapulco is preferred to Honolulu, and so society's preferences are not transitive.

But things can get even more confusing! We haven't yet considered what happens if players vote in ways that seem to go against their own self-interest. To be specific, suppose the convention site is to be chosen by means of principle 3 and Acapulco is first pitted against Honolulu (and the winner against San Francisco). Given the preferences of the committee members, the final outcome is easy to work out: If events run their normal course, Acapulco will win the first contest and lose the second to San Francisco.

This is all very well for Klein, who prefers San Francisco most. But this impending outcome should give Green pause, since San Francisco is his last choice, and there is a counterstrategy he can employ. Observe that Green's vote for Acapulco on the first round is a futile gesture, since Acapulco is destined to be knocked out on the next round by San Francisco anyway. If Green departs from his own preference order on the first round and votes for Honolulu instead of Acapulco, Honolulu will not only carry the first vote but will go on to defeat San Francisco in the second vote as well. So by voting insincerely, or strategically, Green seems to have the power to change the ultimate winner from San Francisco, his last choice, to Honolulu, his second choice (see "fact" 1).

And what of the others? Gomez will not be motivated to change anything, since he is now getting his first choice of Honolulu rather than his second choice of San Francisco. Honolulu will not sit well with Klein, of course, since it is his last choice, but there is nothing he can do about it. He can shift his vote on either election as he pleases, but the outcome will always be the same: Honolulu is destined to win both votes. So once again appearances have proved to be deceptive.

exercises

1 In the 1970 New York senatorial election there were three candidates: Ottinger, Buckley, and Goodell. According to most political commentators, voters who supported either Goodell or Ottinger as their first choice tended to support the other as their second choice, since both of them were substantially less conservative than Buckley. In the actual election Buckley won with a plurality of 39% over Ottinger's 37% and Goodell's

24%. If the Ottinger and Goodell first preference voters supported one another as suspected, what would be the result of a two-way match between Buckley and Goodell? Between Buckley and Ottinger? Do you feel that the outcome of this election accurately reflected "the general will"?

2 In the 1969 New York mayoralty election the shoe was on the other foot: There was a single liberal candidate, Lindsay, and two relatively conservative candidates, Marchi and Biaggi. Assume (for simplicity) that everyone who preferred Biaggi as a candidate had Marchi as a second choice and everyone who most preferred Marchi had Biaggi as a second choice. Show that if the outcome was to be determined by a plurality vote (as it was), Lindsay might win with as little as 34% of the total vote; and yet in this case he would lose in a two-way race to *either* of his rivals by a margin of almost two to one. Would you say this outcome is consistent with "the general will"?

3 Consider again the problem of choosing a convention site, in which the preferences of the three committee members are as indicated on page 210. Assume that the selection of the site is to be made by two two-way elections in the way we described earlier. Show that if each committee member knows the others' preferences and votes in the most sophisticated way, the ultimate winner will be the city that would have lost in the initial election had everyone voted sincerely.

4 It has been suggested that one way of settling a three-way (or larger) race is to have a preliminary vote in which everyone votes honestly and then knock out the candidate with the lowest vote (if no single candidate receives a majority). Then have all but the least popular candidate have a second go, and continue in this way, eliminating the least popular candidate at each turn, until some candidate receives a majority vote. This must inevitably happen because eventually there will be only one candidate left.

Suppose a convention site is to be selected and the percentages of voters who hold various preference orders are as follows:

Percentage of voters	Preferences		
	First	Second	Third
34	Honolulu	Acapulco	San Francisco
34	San Francisco	Acapulco	Honolulu
32	Acapulco	Honolulu	San Francisco

(a) If everyone votes honestly for their first choice on the first vote, who will be eliminated?

(b) If a vote were taken between Acapulco and Honolulu only, which city would win? By what margin?

(c) If a vote were taken between Acapulco and San Francisco, which city would win? By what margin?

(d) Do you think this method of eliminating the "weakest" candidate at each stage is entirely persuasive as a method of reaching group decisions? (See "fact" 5.)

5 The city council, which consists of five members (Alice, Barbara, Charles, David, and Evelyn), must decide on which one of three alternative transportation systems to use: trolleys, buses, or trains. Their preferences are as follows:

Committee member	Preferences		
	First	Second	Third
Alice	Buses	Trains	Trolleys
Barbara	Buses	Trains	Trolleys
Charles	Trolleys	Trains	Buses
David	Trains	Trolleys	Buses
Evelyn	Trolleys	Buses	Trains

(a) Suppose the method of transportation to be used is to be determined by first comparing two and then matching the more popular against the third. What will the outcome be if all members vote their true preferences and the first two compared are
   (i) buses and trains?
   (ii) buses and trolleys?
   (iii) trains and trolleys?

(b) In which of the three cases can someone gain by *not* voting his or her true preferences?

6 A committee of 54 must award first, second, and third prizes to three finalists, Lamont, Karp, and Brown. Their preferences are as follows:

Preference ordering of candidates (most preferred listed first)	Number of committee members holding preference
Lamont, Karp, Brown	8
Lamont, Brown, Karp	11
Karp, Lamont, Brown	3
Karp, Brown, Lamont	15
Brown, Lamont, Karp	1
Brown, Karp, Lamont	16
	54

(a) If the prizes are to be awarded by taking a single (honest) vote—the largest vote getter receiving first prize, the next second prize, and so on—predict the outcome.

(b) Suppose only one prize is to be awarded, by comparing the voters two at a time. If the voting is honest, show that Brown [who would get third prize in part (a)] would defeat both Lamont and Karp, while Lamont [who would win first prize in part (a)] would lose to both Brown and Karp.

(c) If the voters act according to their preferences in part (b), without voting strategically, Brown will win any knockout tournament no matter how the pairing is done. Show that he will win even if the committee votes strategically. Calculate who should shift their votes for each of the possible pairings.

7   Repeat Exercise 6, assuming that only a single prize will be awarded and that the preferences of the committee members (now numbering 52) are as follows:

Preference ordering of candidates	Number of committee members holding preference
Lamont, Karp, Brown	15
Lamont, Brown, Karp	4
Karp, Lamont, Brown	2
Karp, Brown, Lamont	13
Brown, Lamont, Karp	12
Brown, Karp, Lamont	6
	52

(a) Show that in a single match with two candidates, Brown will beat Lamont, Lamont will beat Karp, and Karp will beat Brown. Show that in a two-step tournament in which the voting is honest, the final winner will be the candidate who does not participate in the first vote.

(b) Show that if the committee members vote strategically and the pair in the initial contest is

  (i) Lamont and Brown, then Lamont will win in both votes. Two groups of voters will not vote according to their true preferences; who are they?

  (ii) Lamont and Karp, then Karp will win twice. By how much? Which two groups will distort their own preferences?

  (iii) Brown and Karp, then Brown will win both votes. By how much? Which groups will vote strategically?

Let's return to the problem of choosing a convention site, where Klein, Gomez, and Green have preference orders as shown on page 210. Suppose that the people who had formed the committee anticipated that the vote might end in a deadlock and so introduced the following wrinkle: If two of the three committee members cast a vote for any one city, that city will be named as the convention site. If there is no clear majority for any one city, Klein (the chairperson of the committee) will cast the deciding vote.

Common sense and "fact" 3 seem to dictate that Klein's power is distinctly greater than that of the two other members, and since the preferences of the three committee members are symmetric, it seems that Klein's preferences should prevail. And yet there is an argument that arrives at precisely the opposite conclusion. Examine each step in the following argument and see whether and where you disagree:

(a)   If Gomez and Green both vote for the same city, that city will be the convention site whatever Klein does; if Gomez and Green vote for different cities, then Klein's choice will prevail. Therefore *Klein should vote for his first choice: San Francisco.*

(b)   If Klein votes for San Francisco, the final site chosen will depend on the choices of Green and Gomez in the following way:

Gomez's Choice

		San Francisco	Acapulco	Honolulu
Green's Choice	San Francisco	San Francisco	San Francisco	San Francisco
	Honolulu	San Francisco	San Francisco	Honolulu
	Acapulco	San Francisco	Acapulco	San Francisco

(Observe that the outcome will be San Francisco unless Gomez and Green agree on some other city.)

(c)   *Gomez does best to choose Honolulu whatever Green does.* Check this statement for all three of Green's strategies. Notice that you can't make a similar statement if you reverse the roles of Gomez and Green.

(d)   If Klein picks San Francisco and Gomez picks Honolulu, *Green does best to pick Honolulu too.*

(e)   *The outcome will be Honolulu!* The "powerful" chairperson will be stuck with his worst alternative, while the other two committee members get either their first or second alternatives. So much for the power of the chair.

Some Applications

In his book *Paradoxes in Politics,** Steven J. Brams mentions a number of political contests in which there were opportunities for strategic voting. Sometimes this opportunity was exploited; at other times it was not. Three such instances occurred in the presidential elections of 1912, 1948, and 1968; the other two instances occurred in the House of Representatives and the Senate.

EXAMPLE 1

The 1948 and 1968 Presidential Elections

In both the 1948 and 1968 presidential elections there were essentially three candidates running, and in both cases the third-party candidate's name was Wallace. Henry Wallace was the candidate of the Progressive Party in 1948, and his position was to the left of his running mates; George Wallace, the third-party candidate in 1968, took a position to the right of his running mates. Neither candidate was given much of a chance to win the election. It is generally believed that supporters of both Wallaces were induced to vote strategically: Henry Wallace supporters defected to Truman, the Democratic candidate, because he was their second choice and preferable to Republican candidate Dewey, and the supporters of George Wallace also reverted to their second choice, presumably because they felt they couldn't afford the luxury of a wasted first-preference vote.

EXAMPLE 2

The Presidential Election of 1912

In the presidential election of 1912 Wilson, Taft, and Roosevelt were the candidates of the Democratic, Republican, and Progres-

---

* Steven J. Brams, *Paradoxes in Politics* (New York: Free Press, 1976).

sive parties. The probable preference orders of the members of each party and the percentage of the vote they eventually received are shown in Table 2.

Table 2

Name of party and percentage of vote received	Preference ordering		
	First	Second	Third
Democrats: 42%	Wilson	Taft	Roosevelt
Republicans: 24%	Taft	Roosevelt	Wilson
Progressives: 27%	Roosevelt	Taft	Wilson

Apparently the opportunity for strategic voting was not exploited here. Wilson, who would have lost a two-way election against either of the other two candidates, won.

EXAMPLE 3

A School Construction Bill

In 1956 the House of Representatives was considering a bill to finance school construction. An amendment was offered by Representative Powell barring aid to segregated schools. If we denote the original bill by O, the amended bill by A, and the failure to pass any bill by N, we can express the three predominant opinions in the House, which were held by Northern Democrats, Southern Democrats, and Republicans, as shown in Table 3. (Any two of these three predominant groups contained more than a majority of the voters.)

Table 3

Primary political group	Preferences		
	First	Second	Third
Southern Democrats	O	N	A
Northern Democrats	A	O	N
Republicans	N	A	O

It is the procedure in the House to use a "tournament" to determine which alternative will be chosen. A vote is first taken on whether to consider the bill with or without the amendment; then another vote is taken to determine whether the bill, amended or not, is passed.

Once again, the opportunity for strategic voting was not used; the bill was amended and then, predictably, defeated. Apparently most voters acted naively.

EXAMPLE 4

Another Construction Bill

In 1955 a situation similar to the one described in Example 3 arose in the Senate but ended differently. A construction bill with a "fair play" provision for workers was being considered, and an amendment that would have deleted the provision was offered. With the same abbreviations used in Example 3, the Northern Democrats' preference order was O, A, N, that of the Southern Democrats was A, N, O, and the Republicans' was N, O, A. (Any two of these political groups contained enough voters to form a majority.) The voting procedure was the same as in Example 3. The amendment was considered first and then the bill itself.

In this case the Northern Democrats, regarding an amended bill as preferable to no bill at all, voted against their true preferences and backed the amendment. The bill finally passed.

1   For Examples 2 through 4 calculate what the outcome would be if everybody voted for their most preferred alternative.

2   If everyone voted honestly in Examples 2 through 4, which group would be stuck with its lowest preference? Is there a way that group can vote strategically to avoid this outcome?

3   Is there some other group that is both strong enough and motivated to counteract the strategy referred to in Exercise 2? How would they do it?

4   If two parties were to form a coalition in Example 2, which parties would be likely to do so? What counteroffer might the third party make and to whom would it be made?

Logrolling

There is still another variation on the voting theme. So far we have been talking about making decisions when there is only a single issue to be resolved. (There may be several alternative ways of resolving the issue, of course.) Now we will assume that there are a number of different, possibly unrelated, issues to be settled at roughly the same time. This introduces a new strategic element as we shall see.

Imagine that a county composed of three townships is completing its fiscal year with a surplus of $9 million. All three of the townships have been plagued by floods, and constructing a dam in any one (or all) of the townships would yield benefits greatly exceeding the costs. The cost of a single dam is $3 million, so the county could afford to build three dams. The financial benefit of

"WELL, THIS IS WHAT WE GET FOR LETTING THOSE DAM BILLS FAIL!"

the dam to the township in which it is constructed is $20 million. If proposals to build a dam in each township are voted on separately, what is the likely outcome?

If each township takes the narrowest view—voting for proposals that are favorable to it and against all the others—the outcome is easy to predict: All three proposals will fail. And this is the case no matter what (reasonable) voting system is used, because there would be a clear 2-to-1 majority in opposition to each proposal. It seems a pity to lose a potential $51 million in benefits (there is a $17 million difference between the cost of each of the three dams and the value of the dam), and it's not too difficult to find a solution. The solution is vote trading (or *logrolling*), in which one party agrees to vote against its own inclinations on one issue in return for another party's promise of support on an issue considered more important. It shouldn't be too surprising that logrolling—a way of getting all the dams built to everyone's mutual advantage—is frequently observed in legislative bodies.

But logrolling introduces new problems. For example, take the following case, which has been adapted from one mentioned in *Game Theory and Politics*, by Steven J. Brams.* In a legislature there are three groups of voters (those from the Southwest, those from the Southeast, and those from the North), and each group votes as a single bloc. Suppose there are six bills (numbered 1 through 6) being considered, and the position of each bloc (yes or no) on each bill as well as the relative importance of each bill is indicated in Table 4.

Table 4

Region	Importance and position of each bill					
Southeast	Y1	Y2	N3	N4	Y5	Y6
Southwest	Y3	Y6	N5	N2	Y1	Y4
North	Y5	Y4	N1	N6	Y3	Y2

The letter Y means the bloc would normally vote yes on the bill, and N means it would vote no. The numbers indicate which of bills 1 through 6 we are referring to, and the order from left to right indicates the importance of the bill to a bloc. For example, the Southwest bloc regards bill 3 as of the highest importance and wants it to pass. Bill 6 is next in importance for the Southwest

_____

* Adapted with permission of Macmillan Publishing Co., Inc., from *Game Theory and Politics* by Steven J. Brams. Copyright © 1976 by The Free Press, a Division of Macmillan Publishing Co., Inc.

bloc, which wants it to pass as well. Bill 5 is third in importance, and they hope it will fail. And so on.

A quick inventory shows that every bill has two apparent yes votes and only one no vote, so in the absence of any vote trading every bill will pass. If this happens, each bloc will have its way on both the two bills it considers most important and the two bills it considers least important and will lose out on the two bills it considers of middling importance. But let's see what happens when the parties try to get together to make things "even better."

Clearly, the Southeast and Southwest have the potential for a deal. If the Southeast changes its vote from yes to no on bill 5 in return for the Southwest changing its vote from yes to no on bill 4, the effect will be a gain for both. The southeast will win on the bill it considers fourth in importance and lose on the bill it considers fifth in importance, and the southwest will win on the bill it considers of third importance at a cost of losing out on its least important bill—an apparent profit for both parties.

And there are even more trades in the air; the Southeast and the North can agree to vote no on bills 6 and 3, respectively, once again to their mutual advantage. Finally, the Southwest and the North can make a similar agreement concerning bills 1 and 2.

Vote trading only occurs when each of the parties involved does better after the trade than before; otherwise, a rational voter would refuse to trade. It is reasonable to suppose that after the air clears and the voters take stock, they will all be better off; they each made two trades, and presumably they each gained something in the process. But it is easy to see that the effect of *all* the trades is the defeat of all six bills when they all seemed destined to pass. It is also easy to see that rather than improving their positions, the voters jointly have messed things up; instead of losing on the bills of middling importance, they now will lose on their first two and last two bills (see "fact" 2).

1   State precisely why each bloc in the logrolling problem just discussed is worse off after all the trades than it was before.

2   (a)   How do you account for the fact that the voters are worse off after each completing two "advantageous" trades? Is this a case of "losing

money on each item but making it up in volume"? (See the discussion on the prisoner's dilemma in Section 6 of Chapter 5. Also see "fact" 2.)

(b) How do you explain the fact that in the logrolling problem we just described everyone loses when all the trades are completed, while in the dam construction example described at the beginning of the section everyone came out ahead?

3 Suppose a single player, clever enough to see the eventual outcome of all the trades, unilaterally refused to trade with anyone. Would this avoid the trap?

4 There are three towns in the county of Oz: A, B, and C.* At the monthly meeting six bills are being considered, and each bill has financial consequences for each town. For example, if a road is constructed between A and B, C would lose (since it helps pay for the road but receives no benefit) and A and B would presumably gain. Let us call the six bills X, Y, Z, U, V, and W and assume that the bills are considered separately and that the decision is by majority rule. The effect of each of the bills on each of the towns is shown in the following table:

Towns	Bills					
	X	Y	Z	U	V	W
A	3	3	2	−4	−4	2
B	2	−4	−4	2	3	3
C	−4	2	3	3	2	−4

If a bill is not passed, each town neither gains nor loses anything. Moreover, each town votes in its own self-interest.

(a) Assume that every town votes honestly. Show that every bill will pass with a net gain for each town of 2.

(b) Show that each of the following agreements will lead to benefits to the parties that make them:

(i) B promises C to vote against X if C votes against Y; what do B and C each gain?

(ii) A promises C to vote against W if C votes against V; who gains what?

(iii) A promises B to vote against Z if B votes against U; who gains what?

---

*Based on Eric M. Uslaner, "Vote Trading in Legislative Bodies: Opportunities, Pitfalls, and Paradoxes," Modules in Applied Mathematics (Cornell University, 1976).

(c) How many bills will pass if all these agreements are carried out? How far ahead are each of the players? How do you account for the fact that each player is worse off after the three "advantageous" agreements?

(d) What if, after *B* and *C* make their deal, *A* refuses to agree to anything? What will *A*'s profit be? What about *B*'s and *C*'s profits? (See "fact" 2.)

1 Here is another example of "picking and choosing"—not of political candidates this time but of football players.* Several football teams (*A, B, C*, . . .) alternately make selections from a pool of players. Team *A* starts by picking a player, *B* does the same, and they continue until each team has selected a player. Then *A* starts the next round by selecting a second player, and the teams continue in this way until the number of players specified in advance has been chosen. There are various preference orderings over the players; in fact, there is one for each team, since one team may need an end while another needs a quarterback. We assume that the requirements of each team are common knowledge, so that each team knows the preference orders of the others as well as its own.

The most straightforward strategy for a team, which we will call the *honest* strategy, is always to pick the most desirable player available. For example, suppose there are only two teams, *A* and *B*, and four players, I, II, III, and IV, and the teams' preferences (in descending order from top to bottom) are as follows:

A	*B*
I	II
II	III
III	IV
IV	I

If each player adopts an honest strategy, the outcome will be

* This problem was suggested in a paper by Steven J. Brams and Phillip D. Straffin, "Prisoners' Dilemma and Professional Sports Draft," *American Mathematical Monthly* 86:2 (February 1979), pp. 80–88.

Team	Choice on each round	
	1	2
A	I	III
B	II	IV

So in this case each team gets its first and third choices.

(a) Explore what happens if team *A* picks II, its second choice, instead of I on the first round. Specifically, show that if team *B* acts in its own self-interest it will pick III on the first round, team *A* will pick I on the second round, and IV will go to team *B*. Thus team *A* will get its first and second choices instead of the first and third choices it obtained when it chose naively.

(b) It might seem that there is no intrinsic advantage or disadvantage in having one preference order as opposed to another. But it may happen that because of their preference orderings one team may have an advantage over the other. Suppose there are two teams, *A* and *B*, with the following preferences among the eight players, I through VIII, who are to be chosen:

A	B
I	IV
II	III
III	V
IV	VI
V	VII
VI	VIII
VII	I
VIII	II

Team *B* is to choose first. Without proving anything formally, convince yourself by exploring the various possibilities that if *A* and *B* pick their players strategically, the outcome will be

Team	Round			
	1	2	3	4
B	IV	V	VII	VIII
A	III	VI	I	II

Since team *B* chooses first, it should presumably do better than *A*. But *B* gets its first, third, fifth, and sixth choices, while *A* gets its first three plus its sixth choice: a clear advantage. If *A* were to start, it would get its first three and its fifth choices, while *B* would get its first, fourth, fifth, and sixth choices: an even clearer advantage for *A*.

2   This problem is for the ambitious reader—it's fun to work out but above and beyond the line of duty. It was constructed and analyzed by Phillip Straffin and Steven Brams.*

The preference order of each of the three teams for each of the six available players is as follows:

A	B	C
I	V	III
II	VI	VI
III	II	V
IV	I	IV
V	IV	I
VI	III	II

If all teams make their strategic choices, the outcome will be one of the paths along the tree in Figure 1. At the leading vertex, *A* III means *A* picks III at its first turn, then *B* picks II or V (they are equivalent), and so on. Notice that whatever route is taken, the result is the same: *A* will get I and III, *B* will get II and V, and *C* will get IV and VI.

(a)   By experimenting with the possibilities, convince yourself that these are the proper strategic choices.

(b)   Consider the following three-way trade: *B* gives II to *A*, *A* gives III to *C*, and *C* gives VI to *B*. Notice that all three teams improve their positions despite all the strategic maneuvering that preceded.

(c)   Can you see that it is impossible to have a trade in which every team clearly betters its position after a selection process in which each team picks naively?

---

*Steven J. Brams and Phillip D. Straffin, "Prisoners' Dilemma and Professional Sports Draft," *American Mathematical Monthly* 86:2 (February 1979), pp. 80–88.

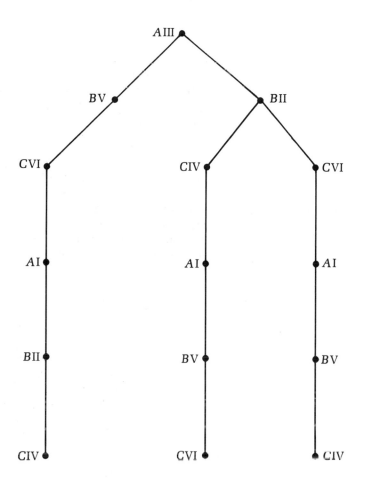

Figure 1

After the census of 1880 Congress sought to reapportion the membership of the House of Representatives in accordance with a formula which was used at that time.* The size of the House, N members, was fixed in advance. The population of the country was divided by N to determine

---

*For a general description see Steven J. Brams, *Paradoxes in Politics* (New York: Free Press, 1976), pp. 137–66.

the ideal number of people in each congressional district, and the representation of a state in the House was determined by the number of districts it contained.

It was not to be expected that a state's population would be an exact multiple of the number of people in a district, nor could a fraction of a representative be sent to congress. So some provision had to be made for the bits and pieces left over. This was the provision: If a state was entitled to a whole number of representatives plus an additional fraction, it would certainly get the whole number. Subsequently, the states with the highest fractions would receive an additional representative so that the total number of representatives would be N, as planned.

To understand this procedure better let's look at a very simple example. Suppose there are five states with populations as shown in the following table and the size of the House is to be 125. Since the total population is 20,000 and there are to be 125 representatives, ideally there should be 160 people in each district.

State	Population	Ideal number of districts	Guaranteed representation
A	3000	$\frac{3000}{160} = 18.75$	18
B	3500	$\frac{3500}{160} = 21.875$	21
C	4000	$\frac{4000}{160} = 25$	25
D	4500	$\frac{4500}{160} = 28.125$	28
E	5000	$\frac{5000}{160} = 31.25$	31
	Total 20,000		Total 123

The third column indicates the ideal number of districts in each state, and the fourth column shows the guaranteed number of representatives.

Since the number of guaranteed representatives is only 123 and there are 125 to be chosen in all, there are 2 representatives to be allocated to two states. The states chosen would be A and B, since they have the largest fractional remainders (.75 and .875). The final representation, then, would be as follows:

State	Representation
A	19
B	22
C	25
D	28
E	31

Now let's consider a hypothetical problem suggested by Brams:

State	Population	I_{25}	N_{25}	I_{26}	N_{26}	I_{27}	N_{27}
A	9061	8.713	9	9.061	9	9.410	9
B	7179	6.903	7	7.179	7	7.455	8
C	5259	5.057	5	5.259	5	5.461	6
D	3319	3.191	3	3.319	4	3.447	3
E	1182	1.137	1	1.182	1	1.227	1
	26,000	25.000	25	26.000	26	27.000	27

In this case it is assumed that there are five states (A, B, C, D, and E) with fixed populations; the table indicates the number of representatives to which they would be entitled if the House size were 25, 26, and 27. The symbol I_{25} denotes the ideal number of representatives a state would have if there were 25 members in the House of Representatives and states were allowed to have fractional representatives in the House. N_{25} is the actual number of representatives that would be allocated to the state according to the formula we described earlier.

(a) For each state and for each of the three sizes of the house, confirm the values of I and use the formula to calculate N.

(b) Notice that *state D has less representation in a House of size 27 than it does if the House size is 26.* This happens without any change in population in any state. What would your reaction be if you were the representative from the fourth congressional district in D and were told, "We're increasing the size of the house from 26 to 27, so your state must lose a representative—and it's you!" (Two states, B and C, gain a representative to balance this, of course.)

The point of this is that an apparently reasonable and simple scheme for allocating votes can lead to absurdities—and in actual fact did. When Colorado lost a seat, the outcome was called a "freak," "atrocity," and a "mathematical impossibility" (see Brams's description). These paradoxes are of more than academic interest (see "fact" 7).

Arrow's Impossibility Theorem   (*Optional*)

The first section of this chapter described a number of ways of answering the question, If you're given the preferences of members of a group concerning certain alternatives, how can you convert these preferences to a single preference order for the group as a whole? We tried a number of different approaches, but they all seemed to break down for one reason or another; the preferences of society didn't seem to reflect the attitudes of its members.

The Nobel Prize winning economist Kenneth Arrow attacked the problem a little differently. He realized different groups might have different decision-making procedures to reflect the greater power or wisdom of some of its members, but he felt that there were certain conditions that *any* decision-making procedure should satisfy, conditions that would be acceptable to every decision-making body whatever its special circumstances were.

Arrow supposed that there were a certain number of alternatives (but in any case at least three) and each member of society (there are at least two) had certain views concerning these alternatives. If *A* and *B* were two alternatives, for example, a member would have to prefer one to the other or be indifferent between them. The preferences of the individual members were also assumed to be consistent in that a member who preferred alternative *A to B* and *B* to *C* would also prefer *A* to *C*.

Starting with these individual preferences, which were assumed to be consistent, Arrow inquired how one could derive an overall preference order for society subject to the following "reasonable conditions":

1. *The decision-making procedure must yield a unique preference order.*
   Whatever the preferences of society's members, the procedure should come up with one and only one preference order for society.
2. *Society should be responsive to its members.*
   The more the individuals in a society like an alternative, the more the society should like it too. Suppose a decision-making procedure yields a preference order for society on the basis of its members' preferences in which alternative X is preferred to Y. If the individual preference orders were changed so that some liked X even better but Y just the same, then in the new preference order society should still prefer X to Y.
3. *Society's choice between two alternatives is based on its members' choices between those two alternatives (and not any others).*
   Suppose society prefers X to Y and people change their minds about other alternatives but not about X and Y. Then X should still be preferred to Y. Society's decision about

whether X is better than Y shouldn't depend on its decision about whether U is better than V.

4. *The decision-making procedure should not prejudge.*
   For any two alternatives X and Y, there must be some possible individual preferences that would allow society to prefer X to Y. Otherwise, Y is automatically preferred to X and the group preferences are unresponsive to those of its members.

5. *There is no prejudgment by an individual.*
   Arrow assumes there is no dictator, that is, society's choices are not identical to the choices of any single individual. If this condition didn't have to be satisfied, it would be easy enough to find a voting mechanism, but Arrow wouldn't consider it representative of the individuals in the whole group.

These five conditions seem to be very reasonable, and it would not be too surprising if there were many decision-making procedures that qualified. But Arrow proved that *no* decision-making procedure can possibly satisfy these conditions. In short, if we adopt Arrow's criteria, our search for a satisfactory voting mechanism is fated to be a wild goose chase.

1 In each of the following situations, which conditions of Arrow are not satisfied?

(a) A town decides whether or not to impose a tax by majority vote.

(b) Three inspectors visit three restaurants, A, B, and C, and each rates the restaurants in order of quality. These ratings are combined into a single overall rating and published in a gourmet journal.

   (i) In the overall rating first place is awarded to the restaurant with the most first choices, second place to the restaurant with the next most first choices, and third place to the remaining restaurant. (Ties are awarded if appropriate.)

   (ii) The overall rating is identical to the rating of the most experienced inspector.

   (iii) Last year's ratings are used whatever the inspectors say.

   (iv) The overall rating is made in a way similar to that in (i) except that the least number of third choices (rather than the most first choices) determines the ordering.

2 You are running in a two-way race for mayor in which there are two salient issues: the passing of a bond issue for a new road and the approval of the school budget.* We abbreviate a position for the bond issue by B, a position against it by $\bar{B}$, and a position for or against the school budget by S or $\bar{S}$, respectively. So, for instance, $B\bar{S}$ is a position for the bond issue and against the school budget. There are three main opinions in town about these issues, and each controls $\frac{1}{3}$ of the vote. The preference orderings of these groups (with the option most preferred listed first) are $(BS, B\bar{S}, \bar{B}S, \bar{B}\bar{S})$, $(B\bar{S}, \bar{B}\bar{S}, BS, \bar{B}S)$, and $(\bar{B}S, \bar{B}\bar{S}, BS, B\bar{S})$.

Assume you must take a position first. Show that whatever position you take on each issue, your rival can take a position that is more popular. (*Hint:* If you take BS, your rival will take $\bar{B}S$; if you take $B\bar{S}$, your rival will take $B\bar{S}$ or $\bar{B}\bar{S}$; and so on.)

Power

In the first part of this chapter we viewed the voting process as a social planner might; our main concern was designing a voting mechanism that would represent the voters fairly, in a way that would reflect their importance, or according to some other general principle. Now let us take another look at the voting procedure but with a different question in mind. Starting with the assumption that the voting procedure has already been established—for better or for worse—let us try to calculate the power of the voters. The term "power," a word used often in everyday language, is roughly synonymous with control and domination. In this section we will try to give a precise, quantitative meaning to this intuitive notion.

The following are four situations in which a clear definition of "power" would be useful.

1. About 200 years ago a number of sovereign states were engaged in the process of forming a union. The larger states were determined to obtain the "power" commensurate with their size; the smaller states were afraid of being overwhelmed by the larger ones. The conflict was resolved, as everyone knows, by forming a Congress composed of two houses. In one house, all states had the same representation; in the other, representation was (roughly) proportional to population.

---

* Adapted with permission of Macmillan Publishing Co., Inc., from *Paradoxes in Politics* by Steven J. Brams. Copyright © 1975 by The Free Press, a Division of Macmillan Publishing Co., Inc.

Suppose you were a representative to the Constitutional Convention and, after negotiating the agreement, were called on to defend it to the people back home. Specifically, how would you compare the "power" of a voter in your state in Congress with the "power" of a voter in some other state?

2. Right now, the direct election of the president of the United States is in the hands of the Electoral College. Traditionally, the members of the College from a state vote as a bloc in accordance with the predominant view expressed by the voters in their state. The number of representatives in the College from a state is the sum of the numbers of senators and House members to which they are entitled.

   (a) Would you say that the small states are overrepresented in the Electoral College? (They do have a disproportionately high representation in the Senate.)

   (b) Does the fact that state voters act as a bloc make a difference in the "power" of the voters in the state? Would it make a difference if each state's representatives voted in the same proportions as the voters in the state? How?

3. You are a financier anxious to gain control of a company. You don't demand absolute control, however; it is sufficient that you have a large enough number of shares to practically ensure the passage of those measures you desire, even if you own less than 51% of the shares. (As you buy shares in the market, the cost goes up steadily, since those willing to sell low are the first to sell. So it behooves you to get control with as few shares as possible.) How do you measure the increase in "power" as the number of shares you control increases? Are the two proportional?

4. You are one of a committee of seven and are secretly approached by two other committee members and presented with the following proposition: Before any measure comes to the floor, the three of you will caucus. After the majority position of your three-person bloc is determined, all three of you will take that position, whatever your true preferences, when the issue comes to the floor. What effect does this agreement have on the "power" of the bloc members?

One thing these examples have in common is that they all deal with "power." The word "power" is used in a variety of situations and takes on a variety of meanings in everyday life. You hear talk about one country's power to defeat another, the power of positive thinking, "power to the people," and "scour power." Intuitively, power is a measure of one's ability to determine what happens by one's own actions; you win a war by sending troops into battle, you change your life by adopting a certain outlook, and you get your utensils clean by using the right soap.

We will try to define power in a way that preserves this intuitive notion but in a narrower context. The power we will be talking about will be the power of a voter or a bloc of voters in some decision-making body. The rules for making decisions may vary in a number of ways. Decisions may be made by majority rule or by unanimous agreement or anything in between. Each voter may have one vote, or different voters may have different numbers of votes. And the decision-making procedure may involve various complications; it may be required (as in fact it is in the U.S. federal government) that for an issue to pass, either (1) one partic-ular member (the President) agree as well as a majority of each of the two houses, or (2) $\frac{2}{3}$ of each house agree. We would like a definition of power that is general enough to allow us to calculate the power of a voter or group of voters, whatever rules are in effect.

On the other hand, there are certain basic assumptions we will make about the decision-making process. These assumptions seem reasonable and are almost invariably true:

1. If the entire voting body opposes an issue, the issue will not pass.
2. If the entire voting body supports an issue, it will pass.
3. Suppose that according to the rules the support of an issue by a group of voters is sufficient to guarantee its passage. Then a larger group of voters that includes the original group will guarantee passage as well; in short, a bill can't be defeated by gaining more votes.

Having set the stage by describing the kinds of rules that may govern our voting body, let's go back to our original question: How much power does a voter or a bloc of voters have? We'll start by analyzing a few examples in which decisions are made by major-ity vote.

1. When the Flugel family goes on a trip, all decisions are made "democratically," by majority vote. The number of votes a member of the family can cast is equal to his or her weight in pounds. Mr. Flugel weighs 220 pounds, his wife weighs 120 pounds, and little Flora and Freddy Flugel

weigh 55 and 95 pounds, respectively. How much power does each family member have to influence family decisions?

2. Suppose Mr. Flugel's weight was only 160 pounds, his wife's weight 150 pounds, and the weights of their son and daughter were 140 and 100 pounds, respectively. How would these new numbers change your answers to question 1?

3. In a state Senate there are four disciplined blocs of 10, 40, 50, and 60 voters. How much power does each bloc have? What is the power of a member of a bloc; specifically, would you rather be a member of a bloc with 40 members or one with 50 members if both blocs have the same power?

4. The City Council has 12 members and a chairperson. All City Council members have one vote; the chairperson votes only if there is a tie vote by the other Council members. What are the powers of the members and the chairperson if there are no absentees and no abstentions?

5. Answer question 4 if there are only 11 members in the City Council.

It seems intuitively clear that the power of a player or bloc of players is somehow related to the vote they control. At first you might even be tempted to define the power of a player as the fraction of the total vote that he controls, but a glance at case 1 and a little thought should convince you that this can't be right. Although Mrs. Flugel has more votes than her son and daughter combined, she has no more power than either of them; in fact, all three of them are absolutely powerless. What Mr. Flugel decides, the family decides. If any family member other than Mr. Flugel changes his or her mind, it *never* makes any difference; whenever he changes his mind, it *always* makes a difference. The person with all the power is generally called a *dictator;* the other family members, who have no power at all, are called *dummies.*

A family member's power is also disproportionate to the number of votes he or she controls in case 2. Mr. Flugel has more votes than his wife, and she has more votes than their son; and yet they all have the same power. A straightforward calculation makes it clear that the approval of two family members—*any* two family members except the daughter—is required for any decision; Mr. Flugel's greater voting power is only an illusion. Although Flora has more than $\frac{2}{3}$ of her brother's vote, she might just as well have no vote at all; her vote will never affect the family's decision, so she is only a dummy.

In case 3 the three largest blocs share power equally and the smallest bloc is a dummy for the reasons we mentioned in case 2. Where two blocs have the same power, it is advantageous to belong to the smaller. Generally speaking, the smaller the bloc, the greater the influence an individual will have on determining the position the bloc will take. A citizen of a small state has an edge over a citizen from a large one if both states have the same power in the electoral college, because a single vote in a small state has a greater effect on the outcome of an election than it does in a large one.

In case 4 it would seem that the chairperson has much less power than other members, since he votes much less frequently. But when the Council takes a 7-to-5 or even more lopsided stand on an issue, the vote of any single member doesn't matter. A single vote is critical only when the Council is divided evenly, and that is precisely when the chairperson gets to vote. This thought may be expressed in another way: Of the 13 members of the Council plus chairperson, you need 7 votes to pass an issue; for this purpose the members' votes and the chairperson's vote are equivalent.

In case 5 the situation is entirely different. If all members are present and voting, the chairperson's vote is worthless.

We said before that the concept of power is very broad and that we would investigate only one aspect of it, power in a voting body. But we must limit ourselves even more. There are undoubtedly a number of factors at work in a real decision-making body that we will ignore. The influence of a voting member may be much greater than one would deduce from the number of votes he directly controls. Personality, tradition, and ideology all play important roles, but we will not try to deal with them here; we are looking for a reasonable definition of power based solely on the ability to pass or block the passage of an issue.

The key to measuring the power of a voter or bloc of voters is to observe the winning coalitions—those coalitions of voters which are sufficient to assure the passage of a measure. A person has power to the extent that his positive vote allows a measure to pass when it would fail if he voted negatively or his negative vote dooms a measure that would pass if he supported it. If he can *never* change the decision of the voting body by changing his vote, he is a dummy and without real power. If he *always* changes the decision of the voting body when he changes his vote, he is a dictator and has all the power.

There is a problem, however. Suppose there is a committee of 7 members that is governed by majority rule. If an issue passes by a vote of 5 to 2, no single voter is crucial; at least 2 voters must change their minds before a committee decision will be reversed. And if a measure passes by a vote of 4 to 3, there are 4 voters whose votes are critical, but which of them is to get credit for being the crucial vote?

One way of answering this question (and it is not the only way) is to assume that there is a roll call vote of all the members. Assume that as the roll is called, each voter votes yes. From our earlier assumption we know that at some point in the roll call someone's vote will be crucial; that is, after he votes yes, the measure will have enough votes for passage for the first time. The critical voter is given credit for passage on that particular roll call. We then calculate all possible orderings for roll calls and determine the critical voter for each of them. The power of a voter (or bloc) is the fraction of all roll calls on which he is crucial (sometimes called *pivotal*).

Since the power of a voter is the fraction of all roll calls in which he is pivotal, it follows that a player's power must be between 0 and 1 and the sum of the powers of all of the players must be 1. This definition of power was originally offered by the mathematician Lloyd Shapley, and the fraction of all roll call votes for which a voter is pivotal is called the *Shapley value*.

EXAMPLE 1

In a legislature with 150 members there are 4 blocs, which vote as a single unit. We will call them *A, B, C,* and *D* and assume that they have 30, 30, 40, and 50 votes, respectively. Assuming that the majority rules, find the power of each of these blocs.

Solution

There are $24 = 4!$ different orders (the number of permutations of 4 objects) in which the blocs can vote. These are listed below, and in each case the bloc that puts the total over the top (over the 76 votes required for passage) is underlined.

$$
\begin{array}{llll}
AB\underline{C}D & BA\underline{C}D & CA\underline{B}D & DA\underline{B}C \\
AB\underline{D}C & BA\underline{D}C & CA\underline{D}B & DA\underline{C}B \\
AC\underline{B}D & BC\underline{A}D & CB\underline{A}D & DB\underline{A}C \\
AC\underline{D}B & BC\underline{D}A & CB\underline{D}A & DB\underline{C}A \\
A\underline{D}BC & BD\underline{A}C & CD\underline{A}B & DC\underline{A}B \\
A\underline{D}CB & BD\underline{C}A & CD\underline{B}A & DC\underline{B}A
\end{array}
$$

On the roll call *ADBC, D* is given credit for passage, since *A* alone has 30 votes, which is insufficient for passage, while *A* and *D* together have 80 votes, which is more than enough for passage. In the roll call *ABCD, C* gets the credit, since *A* and *B* have only 60 votes but *A, B,* and *C* have 100 votes; adding *C* has pushed the supporters over the 76 vote line.

If you count the number of times each bloc is critical, you will find that *D* is critical 12 times and *A, B,* and *C* are each critical 4 times. So *D*'s power is $\frac{1}{2}$ and *A, B,* and *C* each have a power of $\frac{1}{6}$.

EXAMPLE 2

A county that consists of five towns (*A,B, C, D,* and *E*) is governed by a board of supervisors in which ordinances are passed by majority vote. Towns *A* and *B* each have 3 representatives on the board, and *C, D,* and *E* each have only 1. Right now, all the representatives vote independently, but *A* is thinking of having its representatives form a single bloc. Before a bill comes to the floor, the 3 representatives from *A* would caucus and determine their majority position. When the bill finally reaches the floor, the representatives from A would take a united position. Town A would like to calculate in advance its power as things are and its power with a bloc. Specifically, it would like the answers to the following questions:

(a) What is the sum of the powers of all 3 of A's representatives when everybody votes independently?

(b) What would A's total power be if everyone voted independently except A's representatives and they formed a bloc?

(c) What would A's total power be if, after A's representatives formed a block, B's representatives reacted and formed a bloc too?

Solution

(a) If each of the 9 representatives votes independently, they would all have the same power. The total power is 1, so each representative has power $\frac{1}{9}$ and A's total power is $\frac{3}{9}$, or $\frac{1}{3}$.

(b) We can calculate the power of A's bloc of 3 votes among 6 other single independent votes as follows: There are in effect 7 voters, so 7! different roll call sequences can be formed. Bloc A is pivotal when it is the third, fourth, or fifth to be called, that is, when the calling sequence is $\_\_\_A\_\_\_\_\_$, $\_\_\_\_A\_\_\_\_$, or $\_\_\_\_\_A\_\_$ (where the blank spaces may be filled in by the 6 single votes in every possible way). Since there are 6! ways of filling in each of these 3 sequences, A's power is $(3)(6!)/(7!) = 3/7$.

(c) If the 3 voters in A form a bloc and the 3 voters in B form a bloc, there will effectively be 5 players: 2 with 3 votes and 3 with 1 vote. There are $5! = 120$ different sequences that can be formed. Bloc A will be pivotal when the sequence takes one of the following 6 forms:

$$B\,A\_\_\_,\ \_B\,A\_\_,\ B\_A\_\_,\ \_\_A\_B,$$
$$\_\_A\,B\_,\ \text{and}\ \_\_\_\_A\,B,$$

(where, once again, the blanks may be filled in in all possible ways by the single voters). There are 6×6 such sequences, or 36 in all, so A's power is $\frac{36}{120} = \frac{3}{10}$.

So it all comes to this: Initially, before blocs are formed, the total power of both A and B is $\frac{3}{9}$. A forms a bloc and B's representatives continue to vote as individuals, the total power of A's representatives rises to $\frac{3}{7}$ and B's total power drops to $\frac{2}{7}$. This is fine as far as it goes, but the calculations shouldn't end here; A must anticipate that B will form a counterbloc, in which case the total power of both *A* and *B* will drop to $\frac{3}{10}$. So ultimately, after both *A* and *B* form blocs, they are worse off than when they voted independently.

A situation similar to the one in the previous example was analyzed by Phillip Straffin in the January 1977 issue of *Mathematics Magazine*. In Rock County in southern Wisconsin there are two large cities, Janesville and Beloit. Despite a rivalry between the two cities, the 11 supervisors from Beloit and the 14 representatives from Janesville continue to vote independently. Although there have been suggestions that Beloit form a bloc because of the initial advantage to be gained, Straffin points out that the ultimate effect would be a loss for both large towns when Janesville reacted by forming a counterbloc. (See "fact" 6).

exercises

1 There are 3 members (*A*, *B*, and *C*) on a committee. On the left are listed the coalitions that are capable of passing a measure, and on the right are the powers of the members. Confirm these calulations.

	Winning coalitions	*Power of* *A*	*B*	*C*
(i)	*ABC*	$\frac{1}{3}$	$\frac{1}{3}$	$\frac{1}{3}$
(ii)	*ABC, BC*	0	$\frac{1}{2}$	$\frac{1}{2}$
(iii)	*ABC, BC, AC*	$\frac{1}{6}$	$\frac{1}{6}$	$\frac{2}{3}$
(iv)	*ABC, AC, AB, BC*	$\frac{1}{3}$	$\frac{1}{3}$	$\frac{1}{3}$
(v)	*C, AC, BC, ABC*	0	0	1

2 In 1958 Nassau County had a weighted voting system for its Board of Supervisors. The county consisted of six municipalities (Hempstead Number 1, Hempstead Number 2, North Hempstead, Oyster Bay, Glen Cove, and Long Beach), and these municipalities cast 9, 9, 7, 3, 1, and 1 vote, respectively. Show that Oyster Bay, Glen Cove, and Long Beach were dummies and the other towns each had power of $\frac{1}{3}$.

It was successfully argued in court that the smaller communities were essentially disenfranchised. For a more detailed account of this argument by the attorney who made it see "Weighted Voting Doesn't Work: A Mathematical Analysis," by John F. Banzhaf III, in the *Rutgers Law Review*, vol. 19 (1965), pp. 317–43. Also see the *New York Times* for November 17, 1974, p. 43.

Exercises 3–5 were described by William F. Lucas in an article entitled "Measuring Power in Weighted Voting Systems," in *Case Studies in Applied Mathematics,* by the Committee on the Undergraduate Program in Mathematics of the Mathematics Association of America (1976), pp. 42–106.

3 In the Canadian election of 1972 the Liberals won 109 seats, the Tories 107 seats, and the New Democrats 31 seats; 17 seats were won by smaller parties.

(a) Show that the "smaller parties" were dummies and that the Liberals, Tories, and New Democrats, despite their different numbers of seats, each had the same power: $\frac{1}{3}$.

(b) Notice that the "smaller parties" had power proportionately less than their numbers, since they had no power at all. The New Democrats, on the other hand, had power disproportionately larger than their size, since they had $\frac{1}{3}$ of the power but considerably less than that in seats. Obviously, there is no simple relationship between power and bloc size.

4 Formerly, in the New York City Board of Estimate, the Mayor, Controller, and Council President each had 3 votes, the borough presidents of Brooklyn and Manhattan had 2 votes, and the presidents of the other three boroughs had 1 vote each. This was changed, however, so that all five borough presidents had 2 votes and the other members had 4 votes. Calculate the power of each board member under each system. It takes a majority to pass an issue.)

5 After the Tokyo elections of 1973 the Metropolitan assembly was distributed as follows: The Liberal-Democrats had 51 votes, the Komeito had 26 votes, the Communists had 24 votes, the Socialists had 20 votes, and the Democratic-Socialists and Independents each had 2 votes.

(a) Show that both the Democratic-Socialists and the Independents were dummies.

(b) Show that the Liberal-Democrats had $\frac{1}{2}$ of the power and the Komeito, Communists, and Socialists each had $\frac{1}{6}$ of the power.

Some Paradoxes and Applications

The Shapley value sometimes suggests relationships that are not immediately obvious. For example, suppose someone has a fixed number of votes in a legislative body. Suppose further that the size of the body is increased but the individual's number of votes remain the same. Since there are new voters who have been given power and the individual's vote has not changed, it might seem that his power cannot possibly increase. But suppose issues are determined by majority vote and there are 3 voters (A, B, and C) with 5, 5, and 9 votes, respectively. It is easily calculated that all three have power equal to $\frac{1}{3}$. If the votes of A, B, and C are kept constant but a new voter D is added with 2 votes (it now takes 11 votes for a measure to pass instead of 10, of course), then C's power increases from $\frac{1}{3}$ to $\frac{1}{2}$. (This is most easily verified by observing that C is pivotal when he is either second or third in the sequence, which occurs exactly half the time.) The power of each of the other players becomes $\frac{1}{6}$.

This point may be seen even more clearly if you consider a voting body in which there are dummies. Initially, dummies have no power at all. If enough new voters with few votes are added, the former dummy will be pivotal in at least one sequence; whatever power he manages to obtain in this way must be an improvement. (See "fact" 7.)

Another conclusion that one might be tempted to reach is that individuals can never do worse by joining a bloc; in fact, they can. Consider the following example, suggested by Steven Brams: There are 5 voters (A, B, C, D, and E) with 2, 2, 1, 1, and 1 vote, respectively, in a body governed by majority rule. If the voters all act independently, each of C, D, and E will have power $\frac{2}{15}$; their total power would be three times that, or $\frac{2}{5}$. If C, D, and E form a coalition, in effect there will be 3 voters, one with 3 votes and the other two with 2 votes. Each of the 3 voters would have the same power, so the total power of CDE would be $\frac{1}{3}$: a loss of $\frac{1}{15}$ from the power they had when acting independently. (See "fact" 6).

Some Applications

There have been a number of interesting applications of the Shapley value to international, national, and local politics. The United Nations Security Council, for example, has five permanent members and ten nonpermanent members; in order to pass a resolution, all the permanent members and at least four nonpermanent members must agree. The same voting rules would hold if the permanent members each had 7 votes, the nonpermanent members had 1 vote, and 39 votes were required to pass a resolution. It has been calculated that the permanent members, with $\frac{7}{9}$ of the vote, actually have 98% of the power.

As is well known, it is necessary to have a majority in the House and Senate to pass a bill with the president's approval; without his approval you need $\frac{2}{3}$ of the vote in each chamber. It

It has been calculated that in the United Nations Security Council, the permanent members, with $\frac{7}{9}$ of the vote, have 98% of the power.

turns out that the House and Senate each have $\frac{5}{12}$ of the power (the power of an individual senator is, of course, much greater than that of a Representative) and the President has the remaining $\frac{1}{6}$.

Perhaps the most interesting application of the Shapley value is to the formation of bandwagons at political conventions. The general idea is this: Imagine two blocs forming behind two of the candidates for the nomination. As the blocs grow in size (as independents become committed to one side or the other), the power of the independents not yet committed and the power of each of the blocs change. At each stage the independents calculate their power and compare it to how much power they would add to a bloc if they joined it. (This marginal increase presumably would be a measure of their reward for joining.) At a certain point the imbalance becomes so large that the independents flock to the stronger bloc; this is known as the "bandwagon" effect.

In a fascinating mixture of theory and application, Phillip Straffin gives an account of the 1976 Republican primary contest between Ford and Reagan (in the November 1977 issue of the *American Journal of Political Science*). Starting with the theory first, he indicates the circumstances under which one bloc or the other will grow explosively. He describes several situations in the Ford-Reagan primary contest when observers thought the balance had tipped and a bandwagon was in effect, although the mathematical model indicated otherwise. When the mathematical model finally did signal that a bandwagon was in effect, this was confirmed in a number of ways. For instance, Reagan accepted an unlikely Vice-Presidential candidate (it was Ford who held the edge), and Reagan's compaign manager distorted the number of committed delegates. The article is particularly recommended as an artful application of theory and a source for further readings.

Summary

When you first begin to analyze the subject matter of this chapter—voting, political power, logrolling, etc.—it seems to be an area that is inhospitable to mathematical analysis. Unlike the physical sciences (in which you can easily assign numbers to such variables as mass, current, speed, and force), political science and the social sciences in general treat concepts such as "power" and "the general will," which are much harder to define precisely. And yet

we have seen that mathematical analysis could sometimes weave these unpromising elements into a plausible, quantitative theory; and where this was not possible, analysis could discover why and explicitly prove that no general theory was possible.

Consider the concept of "power," for example. An intuitive notion of "power" was translated into a formal mathematical concept, and the power of various groups and individuals were calculated. The intuitive notion and the mathematical values were compared in simple cases to see whether they were consistent, and after a number of such trials, we gained some confidence in the mathematical definition. At this point we relied solely on the mathematical calculations and discarded the intuitive notion, which would be "out of its depth" in such problems as calculating the power of a member of the U.N. Security Council or of a senator in Congress, or the changes in power of members of a voting body when blocs are formed.

We also considered a somewhat different, but equally useful, role that mathematical analysis could play. We examined the problem of setting up a voting procedure and tried out various methods that seemed reasonable to see whether the results were consistent with our sense of fairness. By taking a more general, mathematical approach, Arrow showed that an such procedure must be irrational or inequitable in some respect.

So a mathematical model can serve to reinforce an intuitive idea or extend it when intuition is inadequate, or it can be used to prove that certain intuitively plausible statements (such as the eight "facts" listed at the beginning of the chapter) are invalid. It has performed both functions in this chapter.

ON CONSTANT CHANGE

MARKOV CHAINS

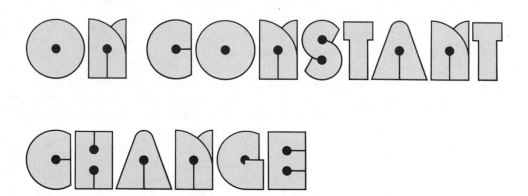

Zlatko Šimunović, "Mobile Sculpture **5AM,**" 1979, ▷
Multicolored painted and varnished wood layers,
10¼" x 8" x 8", ⓒ 28th Street Gallery.

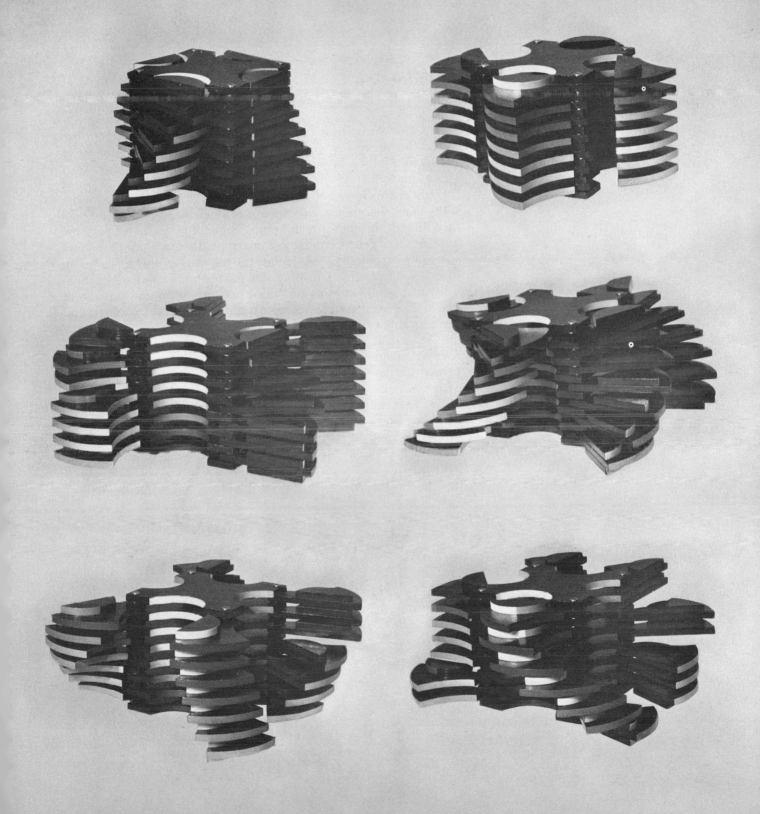

1 Introduction

It is a little surprising and very fortunate when a mathematical model that serves us well in one application also turns out to be useful in another. A particularly versatile model is the Markov chain, named after the Russian mathematician A. A. Markov. Markov chains have been used in business to calculate the probability that a warehouse will run out of stock, in physics as a model of diffusion through a membrane, in genetics to investigate the fluctuation of gene frequency under the influence of mutation and selection, in sociology to calculate the mobility of social classes, in theoretical investigations of casino games to calculate the probability of a gambler going bankrupt because of an unusual run of bad luck, as well as in meteorology, the stock market, ecology, and learning theory.

As an example of a situation that can be analyzed by means of Markov chains, consider the problem facing a certain theater owner. Let us suppose that after many years of experience the owner has observed that the film a customer sees this week influences what the customer will do next week. A typical customer likes 3 of every 5 films he or she sees. Customers who stay home this week have an 80% chance of staying home next week. Customers who go to a movie next week after staying home this week (which happens the remaining 20% of the time), will like it $\frac{3}{5}$ of the time ($\frac{3}{5}$ of 20% = 12%) and dislike it $\frac{2}{5}$ of the time ($\frac{2}{5}$ of 20% = 8%.)

Customers who see a film and like it this week will stay home 70% of the time next week. Of the remaining 30%, they will see a movie they like 18% of the time and will see a movie they don't like 12% of the time.

Finally, customers who see a movie they don't like this week will stay home next week 90% of the time, see a movie they like 6% of the time, and see a movie they don't like 4% of the time.

All this information is summarized in the following table, which is known as a **transition matrix:**

Activity this week

	Stays home	*Likes movie*	*Dislikes movie*
Stays home	.8	.7	.9
Likes movie	.12	.18	.06
Dislikes movie	.08	.12	.04

Activity next week (row label for the left side)

The probability of a customer's activities next week given his activities this week.

If there are 10,000 potential customers, all of this type, and if 20% of them attend on opening night, what will be the size of an average audience over the long run?

This process, in which the size of the audience varies from week to week, is called a **Markov chain.** Though there are many differences in the kinds of problems that make use of Markov chains, all of them have certain properties in common:

1. There is some fixed number of **states** that the world can be in, and at any one time the process will be in exactly one of them.
2. At certain prescribed times—we call them **steps**—the process may change its state in accordance with certain probabilities.
3. The probability of passing from one state to another depends only on the states involved and does not vary with time. These probabilities are called the **transition probabilities.**

In the last example a customer can be in any one of three states in any one week: seeing a film and liking it, seeing a film and disliking it, or staying home. Each week is a new step, which might (or might not) change a customer's state. Condition 3 is particularly important. In effect, it says that in deciding what to do one week, a customer remembers only what happened the week before. So if we knew that Ms. Sanchez would stay home 50 weeks from now, we would also know that she had an 80% chance of staying home 51 weeks from now as well.

You can think of a Markov chain as a game of musical chairs. You're sitting in a chair at the step 0 (the start of the game) and rise when the music starts. You walk until the music stops and then sit down in a new chair, possibly the one you just left. The new chair is your state at the step 1. The music starts again, you rise again, and the process goes on indefinitely. If you're in chair *A* now, your chance of reaching chair *B* next is called the *one-step transition probability,* and this never changes; if you're ever in chair *A* in the future, your chance of reaching chair *B* immediately afterward will be the same as the probability of going from *A* to *B* now.

As we said earlier, this same model may be used in many different applications. To a bettor at the track his state is the state of his finances; if he bets the same amount at fixed odds at each step, he's always ahead or behind by some fixed amount. The steps are the races, and after each race his finances change. To a stockholder the states are the closing prices of the stock on trading days, and the steps may be defined as the close of each trading day. The geneticist might see the states in terms of eye color, and the sociologist might see them in terms of social class; both might define a step as a generation.

1   When a rat passing through a maze arrives at a critical junction, it must either go left or go right. It receives an electric shock in one case and food in the other case. After the rat has performed repeatedly its decision is influenced by its past experience. Why would you *not* expect a Markov chain model to apply here? If the memory of the rat had a certain quirk, you could use Markov chains after all; what would that quirk be?

2   Suppose you toss a fair coin many times and get a long sequence of heads. Since the coin is fair, you should get the same number of heads as tails in "the long run," so it might be argued that a tail is more likely on the next toss to balance the long string of heads. (This is the theory of "the

maturity of chances" and is not to be taken too seriously.) If you accept this argument, could you apply the theory of Markov chains? (Assume there are two states, heads and tails, and the tosses are the steps.) Can Markov chain theory be applied in fact?

3 According to many technicians in the stock market, tomorrow's price for a stock is intimately related to the stock's past history. These technicians talk about "resistance levels," "support levels," "head and shoulders patterns," and "double bottoms," all reflecting a stock's past history and all presumably affecting the prognosis of a stock's future price.

On the other hand, certain economists take a different view; they feel that everything that can be gleaned from past history is already reflected in the current price and all additional information is worthless as a guide for the future. Which of these two groups of analysts would use a Markov chain model?

4 Each of the following situations can be analyzed by using a Markov chain model. In each case state what the steps, states, and transition probabilities are.

(a) If a person has an IQ less than 110 there is a probability of .4 that his or her first child will have an IQ greater than 110. If a person's IQ is greater than 110, there is a probability of .65 that his or her first child will also have an IQ greater than 110.

(b) A person with collision insurance who has no accidents this year has a probability of .85 of having collision insurance next year. A person without collision insurance who has no accidents this year will have collision insurance next year with a probability of .25. Anybody who has an accident this year has a probability of .95 of having collision insurance next year. One tenth of all drivers have an accident in a given year.

(c) People who cheat on their income tax and aren't audited have a .4 probability of cheating next year; if they are audited, their probability of cheating drops to .2. Those who didn't cheat this year have a probability of .1 of cheating next year if they were audited this year and a probability of .25 of cheating next year if they were not audited. One eighth of all taxpayers are audited every year.

5 Describe each of the following situations as a Markov chain:

(a) Each year two cities have a soccer match consisting of two games. The winner is considered the local champion. The teams are evenly matched. The champion of last year remains the champion this year unless it loses both games.

(b) At a carnival I keep betting $1. I win $\frac{1}{5}$ of the time, and when I do I get $3 back. The other $\frac{4}{5}$ of the time I get nothing back.

(c) On a given day the cost of a share of stock increases 2 points with probability .15, increases 1 point with probability .25, remains unchanged with probability .25, drops a point with probability .2, and drops two points with probability .15.

6   In a certain course tests are given weekly. A particular student who cheats on a test has a 25% chance of getting caught. If he is caught, he will cheat on the following week's test 16% of the time; if he's not caught, he'll cheat the next week 80% of the time. If the student doesn't cheat this week, he has a 40% chance of cheating next week. Fill in the entries in the transition matrix below so that it reflects this student's behavior:

Behavior this week

	Cheats and gets caught	Cheats successfully	Does not cheat
Cheats and gets caught			
Cheats successfully			
Does not cheat			

Behavior next week (row label at left)

7   When Joan is asked to lend money, she considers the last experience she had lending money. If the last person who borrowed never repaid, she refuses the next request; if the last loan was repaid, she agrees to the loan. If the last person to ask for a loan was turned down, she grants the next request 48% of the time. Loans are repaid 50% of the time. Construct a transition matrix that reflects her behavior.

8   A cosmetics firm uses three advertising agencies: A, B and C. If the last advertising campaign was successful, the firm retains the agency 84% of the time; if it was unsuccessful, the firm retains the agency 32% of the time. In each case if the firm decides to switch, it is as likely to go to one of the competing agencies as the other. Each agency is successful 75% of the time. Construct an appropriate transition matrix.

9   A newspaper dealer uses the honor system on one day with probability 80% if he used it the day before and was not cheated and with probability 30% if he used it the day before and was cheated. If he didn't use the honor system on the previous day, he uses it on the next 60% of the time. Cheating occurs 50% of the time with the honor system and not at all without it. Construct an appropriate transition matrix.

10 At a roulette table a gambler makes a series of $5 and $10 bets. If he wins, he makes a $10 bet; if he loses, he makes a $5 bet. He has a 40% chance of winning each bet. Construct an appropriate transition matrix.

11 At Siwash college a research study showed that honor students do better when they get lower grades. An honor student who has less than a B average this term has only a 25% chance of getting less than a B average next term; an honor student who has better than a B average this term has a 40% chance of getting less than a B average next term. Construct an appropriate transition matrix.

Transition Probabilities

The basic building blocks of Markov chains are the transition probabilities—the probabilities of passing from one state to another. It is convenient to use the notation P^k_{AB} to designate *the probability of being in state B k steps from now if you are presently in state A*. If no value of k is given, it is assumed to be 1. So if you are at A now, your chance of getting to B at the next step is P_{AB}.

In a real-life application you will generally know the one-step transition probabilities or will estimate them. All the other properties of the Markov chains can be derived from these probabilities. The kinds of things one may want to know about a Markov chain varies from application to application, but these are a few questions that are frequently asked:

1. If I'm at A now, what is the probability of being at B exactly k steps from now? That is, what is P^k_{AB}?
2. Is there a set of states which, once entered, can never be left?
3. Suppose X represents the number of steps it takes to get from A to B for the first time. What is the expected value of X?
4. As the Markov process continues over a long period of time, how frequently can you expect to be in each state?

Before we try to answer these or any other questions, let's look at another example:

EXAMPLE 1

In a certain society there are two classes: aristocrats and peasants. It has been observed that children of peasants become aristocrats $\frac{1}{5}$ of the time (through hard work, luck, or prudent marriages) and remain peasants the other $\frac{4}{5}$ of the time. Children of aristocrats remain aristocrats $\frac{9}{10}$ of the time and become peasants $\frac{1}{10}$ of the time. Calculate the probability that a randomly chosen descendant of generation k of an aristocrat is an aristocrat.

Solution

In the transition matrix we use "a" and "p" to abbreviate "aristocrats" and "peasants":

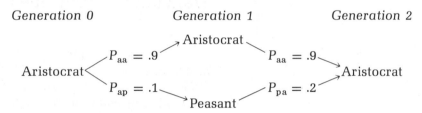

Generation 0   *Generation 1*   *Generation 2*

Aristocrat
$P_{aa} = .9$ → Aristocrat
$P_{ap} = .1$ → Peasant
$P_{aa} = .9$ → Aristocrat
$P_{pa} = .2$

The probabilities for the first generation are easy enough to calculate. If a person is an aristocrat initially (we consider this step 0 or generation 0), then the transition matrix shows that the probability of a generation 1 descendant being an aristocrat is .9. The calculations for the second generation are a bit harder.

A grandchild of an aristocrat can become an aristocrat in either of two ways:

1. Parent, child, and grandchild can all be aristocrats.
2. The aristocratic parent can bear a child that becomes a peasant, who has a child that becomes an aristocrat again. In the diagram below, the upper path corresponds to the first possibility and the lower path to the second:

		Generation k	
		Aristocrats	*Peasants*
Generation k + 1	*Aristocrats*	$P_{aa} = .9$	$P_{pa} = .2$
	Peasants	$P_{ap} = .1$	$P_{pp} = .8$

Since the transition probabilities at each step are independent, the probability of the upper path is $(P_{aa})(P_{aa}) = (.9)(.9) = .81$ and the probability of the lower path is $(P_{ap})(P_{pa}) = (.1)(.2) = .02$. These two possibilities are mutually exclusive, so we can add them and obtain $P_{aa}^2 = .81 + .02 = .83$.

In the same way you can calculate the probability that an aristocrat's descendant of generation $k + 1$ will be an aristocrat if you know the probabilities for generation k. Again there are two possibilities: The probability that generations 0, k, and $k + 1$ are all aristocrats is $P_{aa}^k P_{aa}$, and the probability that generation k is a peasant and generations 0 and $k + 1$ are aristocrats is $P_{ap}^k P_{pa}$. Since these are the only possibilities and they are mutually exclusive, we have the formula

$$P_{aa}^{k+1} = P_{aa}^k P_{aa} + P_{ap}^k P_{pa}$$

In the model that we just discussed we looked at a single individual and calculated the probability that the descendant of generation k will be an aristocrat. It will sometimes be convenient to interpret the model in a slightly different way. We can assume that a population is composed of aristocrats and peasants and ask what the expected percentage of aristocrats will be some number of generations later. The calculations in both cases are exactly the same.

"HI, MOM AND DAD!"

Table 1

| Generation | Probability that the descendant in generation k will be an aristocrat if the ancestor in generation 0 is: | |
	An aristocrat	A peasant
0	1.0	.0
1	.9	.2
2	.83	.34
3	.781	.438
4	.7467	.5066
5	.7227	.5546
6	.7059	.5882
7	.6941	.6118
8	.6859	.6282
9	.6801	.6398
10	.6761	.6478

Table 1 lists the probability of being an aristocrat in each of the first 10 generations. If you look at the original one-step transition probabilities, you will see that there is a great deal of inertia in this model, that is, aristocrats tend to remain aristocrats and peasants tend to remain peasants. The effect of this inertia is most clearly seen in the early generations; the class of an individual is likely to be the same as that of the ancestor. But as more and more generations pass, the probabilities approach one another and the class of the ancestor seems to matter less. If we interpret these probabilities as the expected percentage of a population that is included in each class, it seems that the population is approaching some stable separation into two classes. In the present case it seems that about 65% of the population will be aristocrats. That this is actually the case—that the initial population matters less and less as time passes—will be confirmed later.

EXAMPLE 2

Charlotte buys a new car every year, restricting her purchases to Fords (f), Chevrolets (c), and Plymouths (p). She always buys the type of car she bought the year before unless it's defective, in which case she flips a fair coin to determine which of the other two makes she will buy. Let us assume that 10% of all Fords, 20% of all Plymouths, and 30% of all Chevrolets are defective (of course, these are fictitious figures) and that Charlotte's present car (the 0th step) is a Plymouth.

(a)   Write out all the transition probabilities.

(b)   Calculate the probability that the car after next will be (i) a Ford, (ii) a Plymouth, (iii) a Chevrolet.

(c)   Write a formula for P_{pf}^{k+1}, the probability that the $(k+1)$st-generation car will be a Ford, in terms of P_{pf}^{k}, P_{pp}^{k}, and P_{pc}^{k}, the probabilities that the kth-generation car will be a Ford, a Plymouth, and a Chevrolet.

Solution

(a) The following are the transition probabilities:

Car owned during generation k

	Ford	Plymouth	Chevrolet
Ford	$P_{ff} = .9$	$P_{pf} = .1$	$P_{cf} = .15$
Plymouth	$P_{fp} = .05$	$P_{pp} = .8$	$P_{cp} = .15$
Chevrolet	$P_{fc} = .05$	$P_{pc} = .1$	$P_{cc} = .7$

Car owned during generation $k + 1$

(b) The probability that the car after next will be a Ford is

(i) $P_{pf}^2 = P_{pf}P_{ff} + P_{pp}P_{pf} + P_{pc}P_{cf} = (.1)\,(.9) +$

$$(.8)\,(.1) + (.1)\,(.15) = .185$$

(ii) In a similar way we can calculate that $P_{pc}^2 = .155$ and

(iii) $P_{pp}^2 = .66$

Notice that the answers to (i), (ii), and (iii) sum to 1. Why?

(c) $P_{pf}^{k+1} = P_{pf}^k\,P_{ff} + P_{pp}^k P_{pf} + P_{pc}^k P_{cf}$

exercises

1 For the population of aristocrats and peasants in Illustrative Example 1, derive a formula expressing P_{ap}^{k+1} in terms of P_{aa}^k, P_{ap}^k, P_{ap}, and P_{pp}.

2 Calculate P_{ap}^2 using the formula derived in Exercise 1.

3 Calculate P_{aa}^3.

4 Notice that $P_{ap} + P_{aa} = 1$ and $P_{ap}^2 + P_{aa}^2 = 1$. Why is $P_{ap}^k + P_{aa}^k = 1$ for any k?

5 Calculate P_{pa}^k and P_{pp}^k for $k = 2$ and $k = 3$.

6 If the weather is fair on a certain day, the next day will be fair 60% of the time and rainy 40% of the time. If it's rainy, the next day will be rainy 30% of the time and fair 70% of the time. Write out the transition matrix. What is the probability that Thursday will be fair if the preceding Tuesday was rainy?

7 Because the parking regulations are not systematically enforced, people often park illegally. The drivers in Tall Oaks have short memories; their parking behavior today is only influenced by their parking experience yesterday.

    (i)   If they park legally today, there's an even chance they'll park legally tomorrow.

    (ii) If they park illegally today and are *not* ticketed, they will certainly park illegally tomorrow.

    (iii) If they park illegally today and *are* ticketed, they have a probability of .75 of parking legally tomorrow.

In the town of Tall Oaks, illegally parked cars have one chance in five of getting a ticket.

(a)   Verify the following transition matrix:

Day k

		Parked legally	Parked illegally Ticketed	Parked illegally Not ticketed
Parked legally		.5	.75	0.0
Parked illegally	Ticketed	.1	.05	.2
	Not ticketed	.4	.20	.8

Day k + 1

(b)   How likely is it that someone who parks legally today will park illegally 2 days from today and receive a ticket?

(c)   If there are 20,000 drivers in town, can you estimate the number of tickets given out on the average, during a single day? (We have to estimate this now. Soon we will be able to calculate it exactly.)

8 A country with a two-party system has an electorate that has a very short memory and is very impatient. At every election both parties make promises they cannot fulfill, so the party in power, which inevitably proves to be disappointing, tends to lose. The Freedom Party has 1 chance in 4 of winning an election if it is presently in office, and the Justice Party has only 1 chance in 20 of winning an election if it is in office. Calculate

the probability that the Freedom Party will be in office three terms from now if the party presently in power is
(a)　the Freedom Party.
(b)　the Justice Party.

9　A hospital uses four terms to describe a patient's condition: healthy, serious, critical, and deceased. Both healthy and deceased patients leave the hospital, whereas serious and critical patients move daily into other categories according to the following transition probabilities:

Patient's condition today

		Healthy	Serious	Critical	Deceased
Patient's condition tomorrow	Healthy	1.0	.2	0.0	0.0
	Serious	0.0	.4	.2	0.0
	Critical	0.0	.3	.5	0.0
	Deceased	0.0	.1	.3	1.0

(a)　If there were no new patients entering the hospital, what would happen to the hospital's population? Specifically, if there were 1000 patients who were either serious or critical today, how many would you expect to have tomorrow? How many would you expect to have on the following day?
(b)　In what way are the categories "healthy" and "deceased" different from the categories "serious" and "critical"?

10　(a)　If the transition matrix for a Markov chain is

	A	B	C
A	p	q	r
B	s	t	u
C	v	w	x

then $p + s + v = q + t + w = r + u + x = 1$. Why?

(b) Suppose you have $2 and two fair coins are tossed. For each head that turns up you win $1, and for each tail that turns up you lose $1. You keep playing until you either lose your $2 or you have $4. Show that the transition matrix is

Your fortune now

		$0	$2	$4
	$0	1	$\frac{1}{4}$	0
Your fortune next turn	$2	0	$\frac{1}{2}$	0
	$4	0	$\frac{1}{4}$	1

In what state(s) do you think you will be in the distant future? Can you explain why in terms of the transition matrix?

(c) Suppose a Markov chain has the following transition matrix:

Now

	A	B	C
A	0	$\frac{1}{4}$	1
B	0	$\frac{1}{2}$	0
C	1	$\frac{1}{4}$	0

Can you predict what will happen in the long run?

(d) Answer question (c) for the following matrix:

	A	B	C
A	$\frac{1}{3}$	$\frac{1}{2}$	$\frac{1}{8}$
B	$\frac{2}{3}$	$\frac{1}{2}$	$\frac{1}{8}$
C	0	0	$\frac{3}{4}$

11   From a given transition matrix you may construct a new transition matrix that will show the probability of your being in any state after 2 steps, 3 steps, or N steps rather than at the next step. For example, if the original transition matrix was

Present state

	A	B
State next step A	.8	.3
B	.2	.7

then the 2-step and 3-step transition matrixes would be

Present state

	A	B
State two steps later A	.7	.45
B	.3	.55

Present state

	A	B
State three steps later A	.65	.525
B	.35	.475

(a)   In each of the following examples construct a two-step transition matrix from the given transition matrix.

(b)   In each of the following examples construct a three-step transition matrix.

(i)

Present state

	A	B
State next step A	0	$\frac{1}{4}$
B	1	$\frac{3}{4}$

(ii)

Present state

	A	B
State next step A	$\frac{1}{2}$	$\frac{1}{3}$
B	$\frac{1}{2}$	$\frac{2}{3}$

Stable Probability Distributions

In principle, the formula that expresses transition probabilities at step $k + 1$ in terms of those at step k can be used over and over to calculate the transition probabilities at any step. In practice, these calculations are lengthy and tedious. But it is possible to find out the long-term effects of a Markov chain without all these calculations, and we return to the example of the aristocrats and peasants to see how.

Suppose the Markov chain in the population of peasants and aristocrats has been in effect for many generations and a stable class division has been attained in which x was the fraction of peasants and $(1 - x)$ was the fraction of aristocrats. The peasants of the next generation come from two sources; they are either the children of aristocrats or the children of peasants. And if the distribution is stable, the fraction of peasants in the next generation, on the average, will be the same as in this one: x. Translating this into algebra, we have

$$xP_{pp} + (1 - x)P_{ap} = .8x + .1(1 - x) = x$$

The solution to this equation is $x = \frac{1}{3}$. We can set up a similar equation for the aristocrats:

$$(1 - x)P_{aa} + xP_{pa} = .9(1 - x) + .2x = 1 - x$$

which has the same solution, $x = \frac{1}{3}$. So a stable population is composed of one third peasants and two thirds aristocrats.

Once you have a stable distribution, there is no change *on the average* from one generation to the next; nevertheless, there will be deviations, since the actual outcome need not be the average outcome. The interesting question is this: Will these deviations tend to grow or to die out?

Suppose y is the deviation from the stable distribution, so that the fraction of peasants is $\frac{1}{3} + y$ and the fraction of aristocrats is $\frac{2}{3} - y$. In the next generation there will be

$$\left(\frac{1}{3} + y\right)(.8) + \left(\frac{2}{3} - y\right)(.1) = \frac{1}{3} + .7y$$

peasants, on the average. Notice that the deviation from the stable distribution is only 70% of what it was earlier. So in this example, at any rate, although there may be deviations in any given generation, the tendency will always be to return to the stable probability distributions in later generations.

There is always at least one stable probability distribution for any Markov chain, but there may be more than one. Consider again the transition matrix in Exercise 10(b) of Section 2:

Now

		$0	$2	$4
	$0	1	$\frac{1}{4}$	0
Next turn	$2	0	$\frac{1}{2}$	0
	$4	0	$\frac{1}{4}$	1

Suppose the fractions of players with $0, $2, and $4 are x, $1 - x - y$, and y, respectively, and suppose this is a stable distribution. The fraction of players who will have $0 on the next step will be everyone who has $0 now and $\frac{1}{4}$ of those who have $2 now, so that

$$x + \frac{1}{4}(1 - x - y) = x$$

If we calculate the average number of players who will have $2 and $4 on the next step, we obtain the respective equations

$$\frac{1}{2}(1 - x - y) = (1 - x - y) \quad \text{and} \quad \frac{1}{4}(1 - x - y) + y = y$$

All three equations are satisfied if $x + y = 1$. The meaning of this algebraic solution is simple: If people keep gambling until they either win or lose $2 and then stop, eventually everyone will have lost or won and no one will be left to bet. *Any probability distribution in which there is no one in the $2 state is stable.*

(**EXAMPLE 1**)

Consider again the theatre-going example from Section 1. The transition matrix is

	This week		
	Stayed home	Liked movie	Disliked movie
Stayed home	.8	.7	.9
Liked movie	.12	.18	.06
Disliked movie	.08	.12	.04

Next week labels the rows on the left.

Find the average audience on a given weekend after many weeks have passed.

Solution

Suppose the fractions of people who stayed home, saw a movie and liked it, and saw a movie and disliked it last week were x, y, and $(1 - x - y)$, respectively. If this is the same, on the average, on the following weekend, then we have

$$.8x + .7y + .9(1 - x - y) = x$$
$$.12x + .18y + .06(1 - x - y) = y$$

and

$$.08x + .12y + .04(1 - x - y) = 1 - x - y.$$

There is exactly one solution to these three equations:

$$x = \frac{39}{49}, \qquad y = \frac{6}{49}, \quad \text{and} \quad 1 - x - y = \frac{4}{49}$$

So on any given weekend $\frac{10}{49}$ of the potential audience can be expected to attend the theatre. On the average, $\frac{6}{49}$ of the potential audience will like the picture and $\frac{4}{49}$ will dislike it. If, as we assumed in the original problem, the potential audience is 10,000 people, then the size of the average audience will be about 2041.

1   In Exercise 4(a) of Section 1, what fraction of the population will have IQs greater than 110 in the long run?

2   In Exercise 4(b) of Section 1, what fraction of the population will carry collision insurance at any given time in the long run?

3   In Exercise 4(c) of Section 1, what fraction of people cheat on their income tax, on the average, during any given year?

4   Find the long-term probability of good weather in Exercise 6 of Section 2.

5   In Exercise 7 of Section 2, calculate the average fraction of cars that park illegally.

6   In Exercise 8 of Section 2, calculate how frequently the Freedom Party will be in power in the long run.

7   In Exercise 9 of Section 2, show that a probability distribution is stable if and only if everyone is either healthy or dead.

8   An insurance company finds that it has been losing money in a certain region and makes inquiries to find out why. It starts by classifying each driver into one of two groups: speeders and nonspeeders. After conducting a study it finds that (i) Speeders have accidents 50% of the time during the year in which they are speeding and nonspeeders have accidents 30% of the time. (ii) Motorists who have accidents in one year have only a 20% chance of speeding during the next year, but motorists who have had no accident during the year have a 60% chance of speeding during the next year.

(a)   Assuming that the probabilities given are independent, verify the following transition matrix:

		Current year	
		Speeder	*Nonspeeder*
Next year	*Speeder*	.4	.48
	Nonspeeder	.6	.52

(b)   What fraction of the driving public will be speeders over the long run?

(c)   Construct the transition matrix that shows the probabilities of having or not having an accident this year in terms of whether you had one or did not have one last year.

(d)   What fraction of the driving public will have an accident during any one year, on the average?

9   The owner of a ski resort knows that a skier's behavior on one weekend depends on his experience on the previous weekend. If a skier had good weather, fair weather, or poor weather on one weekend, he will return with a probability of .4, .3, and .1, respectively, on the following weekend. If he stayed home one weekend, his chance of skiing the next weekend is .2 whatever the weather. There are 2000 potential customers, and 10% of the days have good weather, 50% have fair weather, and the remaining 40% have poor weather.

(a)   Set up a transition matrix for the Markov chain.

(b)   Calculate the number of guests the ski resort will have on the average during one weekend.

1 You have $2000 and want to double your money. You flip a fair coin repeatedly, betting $1000 on every toss until you either lose your $2000 or make another $2000.

(a) Verify that the transition probabilities are as follows:

Your assets after n tosses

		$0	$1000	$2000	$3000	$4000
	$0	1.0	.5	0	0	0
	$1000	0	0	.5	0	0
Your assets after n + 1 tosses	$2000	0	.5	0	.5	0
	$3000	0	0	.5	0	0
	$4000	0	0	0	.5	1.0

(b) Verify that the probabilities of having various amounts of money after $2n$ and $2n + 1$ steps are as follows:

Capital	Step $2n$	Step $2n + 1$
$0	$\frac{1}{2} - \left(\frac{1}{2}\right)^{n+1}$	$\frac{1}{2} - \left(\frac{1}{2}\right)^{n+1}$
$1000	0	$\left(\frac{1}{2}\right)^{n+1}$
$2000	$\left(\frac{1}{2}\right)^{n}$	0
$3000	0	$\left(\frac{1}{2}\right)^{n+1}$
$4000	$\frac{1}{2} - \left(\frac{1}{2}\right)^{n+1}$	$\frac{1}{2} - \left(\frac{1}{2}\right)^{n+1}$

(Notice that when $n = 0$, the probability of having $2000 is 1 and the probability of having any other amount is 0, which is as it should be.)

(c) Notice that if you have $2000 at step $2n$ you must have had either $1000 or $3000 at step $2n - 1$. In fact, the probability of having $2000 at step $2n$ is half the probability that you had $3000 at step $2n - 1$ plus half the probability that you had $1000 at step $2n - 1$. Verify this algebraically. Verify a similar statement for each of the other four states.

2 In a certain factory in which work is done around the clock, there are three shifts: morning, evening, and graveyard. Each week a worker on the morning shift can either remain on the morning shift or switch to the evening shift, an evening shift worker can go the graveyard shift or keep working evenings, and a graveyard shift worker can continue on that shift or take the morning shift. Assume that the workers' preferences are alike and the transition matrix is as follows:

Shift this week

		Morning	Evening	Graveyard
	Morning	$1 - a$	0	c
Shift next week	Evening	a	$1 - b$	0
	Graveyard	0	b	$1 - c$

How many workers will be on each shift in the long run, on the average? If $c = 1$, show that there cannot be more graveyard workers than there are either morning or evening workers in the long run. Can you see why this is so directly from the matrix without any calculations?

3 You and a friend have a total of N dollars. You repeatedly bet one dollar with your friend until one or the other of you has no money left. (This game is called "The Gambler's Ruin.") Suppose your chance of winning a bet is p and your friend's chance of winning a bet is $q = 1 - p$, and you start with i dollars and your friend starts with $(N - i)$ dollars.

It can be shown that the probability that you will eventually win the game, r_i, is given by the formula

$$r_i = \frac{\left(\frac{q}{p}\right)^i - 1}{\left(\frac{q}{p}\right)^N - 1} \quad \text{if } p \neq q$$

and

$$r_i = \frac{i}{N} \quad \text{if } p = q = \frac{1}{2}$$

(a) Using the same reasoning we used in Question 1, state in words why r_i should be equal to $pr_{i+1} + qr_{i-1}$. Then show that this equality does hold.

(b) Explain why r_0 should be 0 and r_N should be 1, and confirm that they have these values.

(c) If you have a probability of $\frac{2}{3}$ of winning each bet and $N = 100$, show that your chance of winning all your friend's money is

   (i) about $\frac{1}{2}$ if you start with $1.

   (ii) about $\frac{3}{4}$ if you start with $2.

   (iii) about $1 - \left(\frac{1}{2}\right)^K$ if you start with k dollars.

(d) Suppose a gambling house has a 1% edge, that is, your chance of winning each bet is .49. You start with i dollars and quit when you either lose it or accumulate $100. Use a hand calculator to show that the chance of winning is as shown in the following table:

Number of dollars you start with, i	Your chance of accumulating $100, r_i
$50	11.9%
60	18.7%
70	28.8%
80	43.9%
82	47.7%
83	49.7%
84	51.8%
85	54.0%

Notice that you need between $83 and $84 to have an even chance of accumulating $100 before losing your capital.

(e) Suppose you want to accumulate $100 but switch to $5 bets with the same .49 probability of winning. Show that the probability of accumulating 20 $5 bills before you go broke, r_i, corresponds to your starting capital of i $5 bills as shown in the following table:

You start with i $5 bills	Your chance of accumulating 20 $5 bills
10	40.1%
12	50.3%
14	61.3%
16	73.1%
17	79.5%

From this table and the previous one, can you reach some conclusion about betting against the odds if you have a fixed capital and you want to win a fixed amount? Specifically, should you bet large or small amounts of money?

4   In the following Markov chain find the stable probability distribution and show that the deviations from the stable probability distribution get smaller in successive steps, on the average, if a and b are both greater than 0 and less than 1.

	S	T
S	$1 - a$	b
T	a	$1 - b$

Giacomo Balla,
"La Bambina che corre
sul balcone,"
1912, Milan,
Galleria d'Arte Moderna.

8

NUTS AND BOLTS

COMPUTERS

The hardware of modern computers is ▷
composed of tiny chips of circuitry,
here shown surrounded by grains of
table salt. Each chip operates in a
few billionths of a second.

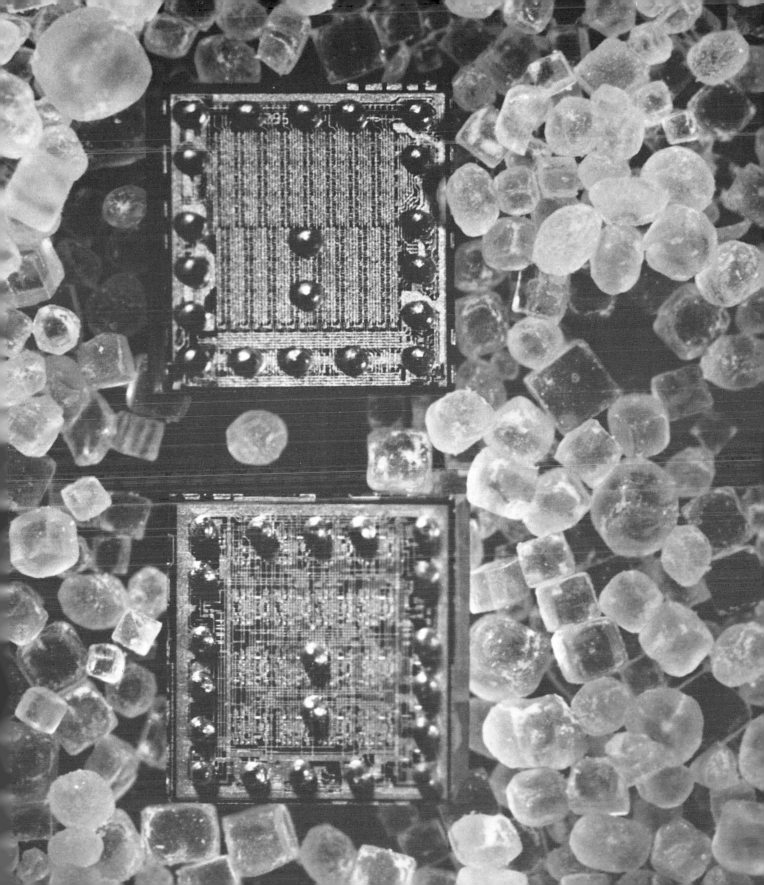

Introduction

For thousands of years humans have been making mathematical calculations. The Egyptians resurveyed the land each year after the flooding of the Nile, the pre-Incans and Druids charted the course of the heavenly bodies, any number of governments have estimated the returns from taxes, businesses have sought the best way to maximize their profits, and millions of people have balanced their budgets. To simplify their calculations people have fashioned a variety of aids. Some were mechanical devices that moved, such as clocks, hand calculators, slide rules, and abacuses; others, such as calendars, logarithm tables, and Stonehenge (a configuration of rock columns that predicted eclipses, among other things), were simply "read" by their users. The modern electronic computer is a relatively recent invention, however, dating back only to about the end of World War II.

Some properties of computers are widely known: They are fast, and they are big. What may not be so well known is that they are not terribly clever. They have absolutely no sense of humor and do not suffer even trivial faults gladly. Except for an occasional machine error, they will do what they are told—but only what they are told; they must have all the i's dotted and the t's crossed, and there is no allowance made for misplaced commas or parentheses. What may be even more surprising is the small repertoire in the computer's bag of basic tricks. If you consider the wide variety of problems that computers have tackled, it would seem that they would have to be very versatile. But actually, though computers move very quickly and can perform a wide variety of tasks, they take very small steps. They can do the ordinary arithmetic operations, of course, and some logical operations as well; they can take absolute values; they can compare two numbers to determine which is larger and base decisions on the result; and that's about all. Of such simple stuff is the computer composed.

Since the basic skills of the computer are so few and so elementary and since instructions have to be spelled out in painful detail (so that no room is left for innovation by the computer), it seems unreasonable to expect the computer to do more than follow orders in a simple-minded, straightforward way. Yet it is common knowledge that we both expect and receive a great deal more. How do you convert a plodding, unimaginative piece of

This picture of a fictitious cathedral was photographed off a color television monitor. The picture was generated by a "paint" program for use by artists in creating backgrounds for a cartoon animation system. It demonstrates the realism one can obtain using computer-generated images.

metal into a mechanism which at times appears to be clever enough to imitate human thought? We'll postpone the answer to this question until the next chapter; for now, let's look at some of the computer's more spectacular achievements:

1. A programmer—a poor amateur at checkers by his own evaluation—wrote a "checker-learning" program that was good enough to win a game from the Connecticut state champion. The experimenter observed that the computer actually "learned" faster than a human novice.*

   In the past few decades computers have been programmed, with steadily increasing skill, to play a variety of parlor games, including chess, go, Kalah, Go-Moku, Hearts, bridge, and poker.

2. In The New York Times for January 4, 1976 (page 10), there is a description of a machine that converts the printed word into computerized speech. When you consider the large number of ways a letter may be pronounced in English, this is no mean feat; in fact, the technique used exploits some recent advances in linguistic theory.

   This ability of the computer to translate and/or analyze patterns is manifested in a number of other ways:

   (a) Computers have been programmed to recognize the spoken word; they have also transformed verbal problems in mathematics into equations. Because the computer's mastery of grammar and syntax was limited, it made errors, but nevertheless it outperformed human beings, on the average.

   (b) A computer has been programmed to analyze three-dimensional objects using two-dimensional pictures. It used this analysis to reconstruct the same figure in three dimensions but from a different view; in this new construction, it suppressed those lines which had been visible earlier but were no longer in sight and added new lines which were now visible but had not been earlier.

   (c) Computers have diagnosed patients' illnesses on the basis of their symptoms, and under certain conditions their diagnoses compared favorably with that of experienced physicians. Computers have also interpreted electroencephalograms and electrocardiograms.

* See A. L. Samuel, "Some Studies in Machine Learning Using the Game of Checkers," *IBM Journal of Research and Development* 3:3 (1959), pp. 210–29.

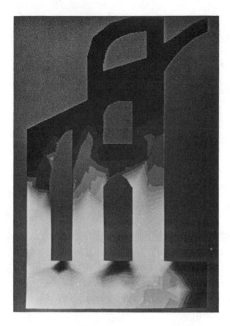

This picture represents the output from a stress analysis computer program. The structure obtained is symbolic of a Gothic church that might have been subjected to vertical dead loads and horizontal wind loads. The increased intensity of the stress contours represents an increase in stresses toward the base of the structure.

(d) Computers have interpreted handwriting, read X-rays, and matched fingerprints.

(e) Computers have translated foreign languages and have parsed and syntactically analyzed sentences.

3. It is well known that music has been composed for the computer, that is, the computer has been used as a musical instrument. What is not so well known is that music has been composed *by* the computer as well. (To be more precise, the computer was used as a tool in the composition.) In the judgment of one composer, not only can the computer take over some of the composer's menial, combinatorial tasks, but as the human composer becomes familiar with its capabilities, the computer actually creates new musical ideas.[*] A program was written that enabled a computer to act as a "banal tunemaker," creating new nursery rhymes from a synthesis of 39 old ones, and another produced tunes in the style of Stephen Foster in a similar way.

4. The computer has been used as a tool in historical analysis. At one time the authorship of some of the *Federalist* papers was in doubt; it was suspected that they were either written jointly by Madison and Hamilton or written by Madison alone. (The styles of both men were very similar; for example, the average number of words they used in a sentence was 34.59 and 34.55, respectively). After Hamilton's death Madison claimed sole authorship of the twelve disputed papers. By analyzing papers that were known to have been written by one or the other, criteria for discriminating between them were generated, and these were then applied to the disputed papers. It was finally concluded that Madison was indeed the sole author, and this conclusion was confirmed by subsequent independent historical studies.

A similar type of analysis was also made in a different context; criteria were devised that enabled a computer to determine when a piece of music was written. The program also enabled the computer to infer the composer of a piece of music, but in this the computer was less successful.

---

[*] See Herbert Brün, "From Musical Ideas to Computers and Back," in Harry B. Lincoln, ed., *The Computer and Music* (Ithaca, N.Y.: Cornell University Press, 1970), pp. 23–36.

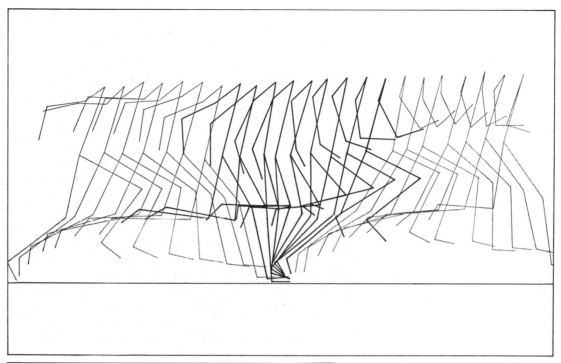

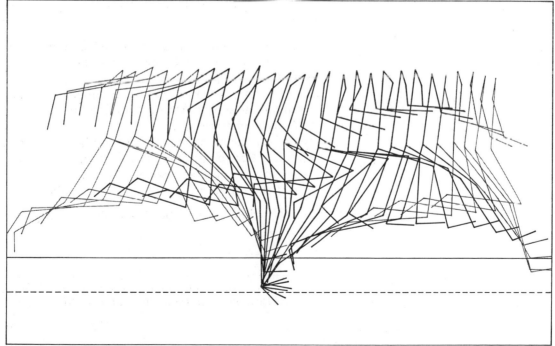

Computer simulations of the motion of a runner on a hard surface (top) and on a softer one (bottom). On the softer surface, the runner takes longer steps and is in contact with the ground longer.

From ''Fast Running Tracks,'' by Thomas A. McMahon and Peter R. Green. Copyright © 1978 by Scientific American, Inc. All rights reserved.

Basically, there are two kinds of computers: analog and digital. Analog computers work in what mathematicians call a continuous way; that is, they measure quantities that need not be expressed in whole numbers. A thermometer may read 5° or 6° or any of the infinite number of readings between these values (in theory). Thermometers, barometers, speedometers, ordinary scales, and slide rules are all crude analog computers. The digital computer, on the other hand, operates in discrete steps—it counts. The abacus, the cash register, a baseball scoreboard, and fingers and toes are all small digital computers. The computers that are popularly thought of as the "giant brains" are also digital computers, and these are the ones we are concerned with here.

It is convenient to think of a digital computer as a huge set of mailboxes (the memory) together with certain other special mailboxes in which logical and arithmetical operations can be performed. *Outside* each of these mailboxes is an address; these addresses serve much the same purpose as addresses on ordinary mailboxes, to allow the user to refer to a particular box. *Inside* each mailbox is contained a number, and these numbers may be interpreted in a variety of ways, as we shall see.

Let's look at an example: Suppose the numbers 5, 6, and 7 are contained in the mailboxes with addresses 500, 510, and 520, respectively (see Figure 1), and the following commands are executed:

1. Take the number in address 500 and place it into the accumulator (one of the special mailboxes in the arithmetic unit).
2. Add the number in 510 to the number in the accumulator and put the sum into the accumulator (deleting the number that was just there).
3. Multiply the number in the accumulator by the number in address 520 and put the product in the accumulator (again deleting the number that was there).
4. Print the number that is in the accumulator.

The number printed would be (5 + 6)(7) = 77 (see Table 1).

Figure 1

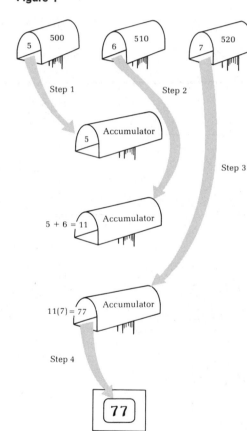

Step 1

Step 2

Accumulator
5

Step 3

5 + 6 = 11   Accumulator

11(7) = 77   Accumulator

Step 4

77

Table 1

	Contents of address 500	Contents of address 510	Contents of address 520	Contents of accumulator
Initially	5	6	7	0
After step 1	5	6	7	5
After step 2	5	6	7	11
After step 3	5	6	7	77
After step 4	The number 77 is printed.			

This short (and unrealistic) program for calculating $(5 + 6)(7)$ raises an interesting question: When the computer finishes executing one command, how does it know what to do next? When you use an ordinary hand calculator, there is no problem. If you want to compute the sum $1 + 2 + 3 + \cdots + 100$, for example, you first add 1 and 2. When the calculator has finished, you add 3 to the sum to get 6. After that you add 4 to the new sum, and then 5, and so on. But on a modern computer in which individual operations are carried out in millionths or billionths of a second, it would obviously be inefficient to have the computer wait for the operator to issue a new command after completing the old one.

To solve this problem the famous mathematician John von Neumann suggested a system of internal programming; a *program* spells out in advance the commands that the computer must perform, and these commands are stored in memory along with the data. When an operation is finished, the computer automatically proceeds to the next command, without human intervention.

The commands in memory are indistinguishable from ordinary numbers; if an engineer were to examine the contents of two locations in memory, it would be impossible to tell one from the other. It is the responsibility of the programmer (the person who writes the program) to do the careful bookkeeping required to guarantee that the two don't get confused. In practice, a master programmer writes a general program which incidental users may call on to avoid much of this tedious work.

We have already said that computers are fast and big, but we haven't said how fast or how big. In the 1940s the time taken to complete a single arithmetic operation was measured in thousandths of a second; by now this time is measured in millionths of a second. A computer's size is a bit more difficult to measure. In a sense a computer has an unlimited size, because information may be stored on tape and recalled at will, virtually without restriction. The drawback in using tape is its low speed; even "high-speed" tape systems are slow compared to the speed with which calculations are made in the computer's core (the "mailbox" system).

It may seem that it would be pointless to increase the speed and capacity of computers above a certain level. In fact, when computers first came into widespread use, there was a general feeling that their technical capacity was all that one could reasonably want. However, it was not very long before demand exceeded their capabilities. To take just one example: At the end of World War II there was an attempt to use computers to get a 24-hour weather prediction. The attempt was successful in that the prediction was reasonably accurate if the appropriate data were supplied. Unfortunately, the calculations themselves took 24 hours, so that what was obtained was in effect a weather "prediction" for the past 24 hours.

In the rest of this chapter we will neither give a short course in programming nor discuss numerical analysis—the mathematics of computers—in a serious way. We will, instead, indicate some of the problems that grow naturally out of the use of computers and describe, in a general way, how they are handled.

The Anatomy of a Number

An essential part of any computer is its *memory*—the numbers and commands that are stored within it and are available on demand. These stored numbers are constructed from certain basic building blocks. These building blocks vary from computer to computer (they may be relays, flip-flops, magnetic core, etc.), but the particular hardware used will not concern us here. What is important is one property that all these basic components have in common: They all have exactly two stable states.

All the different types of components that we just mentioned are basically like light switches; they are either ON or OFF. We will see that, because numbers are represented in memory by these ON or OFF switches, it is convenient to express numbers in certain ways when you are reading them into the computer, having the computer print them out, or storing them in memory. This method of representing numbers suggests the following question: If N switches are assigned to represent numbers, how many different numbers can they represent?

Since N two-way switches can only be in a certain number of different positions, this limits how many different numbers can be represented. A single switch has two possible positions, ON and OFF, so a single switch can represent two numbers. These may be any two numbers at all: 0 and 1, $+1$ and -1, 15 and -7, or whatever, as long as there are only two numbers. If N is 2, there are 4 possible positions the switches can assume:

(ON, ON),   (ON, OFF),   (OFF, ON),   and   (OFF, OFF).

(Note that switch 1 ON and switch 2 OFF is different from switch 1 OFF and switch 2 ON.)

If we let  indicate that the switch is ON and ○ indicate that the switch is OFF, then the 4 positions that the two switches may be in are

Three switches may be in any one of 8 positions:

There is a simple relationship between the number of switches and the number of different positions they can asume. Suppose N light switches can assume M different positions. If you add a new switch, you double the number of possible positions; you can put the original N switches in M different positions with the new switch ON and put them in M different positions with the new switch OFF—a total of $2M$ positions in all. Table 2 shows how the number of positions increases from 4 to 8 when the number of switches increases from 2 to 3. Since one light switch has 2 possible positions, 2 light switches have double that or 4 possible positions, 3 light switches have double that or 8 positions, and in general N light switches have 2^N possible positions.

Table 2

4 different possible positions of 2 switches	Position of third switch	4 different possible positions of 2 switches	Position of third switch
(OFF, OFF)	ON	(OFF, OFF)	OFF
(OFF, ON)	ON	(OFF, ON)	OFF
(ON, OFF)	ON	(ON, OFF)	OFF
(ON, ON)	ON	(ON, ON)	OFF

Numbers in Binary Form

Suppose now that N switches have been assigned to some particular mailbox and that each of the 2^N different positions the switches can take is to represent some number from 0 to $2^N - 1$. Thus all the positions and all the numbers are matched. There is a natural way of translating positions into numbers, and it will be best understood if we first look at a system with which we are familiar: the ordinary decimal system.

When we express a number in our decimal system, we are actually using a kind of shorthand in which the *positions* of the digits as well as the digits themselves play an important role. It is because the positions of the digits have significance that 1234 and 4321 represent different numbers. When we write 1234, we mean

$$1 \times 10^3 + 2 \times 10^2 + 3 \times 10^1 + 4 \times 10^0 \quad (10^0 = 1)$$

or $1000 + 200 + 30 + 4$. A decimal number such as this might be represented by 4 switches—a units switch, a tens switch, a hundreds switch, and a thousands switch—where each switch may assume any of the ten positions $0, 1, 2, 3, \cdots, 8$, and 9 (see Figure 2).

Figure 2

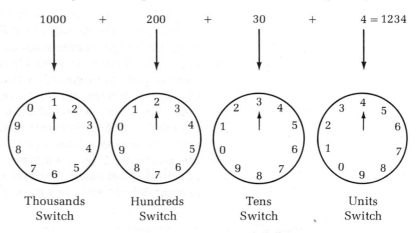

$$1000 \quad + \quad 200 \quad + \quad 30 \quad + \quad 4 = 1234$$

Thousands Switch   Hundreds Switch   Tens Switch   Units Switch

Numbers can also be represented by 2-state, rather than 10-state, switches. Numbers expressed in this form are called *binary numbers*. They can be constructed in a similar way, by letting the OFF state represent 0 and the ON state represent 1. You use one switch as a 2^0 switch, another as a 2^1 switch, another as a 2^2 switch, and so on, and the last switch as the 2^N switch. When the 2^i switch is in the ON position, 2^i is added to the number being represented; if it is in the OFF position, nothing is added. So if six switches are in the positions (ON, OFF, ON, ON, ON, OFF), for example, the binary number represented would be 101110, or

$$1 \times 2^5 + 0 \times 2^4 + 1 \times 2^3 + 1 \times 2^2 + 1 \times 2^1 + 0 \times 2^0$$
$$= 32 + 8 + 4 + 2 = 46$$

EXAMPLE 1

Write each of the following binary numbers in decimal form:
(a) 110010   (b) 101010   (c) 1101

Solution

(a) $110010 = 2^5 + 2^4 + 2^1 = 32 + 16 + 2 = 50$
(b) $101010 = 2^5 + 2^3 + 2^1 = 32 + 8 + 2 = 42$
(c) $001101 = 2^3 + 2^2 + 2^0 = 8 + 4 + 1 = 13$

EXAMPLE 2

How many switches would be needed to represent all the numbers
from 1 to   (a)   50?   (b)   100?   (c)   1000?

Solution

(a) Since 50 is less than $2^6 = 64$, 6 switches are enough; 2^5 is only
     32, so 5 switches won't do.
(b) 100 is between $2^6 = 64$ and $2^7 = 128$, so 7 switches are the
     least number you can use.
(c) 1000 is between $2^9 = 512$ and $2^{10} = 1024$, so 10 switches are
     required.

EXAMPLE 3

Write each of the following decimal numbers in binary form:
(a)   42   (b)   112   (c)   5000   (d)   61

Solution

The following listing of powers of 2 will be helpful in transforming
decimal numbers into binary numbers:

$$2^0 = 1 \qquad 2^4 = 16 \qquad 2^8 = 256 \qquad 2^{12} = 4096$$
$$2^1 = 2 \qquad 2^5 = 32 \qquad 2^9 = 512$$
$$2^2 = 4 \qquad 2^6 = 64 \qquad 1^{10} = 1024$$
$$2^3 = 8 \qquad 2^7 = 128 \qquad 2^{11} = 2048$$

(a) The largest power of 2 that is less than 42 is $2^5 = 32$;
     $42 = 32 + 10$. Since $10 = 8 + 2$, we write

$$42 = 32 + 8 + 2 = 2^5 + 2^3 + 2^1, \quad \text{or} \quad 101010$$

(b) We can break down 112 into a sum of numbers which are
     all powers of 2 as follows: $112 = 64 + 48 = 64 + 32 + 16 =$
     $2^6 + 2^5 + 2^4$, or 1110000 in binary form.

(c) $5000 = 4096 + 904 = 4096 + 512 + 392 = 4096 + 512 +$
256 + 136 = 4096 + 512 + 256 + 128 + 8 = $2^{12} + 2^9 + 2^8$
$+ 2^7 + 2^3$, or 1001110001000 in binary form.

(d) $61 = 32 + 29 = 32 + 16 + 13 = 32 + 16 + 8 + 5 = 32 +$
$16 + 8 + 4 + 1 = 2^5 + 2^4 + 2^3 + 2^2 + 2^0$, or 111101 in binary
form.

Figure 3

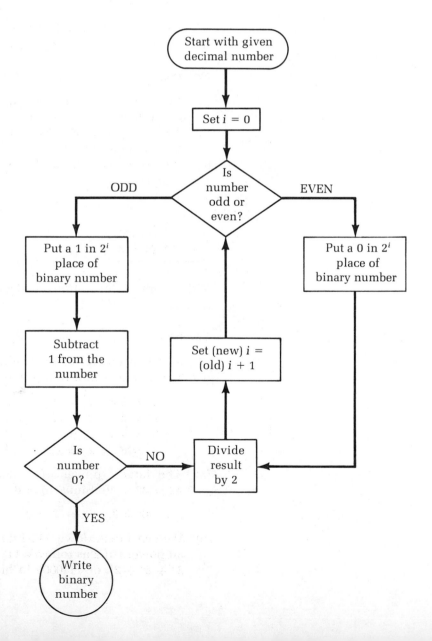

There is a simple, mechanical procedure for converting decimal integers into binary integers. (Such a procedure is called an *algorithm.*)

Step 1: Check to see if the decimal number is odd or even. If odd, put a 1 in the 2^0 place, subtract 1 from the number, divide by 2, and go on to Step 2. If even, put a 0 in the 2^0 place, divide by 2, and go on to Step 2.

Step 2: Check to see if the number obtained in Step 1 is odd or even. If odd, put a 1 in the 2^1 place, subtract 1 from the number, divide by 2, and go on to Step 3. If even, divide by 2, put a 0 in the 2^1 place, and go on to Step 3.

Continue in this way until the decimal number becomes zero—you will then have your binary number. (Figure 3 shows a flowchart for this procedure. You may find it helpful to follow the steps in the algorithm on the chart.)

EXAMPLE 4

Convert 107 to binary.

Solution

		Binary Number
Step 1: $(107 - 1)/2 = 53$	(107 is odd)	1
Step 2: $(53 - 1)/2 = 26$	(53 is odd)	11
Step 3: $(26)/2 = 13$	(26 is even)	011
Step 4: $(13 - 1)/2 = 6$	(13 is odd)	1011
Step 5: $6/2 = 3$	(6 is even)	01011
Step 6: $(3 - 1)/2 = 1$	(3 is odd)	101011
Step 7: $(1 - 1)/2 = 0$	(1 is odd)	1101011

Thus 107 in decimal is equivalent to 1101011 in binary.

Because computers are constructed from elements that have two possible states, it is convenient to use binary arithmetic. Still, since it's easy enough to convert binary numbers to decimal numbers (or to reverse the process), it's worth asking which is better. Humans presumably use decimal numbers because they happen to have ten fingers, but would a binary number system be likely to gain general acceptance under any circumstances?

It is clear that the binary system has one advantage: its simplicity. To learn the binary system you need only master two symbols, 0 and 1, whereas in the decimal system you must master ten. On the other hand, the awkwardness of the binary system makes it impractical; if you compare the decimal number 1,048,576 with its representation in the binary system, which is 100,000,000,000,000,000,000, it becomes obvious that the binary system has its own drawbacks. The length of numbers in binary form—even numbers that are not terribly large in decimal form—makes them unwieldy, and so one is tempted to move back and forth between binary form (the "natural" language of the computer) and decimal form (which is more convenient for people).

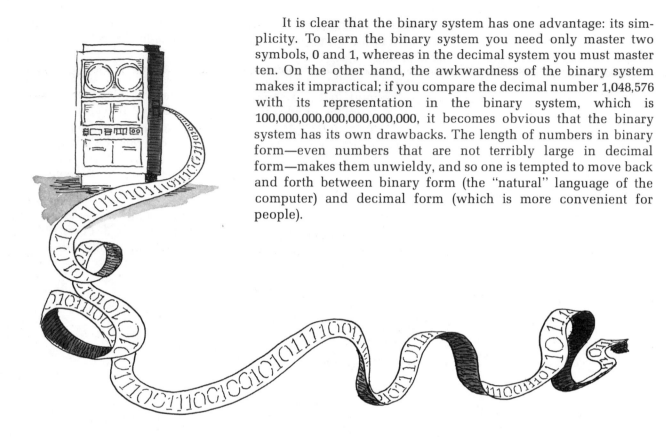

Binary numbers quickly become much longer than decimal numbers of comparable size.

One compromise between binary and decimal representations, which has been tried in practice, is to set aside binary switches to represent each decimal digit and then do everything in decimal. (We know that 4 switches can be in 16 different positions, more than enough for the 10 decimal digits.) This makes the computer and its human operators most compatible, but one pays a heavy price in loss of efficiency, as we will see.

Suppose $4N$ switches have been set aside to form N decimal digits. How would the quantity of numbers that can be represented in this way compare to the quantity of numbers that could be represented in binary form for $N = 1, 2, 4, 8$, and 10?

If $N = 1$, so that there are 4 switches, it is possible to form a single decimal digit; with this digit you can represent all the numbers from 0 to 9, or 10 numbers in all. Using these same 4 switches, one can generate the binary numbers from 0 to 15, or 16 numbers in all. So when $N = 1$, the decimal system has only 62.5% of the efficiency of the binary system; that is, you lose 6 of the 16 possible numbers you can generate when you switch from binary to decimal storage. The figures for higher values of N are shown in Table 3.

Table 3

A Number of switches available	B Number of representations in the binary system	C Number of representations in the decimal system	D Column C as a percentage of column B (the efficiency)
4	16	10	62.5%
8	256	100	39.1%
16	65,536	10,000	15.3%
32	43×10^8 (approximately)	10^8	2.3%
40	11×10^{11} (approximately)	10^{10}	.9%

So if you try to convert a computer from binary to decimal by taking 4 components capable of forming 16 states and using them to form a single decimal digit (thus wasting 6 states), you lose a great deal of storage capacity. This is why computers almost invariably do their arithmetic in binary and only convert to or from decimal when dealing with humans—in the process of reading in or reading out data.

1 Write the following decimal numbers in binary form:
   (a) 10   (b) 100
   (c) 621   (d) 1025

2 Write the following binary numbers in decimal form:
   (a) 100   (b) 1010
   (c) 110011   (d) 11111

3 How many different decimal numbers can be represented using N switches each of which can take on 10 different positions?

4 How many decimal digits would be required to represent all the whole numbers from 0 to

(a) 100?   (b) 600?   (c) 2000?

5 Can you represent more numbers using (a) 3 switches each of which can take 2 positions or (b) 2 switches each of which can take 3 positions?

6 Show that you can represent as many numbers using 6 switches each with 2 different positions as you can using 2 switches with 8 different positions.

1 It can be shown that

$$2^0 + 2^1 + 2^2 + \cdots + 2^K = 2^{K+1} - 1$$

(a) Use this formula to show that the binary number consisting of a 1 followed by N 0's is larger than the binary number consisting of N 1's, that is,

$$\underset{N \ \ 0\text{'s}}{1000\cdots000} > \underset{N \ \ 1\text{'s}}{11\cdots11}$$

(b) Use part (a) to show that two different binary representations for the same decimal number can never be found. (*Hint:* Assume that two different binary representations for a given number exist; focus your attention on that binary digit corresponding to the highest power of 2 where they differ.)

2 (a) Suppose you had 2 switches and each could be in any of 3 positions. How many different numbers could you represent?

(b) Answer part (a) if there were (i) 4 possible positions for each switch, (ii) P possible positions for each switch.

(c) Answer part (a) if there were S different switches and P possible positions for each switch.

3 A student learning to multiply in base 10 first learns the multiplication table for integers from 0 to 9. In base 2 you need only know that $(0)(0) = 0$, $(0)(1) = (1)(0) = 0$, and $(1)(1) = 1$. With this information you can multiply in binary as you would in decimal. To multiply 13 (which is 1101 in binary) by 5 (101 in binary), simply write

$$
\begin{array}{r}
1101 \\
\underline{101} \\
1101 \\
0000 \\
\underline{1101} \\
1000001
\end{array}
$$

This is 65 in binary, as it should be.

(a) Translate each of the following pairs of numbers into binary and multiply them. Then translate back into decimal and check the product.

(i)   4, 6   (ii)   3, 7   (iii)   12, 15

(b) Multiplying by 10 is particularly easy in decimal. Can you think of multiplications that would be as easy in binary? What would be the effect of multiplying by these special numbers?

4 From the equation $2^{10} = 1024$ it follows that approximately 3 decimal places (which cover the numbers from 0 to 999) do the work of 10 binary places (0 to 1111111111, which is 0 to 1023 in decimal).

(a) What can you conclude from the fact that $2^{20} = 1,048,576$?

(b) If it takes 3N decimal places to describe an integer in decimal form, how many binary places (approximately) will it take?

5 Binary numbers are written using only 0's and 1's. What do you think the following would represent in the binary system?

(a) .1   (b) .101   (c) .010101 . . .

(*Hint:* Multiply the binary number .1 by the binary number 10 just as you would in ordinary decimal arithmetic and see what you get; multiply .101 by 1000).

The Representation of Numbers: Fixed-Point and Floating-Point Forms

Let's suppose we have allotted N decimal places to some particular address and wish to use these to represent some set of numbers. (For simplicity we will assume the numbers are in decimal rather than binary form; the principle is the same in both cases, and the decimal form is more familiar.) The simplest and most straightforward way to do this is to use the form we have been using so far: the **fixed-point** form. The term "fixed-point" is used because the decimal point is assumed to be to the right of the lowest decimal digit; fixed-point numbers are always integers. So if N is 8 and the digits located in that address in memory are "12345678," "00874321," or "01020304," these would be interpreted as the integers 12,345,678, 874,321, and 1,020,304, respectively. If there are 8 places and each can contain any of ten digits, the number of different configurations that can be formed is 10^8; if the fixed-point form is used, the configurations will represent the whole numbers from 0 (00000000) to $10^8 - 1$ (99999999).

The fixed-point form will be most useful when we are dealing with nonnegative, whole numbers. If we are just counting—counting votes during an election, counting the number of books sold by a large chain of stores, or counting people in a census—the fixed-point form is ideal.

But if the application is even slightly more complicated, you need greater flexibility. If you are a bookkeeper, you may want to use negative numbers to represent your debts; this can be handled by adding a single ON–OFF switch to indicate the sign of the number. By adding such a switch to the 8 decimal places we can represent a greater range of numbers, from −99,999,999 to 99,999,999.

Even after adding a +/− switch there still remain very serious limitations to the fixed-point form. Since only integers (whole numbers) are represented, the only way to represent a fraction is to approximate it by using the closest integer. Moreover, the range of −100,000,000 to +100,000,000 is inadequate for many applications (in astronomy and biology, in particular), and we don't improve matters much by adding a few more decimal places. In short, we run into problems when we're dealing with very large or very small (in magnitude)* numbers.

Now it is certainly true that you can only represent a finite set of numbers if you only have a finite number of decimal places at your disposal, whatever method you use to represent them. So no matter how we decide to represent numbers in memory, there will inevitably be numbers that are out of range or, if they are within range, that cannot be represented exactly, only approximately. Still, if we have some fixed number of decimal places, one type of representation can be more useful than another. (A representation may be more efficient than another in one kind of application but less efficient in another.)

A particularly useful way of representing numbers is the **floating-point** form. Before we describe what the floating-point representation of a number is, we need to know something about *scientific notation.*

An astronomer who has to make statements such as "A certain star is 473,000,000,000,000,000,000,000,000 miles away" will quickly become disenchanted with the ordinary way of writing numbers. Aside from the waste of paper and ink, it is hard to keep track of the number of zeros in such very large integers. To put such numbers into a more manageable form we make use of exponents

---

* The magnitude of a number is that number with a positive sign. For example, the magnitudes of 1, −2, 0, and −7 are 1, 2, 0, and 7, respectively.

(and, in particular, of powers of 10). So 1000 is written as 10^3, 100,000 is written as 10^5, and a 1 followed by N zeros is written as 10^N. The astronomer would surely prefer to write "A certain star is 473×10^{24} miles away."

Problems also arise when you write very small numbers (in magnitude). If you are measuring the length of a light wave in centimeters, for example, you encounter numbers such as .00000542. Such numbers are more conveniently expressed with the use of *negative* powers of 10. In general, 10^{-K} means $\frac{1}{10}^K$, so that $10^{-1} = \frac{1}{10} = .1$, $10^{-2} = \frac{1}{100} = .01$, $10^{-3} = \frac{1}{1000} = .001$, and so on. We know that .00000542 means $\dfrac{542}{100,000,000}$, so

$$.00000542 = \frac{542}{100,000,000} = \frac{542}{10^8} = 542 \times 10^{-8}$$

EXAMPLE 1

Express each of the following numbers as a product of some power of 10 and the smallest possible whole number:
(a) 21,000,000,000   (b) 1,040,000,000,000,000
(c) .000000078   (d) 6.2104

Solution

(a) $21,000,000,000 = 21 \times 1,000,000,000$
$$= 21 \times 10^9$$

(b) 104×10^{13}

(c) $.000000078 = \dfrac{78}{10^9} = 78 \times 10^{-9}$

(d) $62,104 \times 10^{-4}$

EXAMPLE 2

Express each of the following as an integer:
(a) 7.21×10^8 (b) $.00000046 \times 10^{10}$

Solution

(a) 721,000,000   (b) 4600

The Floating-Point Form

We will describe the floating-point form in the case where we are given 8 decimal places and a $+/-$ switch. When the number of decimal places assigned to a number varies, we must modify the description slightly, but the principle is the same.

Numbers in fixed-point form and in floating-point form look exactly alike in the computer; the only difference between them is in the way they are interpreted. Suppose we have the number $+18276354$ in the computer. If this is a fixed-point number, it would be interpreted as the whole number $+18,276,354$. The floating-point interpretation is more complicated.

The first thing we do is break the number into three parts: the sign $(+)$, the first six digits on the left (182763), and the last two digits on the right (54). We put a decimal point to the left of the first 6 digits to obtain .182763, which we call I. We subtract 50 from the two digits on the right, 54, to obtain 4 and call this J. Then the floating-point number represented by $+18276354$ is $+I \times 10^J$, or $+0.182763 \times 10^4$.

A few questions and observations are in order. For one thing, it should be noted that one number may be represented in several different ways in floating-point form. The number 2 may be represented by $+20000051$, $+02000052$, $+00200053$, and so on. The interpretations for these representations would be $.2 \times 10^1$, $.02 \times 10^2$, and $.002 \times 10^3$, respectively, and all of these mean 2. Generally, the most significant digit of floating-point numbers, the digit farthest to the left, is not zero, so that we would use the form 20000051, rather than 02000052 or 00200053, to represent 2.

It might seem curious that we subtract 50 from the value of the last two digits before using this value as an exponent. We will spell out the reason for doing this in the exercises, but you may want to consider the problem first before reading further. The question to be answered is this: What is the range of numbers that can be represented when we subtract 50, and what is this range when we don't?

In order to decide whether the fixed-point or floating-point form is more appropriate for a particular application, it is necessary to calculate the set of *machine numbers* that are generated in each case. For any particular representation form, a **machine number** is a number that can be represented exactly by the computer. In any finite computer, whatever the form of representation, there can be only a finite number of machine numbers. In our comparison we will suppose as we did earlier that we have at our disposal a $+/-$ switch and 8 decimal digits for both kinds of representation.

"CONTROL TOWER, WE ARE FLYING AT 23,000 ft."

An error of 1 foot in measuring the height at which a plane flies is probably much less important than an error of 1 foot in measuring the height of a bridge.

If you use a fixed-point representation, the machine numbers are all the whole numbers from −99,999,999 to +99,999,999, as we observed earlier. In floating-point the range is much greater; the number 10000000 in memory would be interpreted as 10^{-51}, while the number 99999999 would be interpreted as $999,999 \times 10^{43}$, or just under 10^{49}. This range of magnitude, from 10^{-51} to 10^{49}, is enormous and generally sufficient for any application.

Having established that floating-point machine numbers have a much larger range than fixed-point machine numbers, we now inquire how accurately each form approximates the numbers that fall within this range. For numbers in the fixed-point range there is always a fixed-point machine number no more than $\frac{1}{2}$ away. If you wanted to add 1000.2 and 2412.1 in fixed point, you would have to approximate these numbers by 1000 + 2412, and you would get 3412.

Is an error of $\frac{1}{2}$ acceptable? It depends on the application and the size of the numbers with which you are dealing.

Generally speaking, you are more concerned with the *percentage* of error than with the error itself. If someone says his house is a mile away but it's actually a mile and a half away, this would be considered a more significant error than if he had said the sun was 93,000,000 miles away when it was actually 93,000,000.5 miles away. So if you are dealing with the larger numbers in the range, which are approximately 100,000,000, the maximum error of $\frac{1}{2}$ is only one part in 200 million—harmless enough for most applications. If you are dealing with numbers that are much smaller, such as the heights of adults in feet, and you approximate a height of 5.6 feet by 6 feet, the error would be considered intolerable for most applications. In general, the error introduced in using fixed-point representations for numbers that are not integers varies; the errors are smallest when the numbers are large. As we said earlier, the ideal application of the fixed-point form is to areas in which fractions cannot occur, such as adding up the night's receipts (in cents) at the local restaurant.

The actual error in floating-point will vary considerably with the size of the numbers you are handling, but the percentage error will vary much less. For numbers not too near the extremes (say, from 10^{-43} to about 10^{43}), the maximum error will *never* exceed 1 part in 200,000. This is less accurate than fixed-point at its best but considerably better than fixed-point at its worst. Generally speaking, it is almost invariably worth while to sacrifice two decimal places of accuracy for a greater range and a finer mesh, unless you are working with integers—and even then they can't be too large in magnitude.

exercises

1　Express each of the following numbers as a product of some power of 10 and the smallest possible whole number:
(a)　462,000,000,000　　　(b)　52,000,098,000,000
(c)　40,000

2　Express each of the following numbers as an integer:
(a)　$.462 \times 10^7$　　　(b)　621.1×10^3　　　(c)　21.123×10^6

3　Convert the following decimals into the product of the smallest possible integer and some power of 10:
(a)　.000345　　　(b)　.0000000000000001765
(c)　.00002100002

4　Write each number in decimal form:
(a)　423.2×10^{-6}　　　(b)　$.163 \times 10^{-6}$　　　(c)　5000×10^{-8}

5　In the three columns below we have listed a number as it appears in memory, the fixed-point interpretation of the number, and the floating-point interpretation of the number. Fill in all the blank spaces.

Memory	Number in fixed-point	Number in floating-point
$+12356258$	$+12,356,258$	$+0.123562 \times 10^8$
-00001235	$-1,235$	$-0.000012 \times 10^{-15}$*
-56123453		
$+23000048$		
	$+1,000,000$	
	$-21,000,046$	
		$+0.12 \times 10^6$
		$+0.231221 \times 10^{-5}$

*The leftmost digit should not be a 0 so this would not normally be a floating-point representation; the number would normally appear in the computer's memory as -12000031.

6　Assume that the numbers below are in floating-point form; add them and put your answer in floating-point form.
(a)　50000050 and 40000049
(b)　10000049 and 20000040
(c)　40000052 and 80000048

7 Repeat Exercise 6, but multiply the numbers rather than adding them.

8 Repeat Exercise 6, but assume that the numbers are in fixed-point rather than floating point form.

9 (a) Can you think of a reason for subtracting 50 from the last two digits before using it as an exponent for 10 in floating-point numbers?
  (b) Suppose you didn't subtract 50; what would the range of machine numbers be? What would the disadvantages of this range be?

10 With 8 decimal places, using a fixed-point representation, you can represent all the numbers from 0 to 99,999,999, or 100,000,000 numbers in all. If you add a sign switch, you should be able to double the quantity of numbers you can represent, but in fact you can only represent the numbers from −99,999,999 to 99,999,999, or 199,999,999 numbers in all. What happened to the missing number?

11 Suppose 46560050 represents the number P in floating-point form, 00000010 represents the number Q in fixed-point form, 23124350 represents the number R in floating-point form, and 00000002 represents the number S in fixed-point form.
  (a) Add P and Q and express your answer in fixed-point and in floating-point form as accurately as possible.
  (b) Add P and R and express your answer in fixed-point and in floating-point form as accurately as possible.
  (c) Multiply P and Q and express your answer in fixed-point and in floating-point form.
  (d) Multiply Q and S and express your answer in floating-point form.

1 Suppose a computer is given two numbers in floating-point form. Can you describe a technique for finding the product of these numbers and putting the answer into floating-point form?

2 Suppose a computer is given two numbers in floating-point form. Devise a technique for adding the two numbers and putting the answer into floating-point form.

3 Suppose we broke down an 8-digit number in memory, "abcdefgh," into two parts: $I = .abcdefg$ and $J = h - 5$, and let the 8 digits represent $I \times 10^J$. What would be the advantages and disadvantages of using this modified floating-point system over the one described in the text?

Errors

When scientists, engineers, or planners want to solve a difficult problem that requires a numerical solution (such as deciding how many booths to install on a toll bridge, calculating the temperature distribution of a metal plate, or predicting tomorrow's weather for a certain town), there is a general procedure they often follow: First they find or construct a general mathematical model that reflects the physical laws governing the problem. Next, a computer program must be written to direct the computer to process data in accordance with these physical laws. They then "tell" the computer the parameters of the particular problem to be solved (the traffic expected on the bridge, the temperatures at the edge of the plate, or today's winds, air pressure, and temperature in the vicinity of the the town), and this information is processed in accordance with the model. If no clerical errors are made, the computer will eventually reach a solution. But, the solution obtained may differ from the ideal solution because of the introduction of a variety of errors along the way.

We have already discussed one kind of error: the **round-off error**. A number that cannot be represented exactly by the computer must be approximated, and the difference between the exact value and the approximation is called the round-off error. And even if the original numbers are represented exactly, they will almost invariably become inexact after you have calculated with them; when two machine numbers are used in an arithmetical operation, the result need not be a machine number. For example, when 1 is divided by 3, we obtain .3333 . . . ; this is a nonterminating decimal that is best approximated in fixed-point by 0. The most accurate floating-point approximation is 33333350, which is considerably better but still leaves us with an error of 1 part in 300,000,000. Errors also result when we perform some mathematical operation incompletely, such as summing an infinite number of terms.

We are interested not only in how errors can arise but also in how they grow. It might seem that an error of one part in a billion is not worth worrying about, and often it may not be. It takes just a few examples, however, to show that seemingly negligible errors under the right circumstances can render a solution worthless. We won't try to analyze errors systematically or in depth, but we will point out a few of the pitfalls.

Since errors are a fact of life, we have to accept them; but we try to avoid letting them grow beyond acceptable levels. Even before we talk about "acceptable levels," we must first decide which errors we will consider large and which we will consider small. For example, is an error of 10,000 to be considered large or small? As we have seen, the answer depends on the context, that is, on the size of the numbers with which you are dealing. Generally, it is more meaningful to express an error as a fraction or a percentage of the true value of a number than as an absolute number.

Suppose that x is the true value of a number and $\bar{x}$ is an approximation of x. We make the following definitions:

The **absolute error** of $\bar{x}$ is the approximate value minus the true value, that is, $\bar{x} - x$.

The **relative error** of $\bar{x}$ is the absolute error divided by the true value of x, that is, $(\bar{x} - x)/x$.

The **percentage error** of $\bar{x}$ is the relative error multiplied by 100, that is, $(100)(\bar{x} - x)/x$.

> ### EXAMPLE 1

Suppose x is 10 but is approximated by 9.8 and y is 20 and is approximated by 19.7.

(a) Which of these two numbers has the greatest absolute error in magnitude?

(b) Which of the two numbers has the greatest relative error in magnitude?

(c) What is the absolute error in $\bar{x}\,\bar{y}$?

(d) What is the relative error in $\bar{x}\,\bar{y}$?

We are interested not only in how errors arise but also in how they grow.

"WHAT'S AN ⅛"   "OH, THIS IS CLOSE ENOUGH"   "WHAT'S A 1/16"

Solution

Since $x = 10$ and $\bar{x} = 9.8$, the absolute error in x is $9.8 - 10 = -0.2$. The relative error in x is $-0.2/10 = -0.02$. The percentage error in x is $(-0.02)(100) = -2\%$. Since $y = 20$ and $\bar{y} = 19.7$, the absolute error in y is -0.3 and the relative error in y is $-0.3/20 = -0.015$. The percentage error in y is $(-0.015)(100) = -1.5\%$. Therefore:

(a) y has the greatest absolute error in magnitude.

(b) x has the greatest relative error in magnitude.

(c) Since $xy = 200$ and $\bar{x}\bar{y} = 193.06$, the absolute error in xy is -6.94.

(d) What is the relative error in $\bar{x}\,\bar{y}$?

Since all measurements in this world are approximations, questions like the following often arise. Suppose $\bar{x}$, an approximation to x, is squared. Will the relative error increase, decrease or can you tell? If there is an error in x, what will be the error in $\sin x$, $\sqrt{x}$, and $1/x$? If there are errors in both x and y, what will be the error in their sum? Their product? If $x^2 + qx - 1 = 0$ and there is an error in q, what will be the effect on the roots (solutions) of the equation?

There is a single, common thread in all of these questions: If the numbers we are working with turn out to be incorrect, what effect will this have on the results of our calculations? The answer depends on what specific calculations we are making. As we have said, a systematic study of how errors propagate is beyond our scope, but we will look at a few special cases to see what can go wrong. Let's start with a simple case first: the addition of positive numbers.

Suppose a mining conglomerate receives annual reports from each of its thousands of subsidiaries which estimate the amount of ore processed by them during the previous year. If each estimate is correct to within 2%, how inaccurate will the total be at the worst?

Your first guess might be that the 2% errors will accumulate and, after thousands of additions, the resulting error will be so large as to make the sum meaningless. But in fact, the total is very likely to be quite accurate; in any case, the error cannot exceed 2%. Let's see why.

For simplicity, suppose there are only 4 subsidiaries and the amounts they mined were A_1, A_2, A_3, and A_4 (in pounds) in the last year. If each subsidiary overestimated its production by the maximum of 2%, they would report diggings of $(1.02)A_1$, $(1.02)A_2$, $(1.02)A_3$, and $(1.02)A_4$. The absolute error of the sum would be

$$(.02)A_1 + (.02)A_2 + (.02)A_3 + (.02)A_4$$

and the relative error would be

$$\frac{(.02)A_1 + (.02)A_2 + (.02)A_3 + (.02)A_4}{A_1 + A_2 + A_3 + A_4} = \frac{(.02)(A_1 + A_2 + A_3 + A_4)}{A_1 + A_2 + A_3 + A_4}$$

$$= .02$$

or equivalently, 2%.

In exactly the same way we can easily show that if production is underestimated by 2% in each subsidiary, the total will also be exactly 2% too low. These two cases are the extremes. If (as is more likely) some subsidiaries overestimate their production while others underestimate it, the two types of error will tend to cancel one another and the error in the sum will be somewhere between the two extreme errors.

In short, if we are doing nothing but adding positive numbers—taking a census, counting votes, adding up donations to a charity—the error in the sum will be greater than the smallest but less than the largest of the errors in the individual terms being added; the relative error of the sum will never be larger than the largest relative error in the terms being added. And this is the case no matter how many terms there are. (Of course, the *absolute* error may grow as the number of terms increases).

We just observed that the relative error of a sum of positive numbers—of however many terms—can be kept within bounds by controlling the individual terms of which it is composed. If we take a product of two approximations, the problem becomes more troublesome.

Suppose that $x = 200$, $\bar{x} = 201$, $y = 800$, and $\bar{y} = 802$. The relative error in $\bar{x}$ is $\frac{1}{200}$ or .005, and the relative error in $\bar{y}$ is $\frac{1}{400}$ or .0025. The real product xy is 160,000, while the approximate product $\bar{x}\bar{y}$ is 161,202, so the relative error of the product is $\frac{1202}{160,000} = .007515$. Notice that the sum of the relative errors in $\bar{x}$ and $\bar{y}$ (.005 + .0025 = .0075) is very close to the relative error in the product $\bar{x}\bar{y}$ (.007515); this is not a coincidence.

The relative error in a product is approximately the sum of the relative errors in its individual factors.

Because of the way errors accumulate in multiplication, it is not hard to imagine conditions under which a large number of small and apparently harmless errors transform an otherwise legitimate calculation into pure rubbish. One hundred factors, each 1% too large, can result in a product that is twice as large as it ought to be. (In fact, the situation is even worse than it seems; if the number of calculations is large enough and the errors aren't sufficiently small, the higher-order errors that we neglect may come back to haunt us. For example, if there are 100 factors in a product and each is too large by 1%, the final product will be almost triple, not double, what it ought to be!)

But the fact is that this very worst situation rarely materializes. If you multiply a large number of factors, they generally won't be all too large or all too small; more often they will be too large about as often as they are too small, so that the factors that are too large will tend to balance the factors that are too small. Thus the final result is usually considerably better than what a pessimist might predict.

The relative error of a quotient can be approximated in terms of the relative errors of the divisor and dividend. *If $\bar{x}$ has a relative error of a and $\bar{y}$ has a relative error of b, then the quotient $\bar{x}\,\bar{y}$ has a relative error of approximately $a - b$.* Notice that if you are calculating $\bar{x}/\bar{y}$ and the relative error in $\bar{y}$ is b, the relative error in the quotient is almost the same as would occur if you were finding the product $\bar{x}\,\bar{y}$ and the relative error in $\bar{y}$ were $-b$.

EXAMPLE 2

Hal travels 60 miles in 80 minutes. His inaccurate odometer indicates that he has traveled 63 miles, and his faulty watch leads him to believe he has only been traveling for 78 minutes.
(a) What are the relative errors in his traveling time and distance?
(b) What is the exact relative error in his average speed?
(c) Find the approximate relative error in his average speed and compare the answer to the one obtained in part (b).

Solution

(a) We calculate the relative error in Hal's traveling time, which is $(78 - 80)/80 = -\frac{1}{40}$. The relative error in the distance traveled is $(63 - 60)/60 = \frac{1}{20}$.

(b) The average speed (in miles per minute) is (distance traveled/ time). Hal's exact average speed is $\frac{60}{80} = \frac{3}{4}$, and the average speed he calculates (because of faulty measurements) is

$\frac{63}{78} = \frac{21}{26}$. So the relative error in his average speed is

$\left(\frac{21}{26} - \frac{3}{4}\right)/\frac{3}{4} = \frac{1}{13}$.

(c) The approximate error in the average speed is $\frac{1}{20} - \left(-\frac{1}{40}\right) = \frac{3}{40}$. Notice that $\frac{1}{13} - \frac{3}{40} = \frac{1}{520}$.

EXAMPLE 3

Suppose $x = 50$, $\bar{x} = 52$, $y = 100$, and $\bar{y} = 99$.
(a) Find the relative error and the absolute error in $\bar{x} + \bar{y}$.
(b) Find the exact value of the relative error in (i) $\bar{x}\,\bar{y}$, (ii) $\bar{x}/\bar{y}$. Then find the approximate values of the relative errors obtained by adding or subtracting the relative errors in $\bar{x}$ and $\bar{y}$.

Solution

(a) Since $x + y = 150$ and $\bar{x} + \bar{y} = 151$, the absolute error is 1 and the relative error is $\frac{1}{150}$. Notice that the *magnitude* of relative error of the sum is less than either the relative error in $x \left(\frac{1}{25}\right)$ or that in $y \left(-\frac{1}{100}\right)$; this is because compensating errors introduced by numbers that are both too large and too small tended to balance out.

(b) (i) Since $xy = 5000$ and $\bar{x}\,\bar{y} = 5148$, the exact value of the relative error in xy is $\frac{148}{5000} = .0296$. If we neglect higher-order errors, we may simply add the relative errors in x and y to obtain $.04 + (-0.01) = .03$.
(ii) Since $x/y = .5$ and $\bar{x}/\bar{y} = .5252525$, the relative error in x/y is $.050505$. If we neglect higher-order errors, we simply subtract the relative error in y from that in x to obtain $.04 - (-0.01) = .05$.

So far, we have seen what happens when three of the four basic arithmetic operations—addition, multiplication, and division—are applied to imperfect data. When positive numbers are added, the relative error in the sum is never greater than the largest error in the terms added, so it doesn't matter how many additions there are; the relative error cannot grow. When you multiply many factors, there is some danger in principle because the errors are added; conceivably, even small errors may become large when they are repeated often (as they may well be on a computer). Still, unless there is a reason for the errors to deviate systematically in one direction or the other, the errors tend to balance out and the actual accuracy is much better than it might be. Division is really multiplication in a thin disguise, and the same remarks apply.

It is in the apparently harmless operation of subtraction of positive numbers that the real danger lies;[*] in a single operation a tiny error may become enormous. Consider the following example:

Suppose $x = 990$, $y = 1010$, $\bar{x} = 985$, and $\bar{y} = 1015$. If we calculate the relative error in the difference $\bar{y} - \bar{x}$, we obtain $(30 - 20)/20 = \frac{1}{2}$ or 50%. The initial relative errors in $\bar{x}$ and $\bar{y}$ were about -0.005 and $+0.005$, respectively. So by a single subtraction we have transformed two numbers that were correct to within $\frac{1}{2}$ of 1% to one with a 50% error.

The reason for this difficulty is not hard to see. The difference between two numbers that are roughly the same size is small. When calculating the relative error in this difference, we divide by this small number, and the quotient becomes correspondingly large.

EXAMPLE 4

Suppose $x = 1000$, $\bar{x} = 1010$, $y = 990$, and $\bar{y} = 995$. Calculate
(a)   the relative error in x.
(b)   the relative error in y.
(c)   the relative error in $x - y$.

Solution

(a)   The relative error in x is .01.
(b)   The relative error in y is .00505.
(c)   The relative error in $x - y$ is $(15 - 10)/10 = .5$.

---

[*] When we discuss addition or subtraction, we assume that we are dealing only with positive numbers. Adding a negative number and subtracting a positive one are equivalent in terms of error analysis. Similarly, adding a positive number and subtracting a negative one are equivalent operations.

1 For each pair of values X and $\bar{X}$, give the relative and absolute errors:
  (a) X = 10, $\bar{X}$ = 10.2
  (b) X = $\bar{X}$ = 10
  (c) X = 4, $\bar{X}$ = 4.1

2 Given X and the relative error, find $\bar{X}$ (the approximation) in each case:
  (a) X = 20, relative error = .1
  (b) X = 10, relative error = −0.01
  (c) X = 100, relative error = −0.01

3 If you have a rectangular yard that is actually 200 feet × 100 feet and relative errors of 1% in the length and 2% in the width, what is the largest relative error you can make if you want to
  (a) fence it?
  (b) paint it?

4 Repeat Exercise 3 for a square yard with sides 50 feet long and a relative error of $\frac{1}{2}$% in the length of the sides.

5 If x = 10 and $\bar{x}$ is within 1% of x, y = 2 and $\bar{y}$ is within 7% of y, find the largest possible absolute error and relative error in the sum, $\bar{x} + \bar{y}$.

6 Find the same errors as you did in Exercise 5 if x and y are within 7% and 1% of x and y, respectively, where x = 10 and y = 2.

7 An oil pipe line from Alaska to a refinery in Chicago is to be constructed by welding 50,000 pipe sections together. The pipe's total length is 4000 miles, and the length of each section is correct to within $\frac{1}{100}$ of 1%.
  (a) What is the largest possible error in the total length?
  (b) What is the largest possible relative error in the total length?
  (c) If there were 10,000 sections, how would this affect your answers?

8 A tunnel is to be built by joining two sections, one $\frac{1}{4}$ mile long and the other $\frac{3}{4}$ mile long, from opposite banks. There is a 1% error in length in one of the parts and a 2% error in the other. Calculate the absolute error and relative error in the total length if the 1% error is in
  (a) the $\frac{1}{4}$ mile pipe
  (b) the $\frac{3}{4}$ mile pipe

9 Your watch is correct to within 1%, and your speedometer is correct to within 2%. You travel along a road for 5 hours (according to your watch) at a steady rate of 20 miles per hour (according to your speedometer).

(a) How much can the actual distance traveled differ from your estimate based on these instruments? What is your maximum possible relative error?

(b) Would your answer to either of these questions be different if it was the speedometer that was correct to within 1% and the watch correct to within 2%?

10   Approximate the maximum possible percentage error you can make in estimating the volume of a box whose length, width, and height are correct to within $\frac{1}{2}$%, $\frac{1}{4}$%, and 1%, respectively.

11   Approximate the maximum relative error in
(a) the area of a circle of radius r (area $= \pi r^2$)
(b) the volume of a sphere of radius r (volume $= \frac{4}{3}\pi r^3$)

12   Suppose that $T = AB + C$. Find the maximum possible error in T (approximately) if A, B, and C all are positive and the percentage errors in A, B, and C are 4%, 5%, and 3%, respectively, for
(a) $C = 900$, $B = 10$, $A = 30$ (b) $C = 900$, $B = 30$, $A = 10$
(c) $A = 5$, $B = 10$, and $C = 900$

13   A manufacturer produces metal squares and has them inspected by two different inspectors. One inspector makes sure that the length of each side is correct to within 2%, and the other makes sure the area is correct to within 2%. (The lengths of the sides are always the same even if they are inaccurate.)
(a) Which inspector has a stricter standard? What is the relationship between the degrees of accuracy required by each?
(b) Answer part (a) if the product were metal cubes and one inspector checked the length of a side to within 2% and the other checked the volume to within 2%.

14   Suppose a computer only has a capacity for storing numbers to 3 decimal places, that is, every number is expressed as an integer plus some whole number of thousandths. If as a result of a calculation we obtain a number such as 24.645z (where z is the integer in the fourth decimal place), we can do either of two things (among others): We can either simply drop the z and convert the number to 26.645, or we can round off, that is, convert to 24.645 if z is 4 or less and convert to 24.646 otherwise. If we had a problem in which we multiplied 10,000 such numbers together, can you see an advantage to using one method rather than the other?

15   Suppose 5 numbers that should all be equal to 10 are multiplied together. If the maximum error in each is 1%, what is the maximum possible relative error in the product (approximately)? Suppose that two of the factors are 1% too high, two of the factors are 1% too low, and the fifth factor is exact; what would the relative error in the product be then? Are estimates of the maximum relative error sometimes unduly pessimistic?

16 Suppose x = 40 and y = 10. Suppose $\bar{x}$ has a percentage error of at most 2% and $\bar{y}$ has a percentage error of at most 1%. If u = x/y, find the values of $\bar{x}$ and $\bar{y}$ that give $\bar{u} = \bar{x}/\bar{y}$ its greatest and smallest values. What are these values? Do your calculations exactly and by using the appropriate approximation.

17 If a number is squared, the relative error in the square is roughly double that of the relative error of the number. (Why?) If a number is raised to the 8th power, the relative error will be about 8 times that of the original number. Can you guess what will happen if you take the square root of the 8th root of a number? Check the following table to confirm your guess.

Exact Number	Approximation	Relative Error	Actual Square Root	Approximate Square Root	Relative Error in Square Root	Actual 8th Root	Approximate 8th Root	Relative Error in 8th Root
6	6.1	.016667	2.44949	2.46982	.00832	1.25103	1.25362	.00207
21	21.01	.000476	4.58258	4.58367	.0002378	1.46311	1.46320	.00006
.5	.51	.02	.70711	.71414	.00994	.91700	.91928	.00249

extras for experts

1 Suppose x and y are two positive numbers, with y larger than x, and $\bar{x}$ and $\bar{y}$ are overestimates of x and y, respectively.
 (a) Will there be a greater absolute error in the sum when x has the larger relative error or when y has the larger relative error?
 (b) Answer the same question about the relative error in the product.
 (c) What if either or both of $\bar{x}$ and $\bar{y}$ can be smaller than the exact values?

2 The probability of the ith component of a computer failing in a given time period is p_i. The probability of 50 components all failing during the same period is the product $p_1 p_2 p_3 \cdots p_{50}$. If the individual probabilities can be in error by as much as $\frac{1}{100}$ of 1%, how far can the estimate of the failure be off? (That is, find the maximum possible relative error in your estimate of a total breakdown of all components.)

3   To solve the quadratic equation $Ax^2 + Bx + C = 0$ you may use the formula

$$x = \frac{-B + \sqrt{B^2 - 4AC}}{2A}$$

The root $(-B + \sqrt{B^2 - 4AC})/2A$ can be converted algebraically to the form $\dfrac{-2C}{(B + \sqrt{B^2 - 4AC})}$ by multiplying numerator and denominator by $(B + \sqrt{B^2 - 4AC})$. Suppose you want to solve the equation $(.001)x^2 + 10x + .1 = 0$ with a computer that can only carry 4 decimal places. After setting $A = .001$, $B = 10$, and $C = .1$, the first formula yields

$$x = \frac{-10 + \sqrt{100 - .0004}}{.002}$$

But $\sqrt{100 - .0004} = \sqrt{99.9996}$, and this quantity turns out to be 9.9999799. To 4 decimal places this is 10, so the answer turns out to be 0. The algebraically "equivalent" formula $x = -2C/(B + \sqrt{B^2 - 4AC})$ becomes $-0.2/(10 + \sqrt{100 - .0004})$, which becomes (because of the computer's limitations) $-0.2/(10 + 10) = -0.01$. The answer (to this degree of accuracy) is indeed -0.01. Can you see why one form turns out to be more accurate than the other?

4   What is the effect on the roots of an equation if the coefficients are slightly inaccurate? Specifically, find the solution to the equation $x^2 - ax + b = 0$ when
(a) $a = 1000$ and $b = 250{,}000$
(b) $a = 1000$ and $b = 249{,}999$
(c)   Confirm that the roots in (a) are both 500 while the roots in (b) are 499 and 501.
(d)   Observe that a relative error in b of 4 millionths (if $b = 250{,}000$ and $\bar{b} = 249{,}999$) can produce an error of 2/1000 in the roots. That is, the relative error in the roots is 500 times that of the coefficients.

5   Suppose there are two equations in two unknowns:

$$ax + y = 10$$
$$x + by = 10$$

and the real and approximate values of a and b are as follows: $a = 1.01$, $\bar{a} = .99$, $b = .99$, and $\bar{b} = 1.01$.
(a)   Confirm that the relative error is about .02 in both $\bar{a}$ and $\bar{b}$.
(b)   Show that the solution to the set of the equations with the exact a and b is $x = 1000$, $y = -1000$ and the solution to the set of equations where a and b are approximated is $x = -1000$, $y = 1000$. Would you have anticipated this?

6   Solve the set of equations
$$ax + 2y - z = 2$$
$$-x + y - z = -2$$
$$3x + z = 4$$

for
(a) $a = 1.1$ to obtain $x = 20$, $y = -38$, $z = -56$
(b) $\bar{a} = .9$ to obtain $x = -20$, $y = 42$, $z = 64$
(c) How do the differences in solutions compare to the differences in the
values of a?

7 Suppose A and B are two numbers you obtain experimentally. By taking
certain pains you can obtain either number more accurately, but only at
the expense of the accuracy of the other number. Specifically, the sum of
the two percentage errors in A and B must be 2%, but you can distribute
this error as you like between the two numbers. Assume that A and B are
positive.
(a) If A and B are to be multiplied, would it make any difference how
accurate A and B are (if the sum of the percentage errors remains
2%)? Do the magnitude of A and B matter?
(b) Answer part (a) assuming that you are adding, rather than multiplying, A and B.

Iterative Methods (Loops)

We saw in Section 1 that every step the computer takes must be
spelled out for it in advance. Of what use is it, then, to have a
computer that can add two numbers in millionths of a second
when the time needed to punch the command card and transmit it
to the computer is measured in seconds? The designers of high-
speed computers were aware of this question, and the following
example will show how they answered it.

Suppose you want to add up the integers from 1 to 100 to
obtain the sum $1 + 2 + \cdots + 99 + 100$. Suppose also, that you
were too busy to do the actual addition yourself so you had to leave
the job with your assistant. We will assume your assistant is
incapable of doing anything original, so you have to write out
explicit, step-by-step instructions.

It might seem that you would have to write out about 100
instructions, which might look something like this:

Step 1:   Add $1 + 2$.
Step 2:   Add 3 to the sum calculated in Step 1.
Step 3:   Add 4 to the sum calculated in Step 2.
⋮
Step 99:   Add 100 to the sum calculated in Step 98.
Step 100: Print the sum calculated in Step 99.

Let's suppose that your assistant has an adding machine that can add any two numbers and that he has a number of pads upon which he can write his answers. He's also capable of keeping an eye on the size of the numbers and stopping at the right moment if he's properly coached in advance; in this case that means he can stop adding when he gets to 100.

Now consider this way of calculating our sum: We start by having the assistant place a pad with a 1 on it in his front pocket (*F*), a pad with a 101 on it in his hip pocket (*H*), and a pad with a zero on it in his breast pocket (*B*). Having taken these initial steps, we proceed to our "loop." In each of the steps listed below a command is to be performed and the location of the next command is indicated.

> *Step 1:* Find the sum of the numbers in *B* and *F* (the numbers on the pads in the breast and front pockets, respectively.) Delete the number now in *B* and replace it by this sum. Go to Step 2.
>
> *Step 2:* Delete the number now in *F* and replace it with the number that is larger by 1. Go to Step 3.
>
> *Step 3:* Compare the numbers in *F* and *H*; if the number in *F* is lower, go to Step 1. Otherwise, halt; the number in *B* is the answer.

To get a clear idea of what is happening, let's look at a step-by-step history of the three pads—how they look initially and how they change after each step. This is shown in Table 4.

The important thing to notice about this program is the flexibility of Steps 1 to 3; although they are only written once, they are performed 100 times. And, as we shall see in the exercises, it would only require a modest revision in the program to make it add up the numbers from 1 to 1000 or 1,000,000. (The number that we put in *H* initially would be 1001 or 1,000,001 rather than 101.) Of course, the computer would perform ten times or ten thousand times as many steps, and this is why it is desirable to have high-speed machines. A set of directions in a program that is performed many times by the computer is called a **loop**.

(**EXAMPLE 1**)

A well-known procedure for calculating square roots is the technique called *Newton's method*. If you want to find the square root of a number *A*, you take a first estimate of the square root

Table 4

Step	Pad F	Pad H	Pad B
Initial reading	1	101	0
1	1	101	1
2	2	101	1
3	2	101	1
1	2	101	$3\,(=1+2)$
2	3	101	3
3	3	101	3
1	3	101	$6\,(=1+2+3)$
2	4	101	6
3	4	101	6
1	4	101	10
2	5	101	10
.	.	.	.
.	.	.	.
.	.	.	.
3	100	101	4950
1	100	101	5050
2	101	101	5050
3			ANSWER = 5050

(possibly just a guess) and label it x_0. You then use the formula

$$x_{n+1} = \tfrac{1}{2}\left(x_n + \frac{A}{x_n}\right)$$

to calculate $x_1, x_2, x_3, \ldots$. It can be shown that the values $x_1, x_2,$ $x_3, \ldots$ are better and better approximations (after the first step) to the actual value of the square root. For example, if $x_0 = 3$ and $A = 8$, then

$$x_1 = \tfrac{1}{2}\left(3 + \frac{8}{3}\right) = \frac{17}{6}$$

$$x_2 = \tfrac{1}{2}\left[\frac{17}{6} + 8/\left(\frac{17}{6}\right)\right] = \frac{577}{204}$$

and so on. Of course, eventually you must stop the process. There are a number of different rules you can use in determining when to stop. For instance, you can stop after some fixed number of iterations, or when two successive approximations are the same (or very close).

Let us assume that we have at our disposal a hypothetical computer. We will show how, by setting up a loop, we can have the computer execute an arbitrarily large number of iterations. The loop will involve writing only a small number of instructions and using only a small amount of computer storage.

So suppose that our computer is constructed in such a way that

1. There are 10 computer locations, with addresses $A(1)$ to $A(10)$, in which we will store numbers as we choose to.
2. There are 10 computer locations, with addresses $A(11)$ to $A(20)$, in which we will store commands.
3. With one command you can add, subtract, multiply, or divide any of the numbers located in locations $A(1)$ to $A(10)$—but always two numbers at a time—and store the result in any of these 10 locations. (The previous contents of that location are then deleted.)
4. The computer can compare any two numbers in locations $A(1)$ to $A(10)$ and determine the next command it will execute depending on whether one number is bigger, smaller, or equal to the other.
5. We can store such numbers and commands in $A(1)$ to $A(20)$ as we wish.

Using this hypothetical machinery, we will calculate an approximation to the square root of 5 with $x_0 = 3$, stopping when two successive estimates differ by less than .001.

Solution

We start by putting 5 in $A(1)$, 2 in $A(2)$, 3 in $A(3)$, and .001 in $A(4)$. (It will become clear as we work the problem what part these numbers play in the computer's calculations.) Then we go to $A(11)$ for the first command. The commands in $A(11)$ on are as follows:

$A(11)$: Divide the number in $A(1)$ by the number in $A(3)$ and put the quotient in $A(5)$. Go to $A(12)$ for the next command.

$A(12)$: Add the number in $A(3)$ to the number in $A(5)$ and put the sum in $A(5)$ (deleting whatever was there). Go to $A(13)$ for the next command.

$A(13)$: Divide the number in $A(5)$ by the number in $A(2)$ and put the quotient in $A(6)$. Go to $A(14)$ for the next command.

$A(14)$: Subtract the number in $A(6)$ from the number in $A(3)$ and put the magnitude of the difference in $A(7)$. Go to $A(15)$ for the next command.

A(15): Compare the numbers in A(4) and A(7).

  (a) If the number in A(7) is smaller, print the number in A(6) and halt; that is the answer.

  (b) If the number in A(7) is not smaller, put the contents of A(6) into A(3). Go to A(11) for the next command.

Listed in Table 5 are the contents of addresses A(1) through A(7) after each step. If you perform this routine repeatedly, the successive approximations to $\sqrt{5}$ are 3, 2.333, 2.238, and 2.236; the last value is correct to 3 decimal places.

Table 5

Step	A(1)	A(2)	A(3)	A(4)	A(5)	A(6)	A(7)
Initially	5	2	3	.001			
11					$\frac{5}{3}$		
12					$\frac{14}{3}$		
13						$\frac{7}{3}$	
14							$\frac{2}{3}$
15			$\frac{7}{3}$				
11					$\frac{15}{7}$		
12					$\frac{94}{21}$		
13						$\frac{47}{21}$	
14							$\frac{2}{21}$
15			$\frac{47}{21}$				
11					$\frac{105}{47}$		
12					$\frac{4414}{987}$		
13						$\frac{2207}{987}$	
14							$\frac{2}{987}$
				etc.			

Note: Entries are made in the table only when there is a *change*.

exercises

1 In the loop we described for adding the numbers from 1 to 100 we described the contents of pads *F*, *H*, and *B* on each of the three steps for about 3 iterations (see Table 4.) Fill in the table for three more iterations.

2   How would you modify the program that adds the numbers from 1 to 100 for the following purposes?

    (a)   To add the numbers from 1 to 1000 instead of from 1 to 100. To add the numbers from 1 to 1,000,000.

    (b)   To find the sum $2 + 4 + 6 + \cdots + 100$.

Would it take any longer to write out the program that would solve part (a)? Would it take any longer to carry out the steps of the program once they were written?

3   Write a general program to calculate each of the following sums, where a, b, n, and r are any numbers:

    (a) $a + (a + b) + (a + 2b) + (a + 3b) + \cdots + (a + nb)$

    (b) $a + ar + ar^2 + ar^3 + \cdots + ar^n$

4   In Illustrative Example 1, which used Newton's method to calculate the square root of 5, show the contents of each address until the entry in $A(7)$ is less than 10^{-7}.

5 $x_{n-1} = (2x_n^3 + A)/3x_n^2$ is an iterative formula to calculate the cube root of A by Newton's method. Write a program to calculate the cube root of 10 starting with $x_0 = 3$. Show the contents of the memory for two complete cycles of the loop.

6   Suppose you deposit A dollars to be compounded annually at $i\%$ in a bank. The principal will be $A(1 + i)^n$ dollars in n years. Write a program for calculating the value of \$2000 at 6% interest after 20 years. (Assume that you have no capacity for taking exponents on your computer and therefore must make your computations year by year.)

7   During a 24-hour period 4 police officers (A, B, C, and D) each spend 6 hours on duty. A break-in is expected, so A, the officer who goes on duty first, is told, "Watch the corner house. When you go off duty, tell B to watch the corner house. And tell B to tell C to watch the corner house when B goes off duty. Finally, tell B to tell C to tell D to watch the corner house when C goes off duty." Can you replace these instructions with a loop? If there were 100 officers instead of just 4, compare the length of the loop with the length of direct instructions (such as those written out above.)

8   Alice, Barbara, Catherine, Dorothy, Edna, and Francine are sitting in a theater in the order indicated. Alice says to Barbara, "We're going to lunch at Alfred's; pass it on." In effect, Alice has created a loop. Restate the same request, writing it out in detail.

Some Comments About Hand Calculators

In the last few years hand calculators have become increasingly popular. They are fast, light, and adaptable to many applications; these applications run the gamut from scientific uses (computing sin x or calculating statistical parameters such as means and variances) through accounting or bookkeeping (computing sales taxes or percentage profits).

Calculators vary considerably in sophistication. The simplest of them, which can be bought for less than $10, have little more than the capacity to do the four basic arithmetical operations. Slightly more sophisticated calculators will have one or two memory locations in which the results of intermediate calculations can be stored. If you want to calculate (.453) (.324) + (2.34) (.067), you would normally calculate the first product, store it, calculate the second product, add to it the stored first product, and obtain your answer. Without storage you must either remember your first product or write it down—a considerable inconvenience. The most sophisticated calculators can be programmed in advance. For example, if you wished to calculate $V = (3 - a)^2 \cos 2b \log 4c$ for various values of a, b, and c, you could write a program that would enable you to obtain V by simply entering the program and then putting in the values of a, b, and c. You would not have to write out the operations each time.

Since sophisticated calculators are much like primitive computers, it is interesting to compare their operations. The internal capacities of computers differ, but almost invariably they have storage capacity for at least tens of thousands of numbers while calculators can only handle very small programs—the length of the program plus the storage available for memory is generally measured in hundreds. The input-output of a computer, which is the slowest facet of its operation, is much more efficient than that of a calculator. One may use high-speed tapes or prepunched cards to get a program into memory, but you rarely punch in data while the computer sits and waits. However, this is what you almost always do on a calculator. Some computers have the capacity to read out the results of one program while they are reading in another and calculating as well; this, of course, is well beyond the calculator's present capacity. But for all the quantitative differences, there is much to be learned from using calculators (especially programmable ones) which carries over to computers.

 BUT
CAN THEY THINK?

COMPUTER LEARNING
AND GAMES

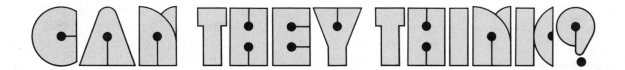

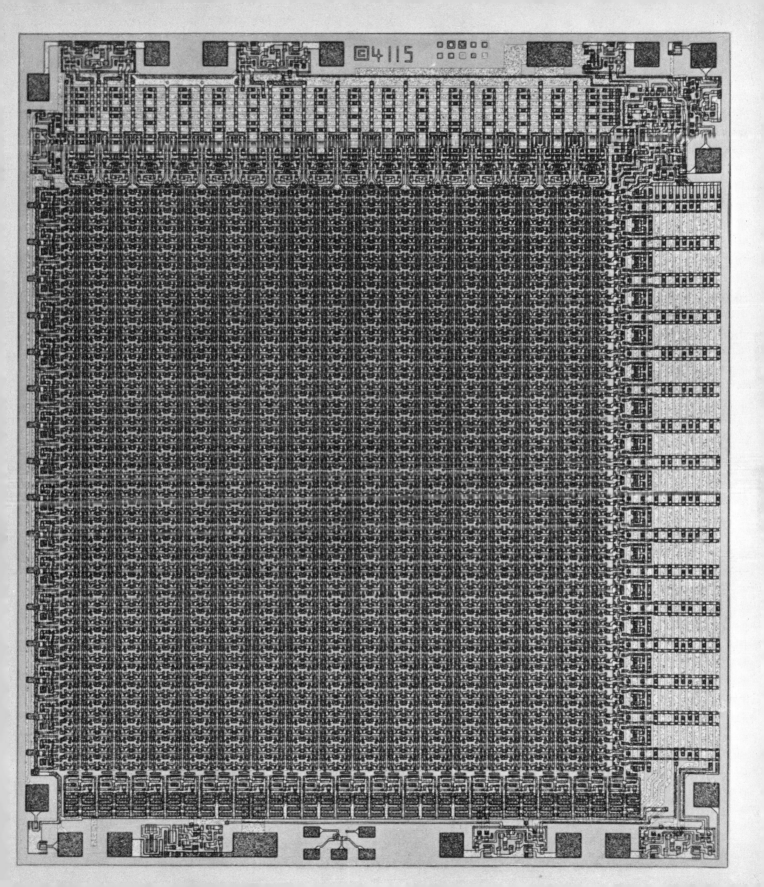

1 Introduction

In 1968, David Levy, a British international chess master, bet a group of computer experts 1250 pounds that in the decade that followed no chess-playing computer could beat him. This year, Mr. Levy expects to collect his wager, but it is turning out to be a much closer race than many had imagined possible.

Chess-playing skill is widely regarded as a legitimate gauge of at least one aspect of human intelligence. Most human chess masters had long believed that although computers could rapidly sift through an enormous number of possibilities, machines lacked certain qualities vital to real mastery of the game.

But new computer programs are proving to be such formidable opponents in tournament play, experts say, that old ideas about the limitations of computer intelligence must be revised. . . .

In 1976, the Paul Masson Chess Tournament in California attracted some 700 contestants, including Mr. Slate's [of Northwestern University] program, which was entered in the class B level against 128 good amateur players. The computer easily won all its games to take a $750 prize, which Northwestern had agreed in advance to relinquish.

In February, 1977, Chess 4.6 [Mr. Slate's chess-playing program] entered the Minnesota Open Tournament, playing against some highly ranked opponents.

Chess 4.6 won the tournament 5 to 1, emerging with an official rating of 2271—a chess master.

FROM The New York *Times*
(2/17/78), Section B, page 1

It is generally known that everything a computer does is a direct or indirect response to a command of its programmer. Computers don't improvise, nor can they modify their instructions (unless they are programmed to). This simple fact may tempt you to draw certain conclusions (such as that computers can never do anything that seems original or creative and can never "outsmart" their users), but they would turn out to be false.

When we first described how computers work, it seemed that every step of every job would have to be spelled out *explicitly* in advance by the programmer. But we later saw that a few judicious commands could generate a loop that would allow the computer to carry out an arbitrarily large number of operations. Once again, you may be tempted to conclude from the computer's absolute dependence on its programmer that it must be incapable of doing anything creative. Yet somehow an impressive list of accomplishments has been compiled, as witness the examples mentioned in the last chapter and the quotations from The New York Times at the beginning of this one.

There has been much discussion about whether computers will ever be able to think. It is not entirely clear what a computer would have to do to establish that it was thinking, since there is little agreement on what thinking is. In any case, we won't discuss the question here; instead we will show how a computer, though completely obedient to its programmer's commands, can "learn" to outperform him.

If we actually wrote such a program in detail, we would become too bogged down in complications to obtain a clear picture of what such program does. So we'll give a general description of one approach to the problem and restrict our attention to the principles involved.

Our main purpose is to show how a computer can be "taught" to do the apparently impossible: "learn" to play a parlor game

Chess master David Levy playing against the CYBER-176 computer, which is programmed with Chess 4.5.

better than its programmer. And our discussion will apply not just to one single game but to an entire class of games: two-person, zero-sum, finite games of perfect information, in the formal language of game theory. This is just a formal way of saying that the game is played by two people, both want to win, each player has only a finite choice at each move, the game lasts only a finite number of moves, and both players know exactly what is happening at all times. (The last provision rules out such games as poker and bridge, in which each player has information unavailable to his opponent.) Chess, checkers, tic-tac-toe, Chinese checkers, *go*, and nim are examples of the type of game we have in mind.

For simplicity it is convenient to assume that there are no draws, that in each game there is a loser and a winner. This assumption is valid for some games, such as nim and hex (which we'll describe later); in others, the rules of the game would have to be modified for it to hold. It is not an essential assumption, but it will make the analysis clearer.

Before going on we should mention a few things we will *not* do. There are some games that are simple enough to analyze completely. In some, such as nim, a mathematical formula prescribes the winning move. In others, such as tic-tac-toe, you can analyze all the possibilities. (The number of variations is considerably reduced because of symmetry.) Although the prospect of a computer defeating a human being may seem impressive, it's really very easy to program a computer to play nim or tic-tac-toe perfectly. It's also easy to write a program that merely comes up with a legal move at the appropriate time even in a complicated game like chess. The real problem is to enable a computer to play much the way a human being does, by using general principles to make choices and not working out every possible sequence of moves.

In the rest of this chapter we will derive certain properties that all games of this type have in common. Then we'll show how to use this information to enable a computer to learn.

puzzles to ponder

1 Think about how you actually make decisions in games of pure skill such as chess and checkers. Suppose that you had to spell out your decision rules in sufficient detail so that a computer could play out a game without coming to you for further clarification. Assuming the computer is as large and fast as you please, can you think of any ways in which the computer might be more efficient than you? Can you think of any reason why the computer might not play as well as you? Is it conceivable that the computer might (a) improve with time? (b) "learn" to beat you at your own game?

2   Would your answers to part (a) be any different if the game involved a mixture of luck and skill such as we find in games like backgammon, poker, and monopoly?

The Extensive Form of a Game

When you discuss the theory of a game, as opposed to actually playing one, it is convenient to use the **extensive form** description of the game. A game in extensive form is like a road map. The positions of the game are represented by points, or vertices, on the map. The moves are represented by directed lines connecting pairs of positions. If a line goes from *A* to *B*, that means it's possible to go from position *A* to position *B* but not necessarily from *B* to *A*.

In this model, playing a game is like taking a trip. Both players begin at the starting position, and one of them (which one is spelled out in the rules) decides on the branch of the road they travel first. They continue on that road until a vertex is reached,

and then the other player decides which of the available sections of the road they will take on the next part of their trip. They go on in this way, alternately making decisions, until they reach an end position and the game is over. The rules of the game determine whether the end position they have reached is a win for player I or for player II. Since the game is of perfect information, we think of each player as having a road map that shows all the vertices, connecting roads, end positions, and the name of the winning player at each end position. Figure 1 illustrates the extensive form of a very simple game.

Figure 1   *Player who has the move*

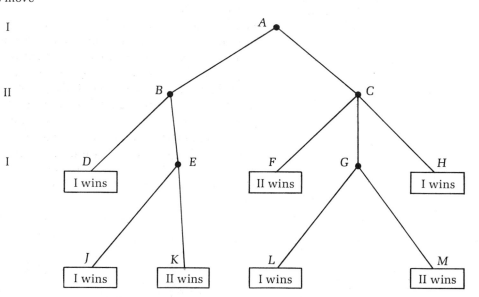

At each of the six end positions (*D, F, H, J, K, L*, and *M*) there is a box that indicates the name of the winning player. The player who has the move at any position is indicated by the Roman numeral at the left; player I moves at positions *A, G,* and *E,* and player II moves at positions *B* and *C*.

The game starts when player I makes his first decision, to move either left to *B* or right to *C*. In either case the next decision is made by II. At *B*, II can move left to *D* (and lose) or move right to *E*. At *C*, II has three choices: to go to *F* (and win), to *G*, or to *H* (and lose). And so on.

In discussing games we will use certain terms repeatedly, so we'll start with a few definitions. When we say that "Position A **precedes** position B," we mean it *immediately* precedes it. Similarly, "Position A **succeeds** position B" means A *immediately* succeeds B. So in the example shown in Figure 1, position A precedes C but not G, and position E succeeds B but not A.

Every position in the game is either an **end position** or a **middle position**. Once an end position is reached, the game is over and the winner is determined by the rules.

In everyday language a strategy is an overall plan, usually a clever one, but we use the term a little differently. By a **strategy** we mean a complete description of what a player will do in every situation that may arise in the game. If for some reason a player were called away in the middle of the game and had to entrust his decisions to an unthinking assistant or a mechanical computer, his strategy would have to be sufficiently detailed to enable the assistant or computer to make decisions exactly the way the player would. Strategies do not have to be clever, but they do have to be complete.

Strategies are mainly of theoretical interest in most parlor games, because it is hopelessly impractical to write out a complete strategy for a game as complicated as, say, chess.

(EXAMPLE 1)

In the game shown in Figure 1, state all the strategies for players I and II. For all possible pairs of strategies, indicate the end position to which they will lead and the winner of the game.

Solution

Player I has 4 strategies. He can go left to B, and then, if II goes to E, he can go left to J. We denote this strategy by LL. If he goes right when II goes to E, we will denote the strategy by LR. If I goes right to C and then left to L or right to M, we denote the strategy by RL or RR, respectively.

Player II has six strategies, which we will denote by LL, LM, LR, RL, RM, and RR. Strategy LL means that if I goes left from A to B, then II will go left from B to D, and if I goes right from A to C, II will go left from C to F. Strategy RM means that if I goes left from A to B, then II will go right from B to E, and if I goes right from A to C, then II will go middle from C to G. The other strategies are defined in the same way. The following table indicates all possible outcomes for each pair of strategies. *The table entries indicate the terminal position reached and the winning player for each pair of strategies.*

Strategy for player II

Strategy for player I

	LL	LM	LR	RL	RM	RR
LL	D(I)	D(I)	D(I)	J(I)	J(I)	J(I)
LR	D(I)	D(I)	D(I)	K(II)	K(II)	K(II)
RL	F(II)	L(I)	H(I)	F(II)	L(I)	H(I)
RR	F(II)	M(II)	H(I)	F(II)	M(II)	H(I)

exercises

1 In the children's game of tic-tac-toe (sometimes called "zeros and crosses") how many different ways are there of filling the 3 × 3 array with zeros and crosses if the game continues even if a player has already won and if you consider the order in which the zeros and crosses are placed?

2 If you assume a player has 20 alternatives at each move in chess (this is generally an underestimate), how many alternative variations are possible if you look three moves ahead (for each player)?

3 Referring to Figure 1 and the strategy matrix in Illustrative Example 1, determine the following:
(a) Does any player have a winning strategy? How can you tell a player has a winning strategy by looking at the matrix? Who has the winning strategy, and what is it?
(b) Could it happen that both players have a winning strategy? Why?
(c) Does either player have a strategy that guarantees he will lose whatever his opponent does? If so, name the player and what the losing strategy is.
(d) Can both players have losing strategies? Is it possible for a player to have both a winning and a losing strategy?
(e) For every strategy that is not necessarily a winning one, indicate the strategy for the player's opponent that will allow him to win.

4 Three games in extensive form are shown in Figure 2. In each case answer the following questions:

(a)   For each middle position list the preceding and succeeding positions.

(b)   Describe two different strategies for each player.

(c)   Choose a strategy for each player and determine the outcome of the game if these strategies are adopted.

(d)   Find a pair of strategies that will lead to a win for player I and another pair of strategies that will lead to a win for player II.

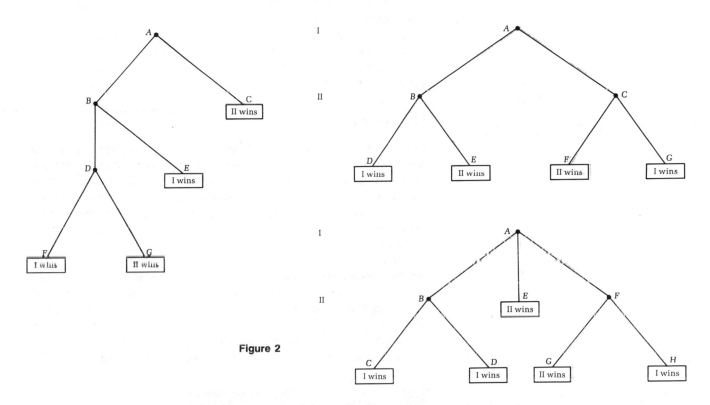

Figure 2

Winning Strategies and Winning Positions

Although only end positions are defined to be "winning" or "losing" for a player, there is a sense in which the middle positions are winning or losing as well. A player has a winning position if he has a strategy that assures him a win whatever his opponent does. Formally, if player X always wins the game whenever he uses a certain strategy S, then S is called a **winning strategy** for X. If whenever a certain position P is reached player X wins by using strategy T, then T is called a **winning positional strategy** at P and P is called a **winning position** for X and a **losing position** for X's opponent.

To identify the winning and losing positions you begin with the end positions and work backwards. First you determine whether the middle positions that come latest in the game are winning or losing. With this information you analyze all the earlier positions in turn in much the same way that a row of falling dominoes knock one another down in succession.

To understand this process better we'll apply it to a particular game, a variation of a very simple game we call **sums**.

The Game of Sums

In this variation of the game of sums player I starts by choosing a whole number from 1 to 5 inclusive. Then player II adds an integer from 1 to 5 to the integer that player I picked. The players alternately add an integer from 1 to 5 to the cumulative sum formed up to that point until the sum reaches or exceeds 100; then the game is over. The first player to form a sum greater than or equal to 100 is the winner.

If you start analyzing this game by trying to find out whether the early positions are winning or losing, the number of possibilities is overwhelming. But if you work backwards from the end positions, life becomes much easier. It is easy to see that the mover (the player who's about to add a number) can win if the cumulative sum up to that point is 95, 96, 97, 98, or 99. In each of these cases he can make the sum 100 at his turn. Therefore, *if the*

A cathode-ray screen display of a stage in a poker game between several computer programs (named Asprate, Caller 2, Exploit, Player 1, Player 4, Player 6, Player 9) and a human player, whose hand is displayed. It is the human player's turn to decide to fold, call, or raise.

cumulative sum is from 95 to 100 inclusive, the player who must move has a winning position.

Now suppose a player must move when the cumulative sum is 94. By adding a number from 1 to 5 he will make this sum something between 95 and 99. But these are all winning positions for his opponent, so he must lose against an intelligent opponent whatever he does. So *if the cumulative sum is 94, the mover has a losing position.*

The rest of the argument that can be used to distinguish winning middle positions from losing ones will be given later in the form of an exercise.

All games in extensive form are analyzed in basically the same way. To see how this is done we will use an extremely simple game of sums in which

1. The players may add either 1 or 2 to the cumulative sum.
2. The first player to reach 4 wins.

The extensive form description of this game is shown in Figure 3.

In this simple game of sums *J* is a winning position for player II, since the cumulative sum is 3 and he has only to add 1 to win.

Figure 3

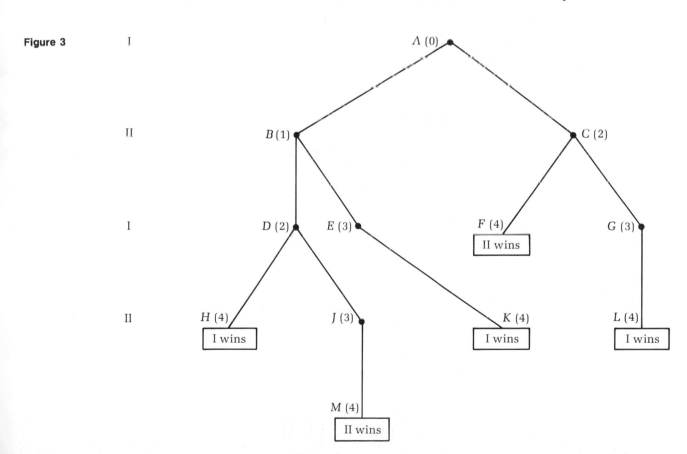

Positions E and G are winning for player I for exactly the same reason that J was for player II. Position D is a winning one for I. He can either add 2 to the cumulative sum of 2 and reach the winning position H or add 1 to 2 to form 3, arriving at J, at which II wins. But the choice is player I's, and his best play is clearly to add 2. Position C is a winning one for player II for the same reason that D was for player I. Position B is a losing one for II, since he must either move to D or E, and in either case I will have a winning positional strategy when he does. Finally, the initial position A is a win for player I. He moves to B and then chooses the appropriate winning strategy when II moves to D or E.

Not only does this kind of analysis sort out the winning positions from the losing ones, but it can also be used to construct a winning strategy. If you have a winning position, just move to where your opponent has a losing position. Specifically, player I should first move to B (pick the number 1); if II goes left to D (by adding 1 to form 2), I should go to H (add 2 to reach the sum of 4). If II goes to E, I goes to K.

When we say a position is a win, we mean it is a win for the player *whose turn it is to move*; similarly, when we say a position is a loss, we mean it is a loss for the player *whose turn it is to move*. If a position is a loss for one player, it is a win for his opponent, and conversely, if it is a win for one player, it is a loss for his opponent.

EXAMPLE 1

The diagram in Figure 4 shows the extensive form of a game. Determine which of the positions are wins, which are losses, and who should win the game. Describe a winning strategy.

Figure 4

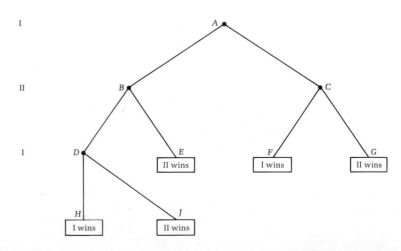

Solution

The diagrams in Figure 5 indicate the steps in the logical process of classifying the positions. We begin by considering the first layer, which consists of C and D, the middle positions that are succeeded only by end positions. Since I moves at D, he will win by going to H, and since II moves at C, he will prefer to win by going to G. We put this new information into the diagram labeled "Step 1," using a dashed box. The next layer consists only of position B; II will clearly prefer to go to E and win rather than go to D and lose, so in the diagram labeled "Step 2" we indicate that B is a win for player

Figure 5

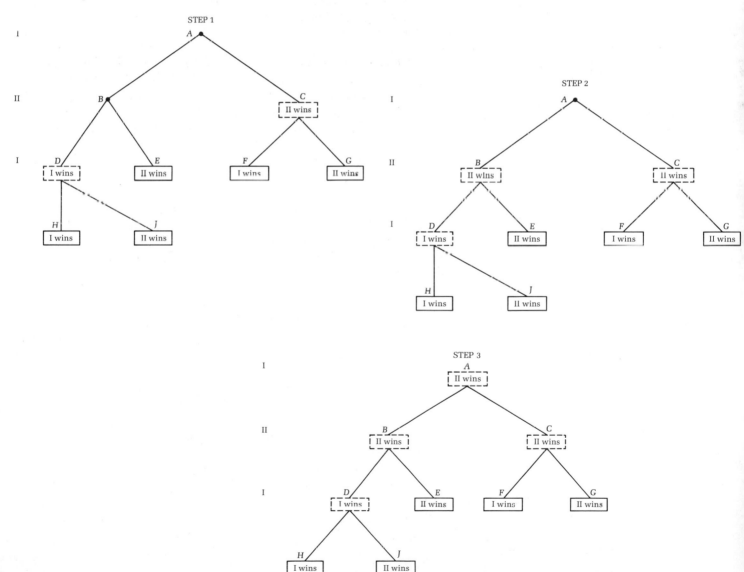

II. Finally, we notice that I has no good choice at *A*; he can go to *B* or *C* but loses in either case. So the game is a win for player II. His winning strategy is:

"If I goes right to *C*, I'll go right to *G*; if I goes left to *B*, I'll go right to *E*."

In the process of "working backwards" that we just described, there is a very simple way to determine whether a position is a losing or winning one; we have already made use of this method, and now we will express it formally. We dignify this rather simple relationship with the name "Basic Principle" because we will want to refer to it later.

> **The Basic Principle** Suppose *S* is a middle position, player I must move at *S*, and all positions that succeed *S* are either end positions or middle positions that have already been established as winning or losing.
> 1. If *every* position succeeding *S* is a win for player II, then *S* is a loss for player I.
> 2. If *at least one* position succeeding *S* is a loss for player II, then *S* is a win for player I.

It is easy to show that the Basic Principle is a reasonable statement without giving a formal proof. (We actually appealed to the Basic Principle in our earlier analysis without formally stating it.) Suppose for a moment that the Basic Principle was false and a player told you that

1. "In the middle of the game I was at a winning position, but no matter what I did I had to put my opponent in a winning position on the following move."

or

2. "I had a losing position, but I found a place to move in which my opponent had a losing position."

With just a little thought you could reply to the player who made remark (1), "If you had to put your opponent in a winning position, you never had a win in the first place." And your response to remark 2 might be, "If you could put your opponent in a losing position, you must have been in a winning position after all."

To sum up: In any game of the type we are discussing, it is always possible to designate a position as either winning or losing. The end positions are winning or losing in accordance with the rules; the middle positions can be diagnosed as winning or losing by a logical process.

(**EXAMPLE 2**)

By applying the Basic Principle to the end positions and working backwards, distinguish the winning middle positions from the losing ones in Figure 6.

Figure 6

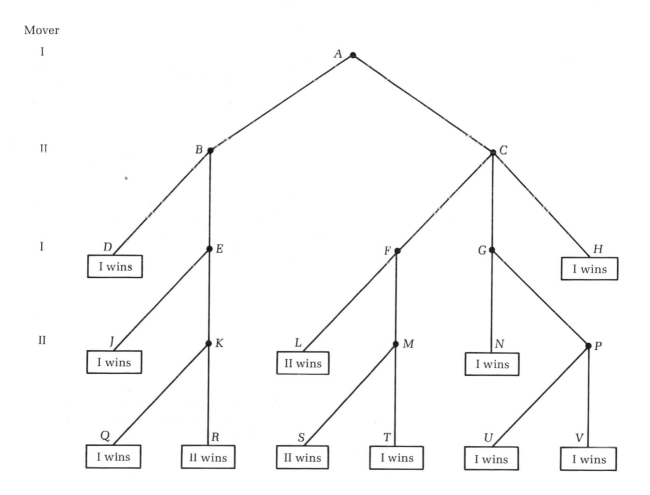

Solution

We first observe that all the end positions (*D, H, J, L, N, Q, R, S, T, U,* and *V*) are winning or losing as determined by the rules of the game and as indicated in the diagram. We can sort out the winning and losing middle positions in four steps:

Step 1: Middle positions *K, M,* and *P* are succeeded only by end positions, which are already designated as winning or losing. Using the Basic Principle, we conclude that *K* and *M* are winning positions and *P* is a losing one.

Step 2: Middle positions *E, G,* and *F* are succeeded only by end positions or middle positions designated as winning or losing in step 1. By the Basic Principle we conclude that *E* and *G* are winning positions and *F* is a losing one.

Step 3: Middle positions *B* and *C* are succeeded only by end positions or middle positions designated as winning or losing in steps 1 and 2. That *B* is a losing position and *C* is a winning one follows from the Basic Principle.

Step 4: The initial position, *A,* is succeeded by 2 positions (*B* and *C*), one of which is losing and the other winning. By the Basic Principle, *A* is a winning position.

With this information it is easy to describe a winning strategy for player I; whenever he moves, I should arrive at a new position that is a loss for II. Such a position must exist by the Basic Principle. In this case, I should move from *A* to *B* initially. If II moves to *E*, I should go to *J*. If I does anything else. II can win.

In Exercises 1 through 3 are shown the "road maps" of three games in extensive form. In each case state which of the middle positions are winning and which are losing, state whether player I or player II has a winning strategy for the game as a whole, and then describe this winning strategy in detail.

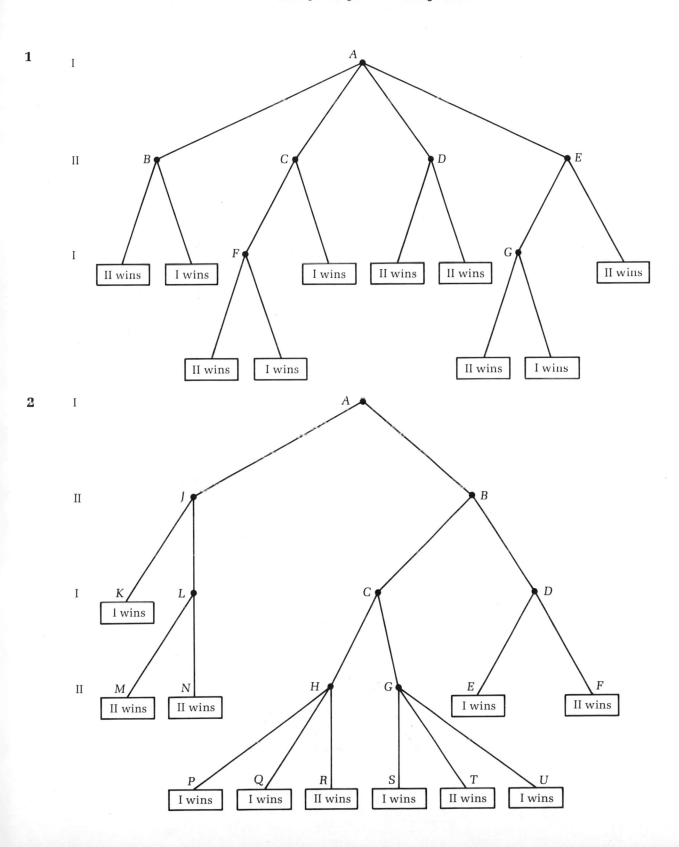

3

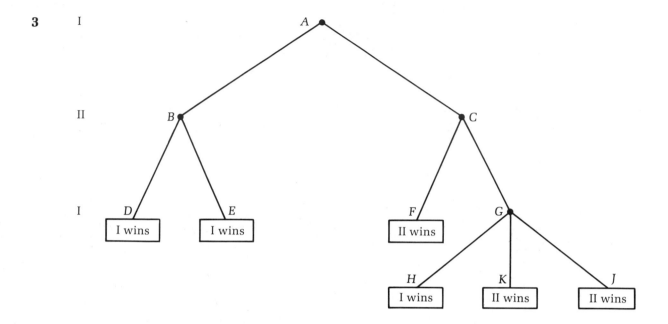

In each of the game trees shown in Exercises 4 through 6, there are 7 terminal positions: *P, Q, R, S, T, U,* and *V.* Whether player I has an initial winning strategy or not depends on how the terminal winning positions are distributed. In each case state which sets of winning positions will guarantee player I a winning strategy and describe the winning strategy.

4

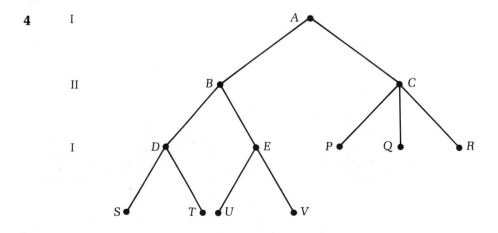

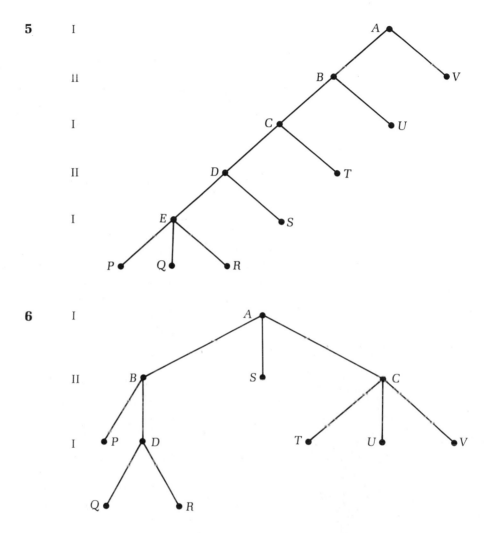

5

6

7 The following questions refer to the game of sums described at the beginning of this section.

(a) Show that if it is your move and the cumulative sum is an integer from 89 to 93, you have a winning position. (*Hint:* Use the fact that the mover has a losing position if the cumulative sum is 94.)

(b) Show that the mover has a losing position if the cumulative sum is 88.

(c) Prove that if it is your move and the cumulative sum is 82, 76, 70, 64, 58, 52, 46, 40, 34, 28, 22, 16, 10, or 4, you have a losing position.

(d) Complete the analysis by stating which are the winning positions for the mover and which are the losing ones and which number you would choose if you moved first.

(e)   Write out a winning strategy for the player who must move when the cumulative sum is 91.

(f)   Write out a winning strategy for the player who moves first.

8   Determine what the outcome of the game shown in Figure 1 (page 320) should be if both players do their best. What should the outcome be if player I initially moves right, to C?

More Examples of Simple Games

Most of the "games" we have discussed so far were not games one is likely to encounter in real life. Except for sums—a children's game—all our illustrative examples were very artificial, because any game complex enough to be interesting to adults generally has an extensive form that is enormous. Still, it would be interesting to look at the extensive form of a game that is actually played, so we will describe a simple version of the game of **nim**.

The Game of Nim

At the start of the game of nim there are a certain number of rows, each containing a certain number of sticks. The players alternately remove at least one stick, but as many as they please, from any one row until there are no sticks remaining in any row. The last player to remove a stick is the winner.

It will be convenient to discuss an identical game but expressed in somewhat different terms:

Initially, player I is presented with a set of nonnegative integers. (These represent the numbers of sticks in the various rows.) He must reduce any one of these to another nonnegative integer. The players continue in this way, alternately taking any

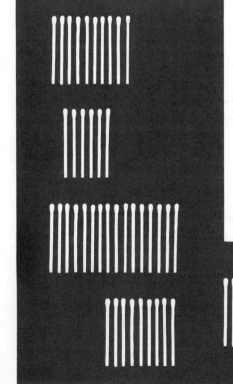

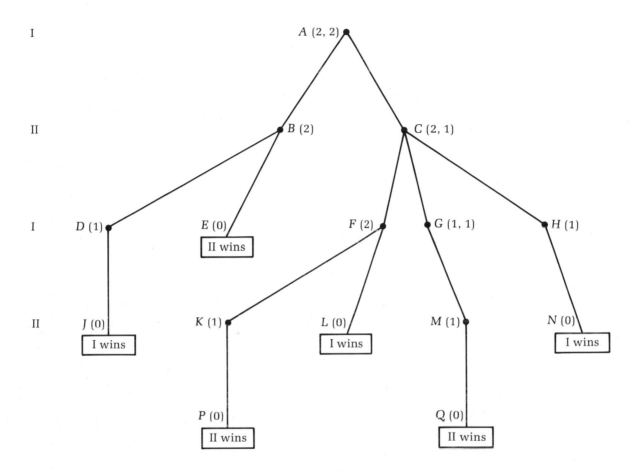

Figure 7

integer and making it smaller (but never less than zero), until all the integers remaining are zero. The last player to reduce an integer wins.

For simplicity, we assume that player I is given the two integers 2 and 2, which we write as (2, 2). The extensive form of the game is shown in Figure 7. Initially, player I has two possible moves; he can reduce one of the 2's to a 1 or a 0.* If II is presented with (2, 1), he has three possible moves; he can reduce the 2 to a 1 or a 0 or reduce the 1 to a 0. And so on.

---

*Technically, there is a distinction between reducing the first 2 and reducing the second, but the rules of this game are such that there is no substantial difference between the two moves; thus we will not distinguish between them.

Another example of a game in extensive form is shown in Figure 8. This is another variation of the game of sums we described earlier.

In this variation each player can add any number from 1 to 3 to the cumulative sum. The first player to add a number that makes the cumulative sum greater than or equal to 5 is the winner.

Figure 8

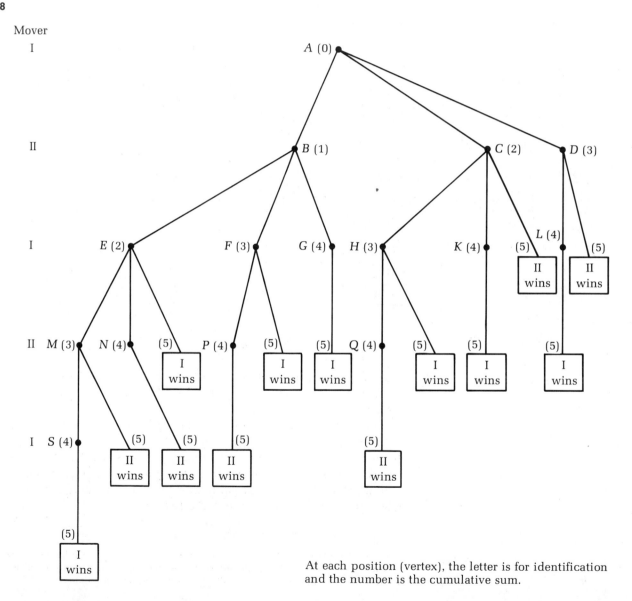

At each position (vertex), the letter is for identification and the number is the cumulative sum.

Exercises 1 through 3 refer to the game of sums shown in Figure 8.

1 Using the diagram in Figure 8, prove the following in the order indicated:
(a) S is a winning position.
(b) M, N, P, and Q are all winning positions.
(c) E, F, G, H, K, and L are all winning positions.
(d) B is a losing position, and C and D are winning ones.
(e) A is a winning position.

2 Describe a winning strategy for player I. Is there more than one?

3 Suppose the rules were changed so that
(a) A player may add any number from 1 to $k - 1$ to the cumulative sum instead of any number from 1 to 3.
(b) The winner is the first person to reach N ($N \geq K$), rather than 5.
Calculate the winning positions and give a verbal description of the winning strategies.

4 Use the diagram of the extensive form of the game of nim shown in Figure 7 to draw the following conclusions in the order indicated:
(a) K and M are winning positions.
(b) D, F, and H are winning positions, and G is a losing position.
(c) B and C are winning positions.
(d) A, the initial position, is a loss. (Therefore player II has a winning strategy; what is it?)

5 In the game of tic-tac-toe, players alternately place zeros and crosses in a 3×3 array, and the first player to place three of his symbols in a row (vertically, horizontally, or diagonally) wins. In the following unfinished game

$$
\begin{array}{ccc}
A & X & X \\
O & B & D \\
C & O & O
\end{array}
$$

it is X's turn to play. The blank squares are those labeled A, B, C, and D. The rest of this game is shown in extensive form in Figure 9 (on the next page), along with the winning and losing final positions. Label the winning and losing middle positions.

6 Assume that players I and II are engaged in the game of nim, that there are two nonzero columns of 1 and 3, and that it is I's turn to move. An extensive form for this game is shown in Figure 10 (on the next page) with the winning and losing final positions. Determine which of the middle positions are losing and which are winning.

Figure 9

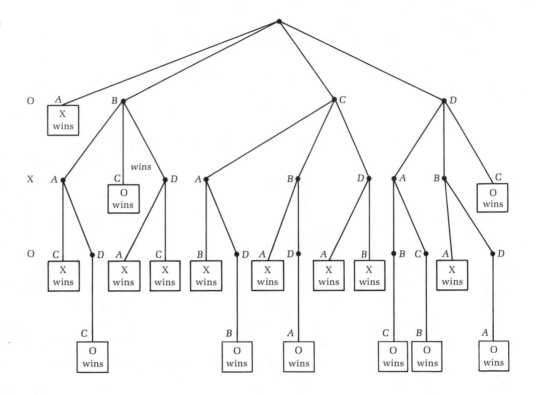

Figure 10

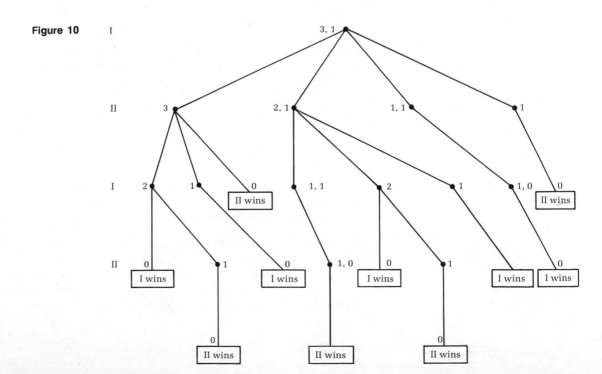

7 In the general game of nim described at the beginning of this section, show that a position at which there is only one nonzero number left is a win.

1 In the general game of nim:
 (a) Show that $(1, 1)$ is a losing position.
 (b) Show that $(2, 2)$ is a losing position.
 (c) Show that $(3, 3)$ is a losing position.
 (d) Suppose you somehow knew that $(1, 1)$, $(2, 2)$, $(3, 3)$, . . . , (N, N) were all losing positions. How would you conclude from this that $(N + 1, N + 1)$ was a losing position as well?
 (e) If you accept (d) as true, prove that (M, N) is a winning position if $M \neq N$.

2 For the general game of nim, decide whether the position described is winning or losing:
 (a) Every nonzero integer is a 1.
 (b) Every integer is a 2.
 (c) All the integers are either 1 or 2.

Evaluating Positions

In theory, for the kind of games we are discussing, you can always work out whether a positon is winning or losing by working backwards from the end positions. We showed how this could be done for nim and sums, although we did not attempt a formal proof.

 If you find yourself in a losing position, there's nothing to do (in theory) but grin and bear it until your opponent makes a mistake. But even if you know you're in a winning position and deduce that you must have a winning strategy, the information is virtually worthless in any real game. In the game of chess, for example, it would take ages to write out *any* strategy in detail, so you are hardly likely to find a winning one in practice.

 Fortunately, you don't have to. A chess player in a real game would never try to plan a completely detailed strategy; it's all he can do to choose his next move. And it is precisely this human thinking process, which enables a person to make a reasonable (if imperfect) choice of moves *without* analyzing the game completely, that we would like to imitate using a computer.

The essential point is this: You don't need to have a complete winning strategy in order to play a perfect game; all you really need to know is the right move to make *next*. If you had a genie who indicated, at each of your turns, which of the immediately following positions were losses for your opponent and which were not, your success would be assured; if you always move from your winning positions to your opponent's losing ones, you'll win. So the problem reduces to this: When it is your turn to move, how do you distinguish between your opponent's winning and losing positions?

You start by analyzing certain "features," which are generally accepted standards for determining whether a position is good or bad. When two chess players discuss the merits of a position, they do it in terms of these "features": "the control of the seventh rank," "the quality and quantity of each player's pieces remaining on the board," "the number of open files and who controls them," "the mobility of the pieces on the board," and so on. We will make the same assumption chess players tacitly make: that by "properly" combining these various features we will fashion a tool capable of distinguishing winning positions from losing ones.

It is the job of the computer, using methods that we will describe later, to work out the "proper" weights. We should not

"KNIGHT TO KING SEVEN.."

Figure 11

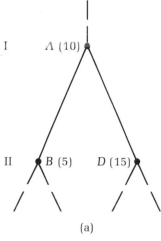

(a)

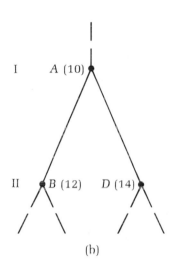

(b)

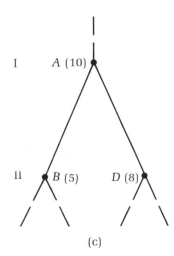

(c)

hope for perfection; it will almost invariably happen that some losing positions will be classified as winning ones and vice versa. But if we can derive a reasonably efficient genie, we will avoid the time-consuming job of looking ahead that plagues computers and humans alike when playing parlor games. Even though we don't expect to construct a perfect genie, our first step will be to discover how one would behave if it did exist. The mechanism that is to play the role of genie is called an *ideal evaluation function*.

In order to sort out the winning positions from the losing ones, we have to develop some new tools: *evaluation functions* and *ideal evaluation functions*. An **evaluation function** G (which we abbreviate by EF) assigns a number to every position in the game. If S is a position, then $G(S)$ denotes the number assigned to it by G. An **ideal evaluation function** (which we abbreviate by IEF) is an evaluation function with the special property that it distinguishes winning positions from losing ones. The number an IEF assigns to any winning position is always greater than the number it assigns to any losing one. (*Reminder:* When we say a position is "winning" or "losing," we mean it is winning or losing for the player whose turn it is to move.)

We can express the special property of an IEF more concisely: If G is an IEF, then for any losing position S and any winning position T we have $G(S) < G(T)$.

As an immediate result of the definition of an IEF we have the following: For any IEF G, there is some number C, called the *cutoff,* with this property: For any position S, $G(S) > C$ means S is a winning position and $G(S) < C$ means S is a losing position. (There are actually many such cutoffs, and any one will do.)

EXAMPLE 1

Each of the three diagrams in Figure 11 shows a portion of a game tree. Next to each vertex is an identifying letter and a number assigned by some IEF. Identify as many winning and losing vertices as possible and estimate the cutoff value.

Solution

In diagram (a), A must be a winning position. If A were a loss, then B, with a smaller IEF value, would be one as well, and according to the Basic Principle two losses cannot occur in succession. Position D is a win, since its value is higher than that of A, and B must be a loss, since the Basic Principle states that every winning position has a losing position immediately following. The cutoff value, C, is somewhere between 5 and 10.

In diagram (b), A cannot be a win. If it were, B and D, with higher values, would be wins as well, and the Basic Principle states that every winning position is followed by at least one losing position. So A is a loss, and B and D must be wins, since only winning positions follow losing ones. The cutoff value C, is somewhere between 10 and 12.

In diagram (c), A is a winning position (why?) and B is a losing position (why?). It is not clear whether D is a winning or losing position. The cutoff value is somewhere between 5 and 10.

EXAMPLE 2

The diagram in Figure 12 shows a small portion of a game tree. Next to each vertex is an identifying letter and a value assigned to that vertex by an IEF, G; that is, $G(A) = x$, $G(B) = y$, and $G(D) = z$. Under what conditions will A be a winning position?

Figure 12

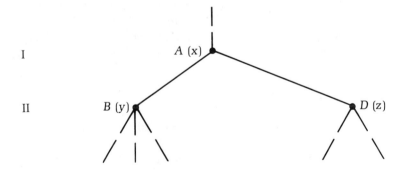

Solution

If x is larger than either y or z, A is a winning position. If A were a loss, all positions following it would have to be wins and have larger IEF values. If x is smaller than both y and z, A would have to be a loss. If it were a win, there would have to be some following position that was a loss, and the IEF value of that position would have to be less than that of A.

If you somehow obtained an IEF and knew the cutoff value C, it would be easy to play a perfect game. At each move you would scan the following positions to see if any had IEF values less than C. If there is at least one such position, move there; it must be a losing one for your opponent. If all the following positions have IEF values greater than C, then you are in a losing position yourself and there is nothing to do but wait for an error by your opponent.

With an IEF you don't even have to know the cutoff value, C, to play a perfect game. All you have to do is to move so that your opponent's position will have the lowest possible IEF value at each turn. If you have a move that puts your opponent in a losing position, you'll be sure to find it.

The difficult question, of course, is, How do you get your hands on an IEF in practice? In any real game it's probably too optimistic to expect to find a perfect IEF; all you can really hope for is a good enough imitation, one that will allow a computer to play reasonably well rather than perfectly. But even if we can't construct a perfect IEF, it is important to be familiar with the properties of one. In order to judge how closely an EF approximates an IEF, we will observe how many properties they have in common. So the next step is to deduce properties of IEFs.

The substance of the Basic Principle is that winning and losing positions are not scattered randomly over the game tree; they are distributed according to a certain pattern. If S is a winning position, there must be a losing position immediately following S; if S is a losing position, every position immediately following it is a winning one. Since the values an IEF assigns to positions are intimately related to whether they are winning or losing, the IEF values should have a pattern also; and indeed they do.

Let's summarize briefly what we have done so far: We first established the Basic Principle, which stated that there are no losing positions following a losing position and there is always at least one losing position following a winning position. We then defined an IEF, which assigns numbers to each position in such a way that any winning position has a higher number than any losing position. Using the Basic Principle and the definition of an IEF, we derived (in the last few Examples) the following result:

If S is any position and G is any IEF, then

(1) if $G(S) < G(T)$ for all positions T following S, then S is a losing position.

(2) if $G(S) \geq G(T)$ for any position T following S, then S is a winning position.

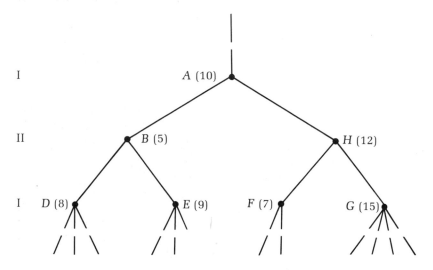

(**EXAMPLE 3**)

Figure 13 shows a portion of a game tree. At each vertex there is a capital letter for identification and the value assigned to that vertex by an IEF. Deduce which of the positions are losing and which are winning. Estimate C, the cutoff, as closely as possible.

Figure 13

Solution

From our previous discussion, A and H are winning positions and B is a loss. (If A were a loss, for example, B would have to be a loss as well, since B's value of 5 is even smaller than A's value of 10; but the Basic Principle states that a losing position cannot immediately succeed another losing position.) From the IEF values at positions B, D, and E it follows that the cutoff value must lie somewhere between 5 and 8; and from the IEF values at H, F, and G it follows that C must lie between 7 and 12. Putting these together, we deduce that C is between 7 and 8.

So far we have looked at our problem from only one direction. We have seen how an IEF, if your ministering genie happened to supply you with one, might be used to construct a winning strategy by distinguishing between your opponent's winning and losing positions. But in fact we are not likely to be given an IEF. What will actually happen in practice is this: We will be presented with a great many EFs, among which there may be a single IEF or, more likely, an EF that resembles an IEF very closely. Our task will be to find this IEF (or quasi-IEF) among the other EFs. The purpose of all the earlier analysis was to derive the properties of an IEF so we could recognize it when in the company of many imposters.

Imagine that you are presented with an EF which you suspect may be an IEF and you must decide if it really is; that is, you must evaluate the evaluation function itself. To test it you might use it in an actual game and see how successful you are. But if you use this approach, you will have to play many games before you can find out if even a single EF is any good, and you may well have many EFs—perhaps even an infinite number—to test.

It turns out that you can eliminate a great many EFs as potential IEFs without playing a single game. In fact, it is often possible to eliminate an EF as a potential IEF by looking at a very small portion of a single game tree. By calculating the values assigned by the EF to each of the vertices, you often find contradictions to the Basic Principle.

EXAMPLE 4

A portion of a game tree is shown in Figure 14. Next to each vertex is a value assigned by an EF. Show that the EF cannot possibly be an IEF.

Figure 14

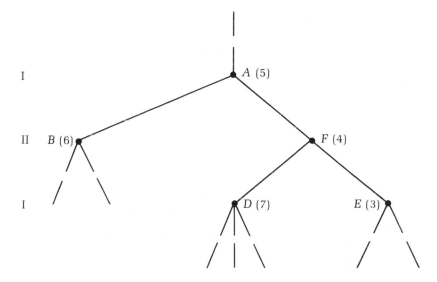

Solution

Since the value at vertex *A* is higher than the value at *F*, it would follow that *A* is a winning position if the EF were an IEF. Also, since *F* has the lowest value of any position following *A*, it must be losing. But *E*'s value is even lower, so *E* must also be a losing position. However, this is impossible; no losing position can immediately follow another losing position (by the Basic Principle). We conclude that the EF cannot be an IEF.

(**EXAMPLE 5**)

The game shown in Figure 15 gives you an opportunity to apply what you know about the relationship between winning and losing positions. Decide which of the vertices *A*, *B*, *F*, *D*, and *E* are winning positions. What are the possible values of x? (As usual, the numbers shown at each vertex are the values assigned by some IEF.)

Figure 15

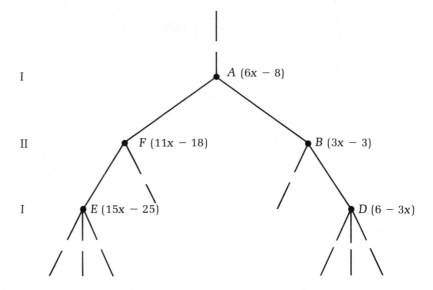

Solution

The solution might proceed along the following lines (you should fill in the details):

1. (a) Show that $G(A) > G(B)$ if and only if $x > \frac{5}{3}$.
   (b) Show that $G(A) > G(F)$ if and onlt if $x < 2$.
   (c) Conclude from (a), (b), and the Basic Principle that A is *a winning position.*

2. (a) Show that $G(B) < G(D)$ if and only if $x < \frac{3}{2}$.
   (b) From (1a) and (2a) deduce that either $G(B) > G(A)$ or $G(B) > G(D)$.
   (c) Use the Basic Principle and (2b) to prove that B *is a winning position.*

3. Use the Basic Principle and the fact that both A and B are winning positions to show that F *is a losing position.*

4. Use the Basic Principle to show that E *is a winning position.*

5. Use (1c) and (3) to deduce that $x < 2$.

6. Use (2) and (3) to deduce that $x < \frac{15}{8}$.

7. Use (3) and (4) to deduce that $x > \frac{7}{4}$. We conclude that $1\frac{3}{4} < x < 1\frac{7}{8}$.

8. Show that $G(D) < G(F)$ if and only if $x > \frac{12}{7}$. Since $\frac{12}{7} < \frac{7}{4}$, deduce from (7) that D *is a losing position.*

exercises

1 Suppose that G is an IEF and that $G(S) \geq G(T)$ for two positions S and T.
(a) Show that if T is a winning position, S must be one too.
(b) Show that if S is a losing position, T must be one too.

2 Suppose that G is an IEF and S is some position. Let M be the smallest value of $G(T)$ for all positions T following S.
(a) Show that if $M < G(S)$, then S is a winning position.
(b) Show that if $M > G(S)$, then S is a losing position.

3 Suppose S is a position, G is an IEF, and M is the smallest value assigned by G to any position that immediately follows S. Show that the cutoff value, C, always lies between $G(S)$ and M. In particular, if S is a losing position, then $G(S) < C < M$; if S is a winning position, then $G(S) > C > M$.

4   Each of the four diagrams below shows a section of a game tree. At each vertex there is a letter for identification and a value assigned by some IEF. In some cases the vertices can be identified as winning or losing, and in other cases they cannot. State which of the vertices are known to be winning or losing and estimate the value of *C* as closely as possible.

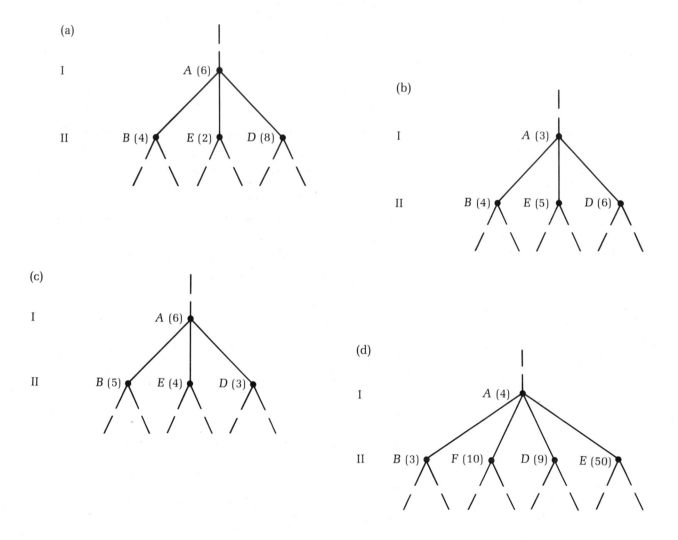

5   In each of the game trees on the next page, the numbers at the vertices are the values assigned to them by some EF. In each case determine whether the EF can possibly be an IEF. In those cases where it might be, assume that the EF actually is an IEF, distinguish the winning and losing positions, and determine the possible range of the cutoff value *C*.

(a)

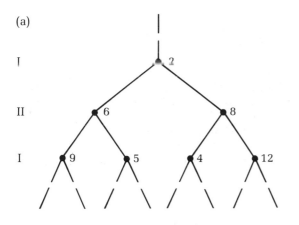

(b)

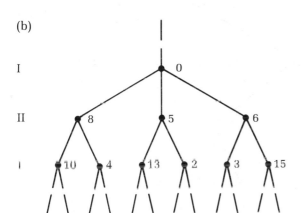

(c)

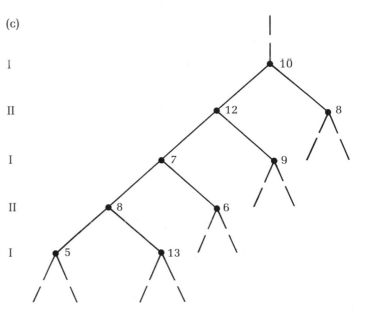

6   The diagrams below show a portion of a game tree. At each vertex is shown the value assigned by some EF. Show that the EF cannot be an IEF.

(a)

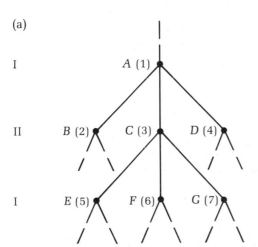

(b)

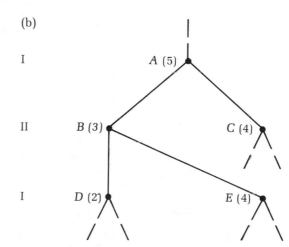

1 The diagram in Figure 16 shows part of the extensive form of a game. At each vertex there is a value expressed in terms of the unknown, x. Use the Basic Principle to derive the following results:

(a) *B* is a winning position. (*Hint:* Show that if *B* were a losing position, *E* would have to be a losing position as well.)

(b) If *A* is a winning position, then *E* must also be a winning position and *F* must be a losing one.

Figure 16

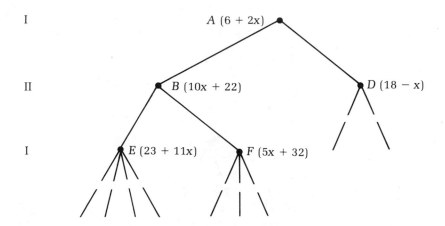

2 Suppose there are 3 possible outcomes to a game: a win, a loss or a draw. A *winning strategy* guarantees that the outcome will be a win for the one who uses it, and a *drawing strategy* will enable a player either to win or to draw (depending on what the opponent does.) A position is a *win* or a *draw* if the mover has a winning or drawing strategy, respectively, and it is a loss if it is not a win or a draw. Find an analogy to the Basic Principle (stated in Section 3) for such games.

3 In the popular children's game of tic-tac-toe, the players alternately fill the nine squares of a 3 × 3 array with zeros and crosses until one of the players has placed three of his symbols in a single row, column, or diagonal. The player who does so wins; if no player does, the game is a draw. For convenience we will number the small squares to identify them as follows:

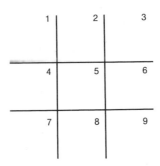

Suppose someone claims that an IEF, G, can be constructed in the following way: If S is any position and it is your turn to move, then

(a) Add 2 points for any row, column, or diagonal (abbreviated RCD) that contains exactly one of your symbols and none of your opponent's, and subtract 2 points for any RCD that contains exactly one of his symbols and none of yours.

(b) Add 7 points for any RCD in which you have exactly two symbols and your opponent has none, and subtract 4 points for any RCD in which your opponent has two symbols and you have none.

(c) If you have two symbols and he has one in any RCD, subtract a point; if he has two symbols in any RCD and you have one, add a point.

In the first position shown in Figure 17, by rule (a), X receives 2 points for diagonal 159, column 147, and row 456 and loses 2 points for row 129, diagonal 357, and column 258. Rules (b) and (c) yield nothing, so the value is 0. The values of the five possibilities for the succeeding position are shown above the diagrams of those positions. Compare the values of each position (for the mover 0) with the strength of each position in the game and decide whether this actually is an IEF. If not, is it a reasonable approximation?

Figure 17

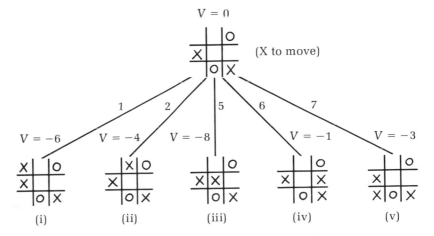

4   Consider two subsequent situations in the game described in Question 3, as shown in Figure 18. On the basis of these positions alone, could the EF be an IEF? If not, does it seem to be a reasonable approximation to one?

Figure 18

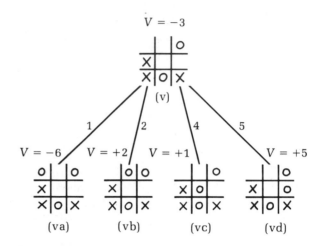

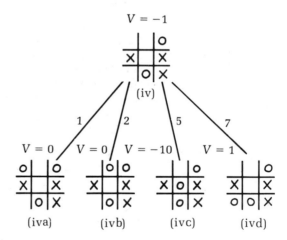

5   In the tic-tac-toe problem described in Questions 3 and 4, someone suggests that an IEF can be defined as follows: With any position S, observe all vacant squares and then take the average of the numbers associated with them (see the numbering of the squares in the array at the beginning of Question 3.) Without calculating show that this cannot be an IEF.

6 Think of any game of the type we are discussing. Can you identify any features for it? Can you roughly predict from this feature whether you're in a good or bad position?

7 Try to identify the significant features in the game of tic-tac-toe. (See Question 3.)

Computer Learning in Practice

Up to now we have been viewing games at close range. From the Basic Principle we know that winning and losing positions are not scattered randomly but fall into certain patterns. Since IEFs assign values to positions based on whether they are winning or losing, these values must have a pattern, as well. In this last section we will indicate how one can harness this information to get a computer to identify an IEF (or something closely resembling an IEF) among a set of EFs. We will discuss the problem in a general way to avoid becoming immersed in detailed calculations.

If you wanted to write a program that would enable a computer to "learn" chess, you might go about it in this way: You get a number of master players to spell out those elements in a position which help determine whether a position is winning or losing. In the game of chess these elements, or features, might be the number, strength, and mobility of the pieces; control of the center, open lines, and seventh rank; the pawn structure; and so on (as we mentioned earlier). Each feature must be defined in a precise, quantitative way so that at any position you could represent a player's control of the center, for example, by a single number.

Your next step is to combine these features in some way and so obtain an overall measure of the strength of the position. One simple way of doing this is to multiply the value of each feature by a constant, or *weights* (which may be thought of as reflecting the feature's importance), and then add up the resulting products.

But how do we get these weights? You start by assigning any weights that seem reasonable and change them as the computer "learns" to play. So suppose you are presented with a set of features along with some corresponding weights. In practice it almost always happens that you will not have an IEF and that you will run into inconsistencies of the type we observed in the exercises. So you define some measure of inconsistency for EFs—a measure of how much the EF differs from an IEF—and then adjust the weights to obtain the least inconsistent EF, that is, the EF that most resembles an IEF.

So the "learning" process is really a searching process. You assign weights to the features and see how positions are evaluated with these weights. If the positional values you get lead to many inconsistencies (that is, if they're very different from the values you would get if you used an IEF), you move on to other weights. If the numbers you get are very similar to what you might have had if you were using an IEF, then the EF becomes a candidate to be chosen as the final evaluation procedure. The weights go through a process of small changes, sometimes raised and sometimes lowered, to minimize inconsistencies. By a process of evolution you hope to arrive at an EF that assigns values in the same way an IEF does. You also hope—and of this there are no guarantees—that if an EF resembles an IEF in one respect (assigning values in a way that is consistent with the Basic Principle), it will resemble it in another one also (distinguishing winning positions from losing ones).

How do these methods work out in practice? We will briefly describe some actual experimental results.* The simplest application of this "learning" procedure was to the game of sums. For this game it was known in advance that the correct weighting procedure would produce an IEF. Using a very crude routine for learning, the computer found the right weights or came very close on 6 out of 11 independent trials.

The learning procedure was given a real test on the game called hex. Hex is generally played on a board composed of hexagons of varying dimensions. (Many a bathroom floor has served as a battleground.) Player I is assigned one pair of opposite sides, and player II is assigned the other pair (see Figure 13). The players alternately place their pieces (white for one player and black for the other) on some vacant hexagon until one player forms a chain from one side to the other; the player who does so is the winner. A chain is simply a set of stones that are connected. Figure 19 shows a winning chain for player I.

In the process of testing the learning procedure for hex, an inexperienced player (the author) defined several features that

---

*For a detailed discussion see Morton Davis, "On Artificial Machine Learning: Some Ideas in Search of a Theory," *International Journal of Computer Mathematics*, Vol. 5, Section B (1976), 315–29.

Figure 19

A GAME OF 5 × 5 HEX

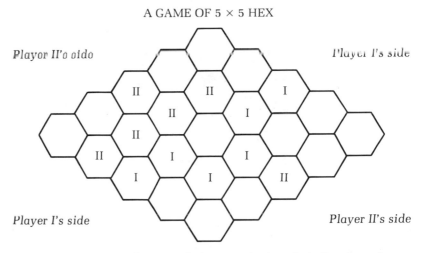

Player II's side   *Player I's side*

Player I's side   *Player II's side*

The hexagons with I's inside form a winning chain for player I

might influence whether a position is winning or losing. A variety of these features were combined in a sequence of trials, with a different combination used each time. In some cases very little happened; if the features are not artfully chosen, if they are irrelevant to the outcome of the game, there may be no weighting procedure that will produce an IEF or even a decent imitation. In other cases the computer pinpointed very effective weightings; at its best it defeated a computer moving randomly several times (in 5 × 5 hex) in the minimum time possible—five moves. This would happen less than 1 in a 1000 times by chance alone.

There are still a great many unanswered questions that are too technical to consider here, and doubtless there are inefficiencies in this approach to computer learning. But this kind of approach has one enormous advantage over one involving a complete, systematic analysis: It avoids the enormous number of possibilities that are encountered by both human and computer players when they try to look ahead. And this advantage alone makes the effort worthwhile.

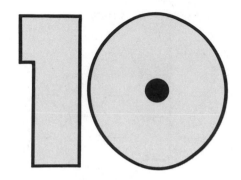

10

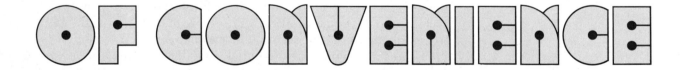

A MARRIAGE OF CONVENIENCE

ANALYTIC GEOMETRY

Introduction

When one considers how many people have contributed to the development of mathematics over several thousand years, it seems that the store of mathematical knowledge must be enormous and no one person could possibly master more than a small part of it. And in a way this is true. Mathematics has become very specialized, and it is not unusual for an expert in one field of mathematics to be almost ignorant of what a specialist in another area is doing. Still, it is remarkable how much mathematics is absorbed even by nonprofessionals. The student who has mastered a good four-year high school mathematics course can routinely solve problems that challenged some of the best minds of the past. And there were very few in the ancient or medieval world who knew a substantial part of the contents of an ordinary freshman course in calculus.

How does it happen that today's student can digest even a small part of this heritage left by the mathematical giants of the past? There are a number of reasons why, but a major one is that apparently different areas of mathematics may sometimes be combined into one. One such subject area, analytic geometry, results from consolidating two other areas into a single structure that is abstract enough to encompass them both. It turns out that algebra and geometry can both be used to describe and draw

Part of an ancient Egyptian papyrus, showing calculations of the area of a rectangle, a circle, a triangle, and a trapezoid.

conclusions about the same concepts (such as "straight line," "distance," "angle," and "circle"). Often a problem that arises in one area is easier to solve in the other.

The parties involved in this marriage of convenience, algebra and geometry, have roots that go back thousands of years. Although there has been some interdependence between the two disciplines, they developed independently for the most part.

Human behavior can be studied from a variety of viewpoints, each of which might be thought of as a separate academic discipline. Listed below are some of these different viewpoints. Comment on how certain pairs of these studies might be combined (in the same way that algebra and geometry were combined) to form a single, unified subject.

(a) Psychology: the study of the traits, emotions, and behavior patterns of the individual.
(b) Physiology: the physical aspect or "chemistry" of the body.
(c) Sociology: the study of the origin and evolution of human society and the laws that govern it.
(d) Economics: the study of the production and distribution of wealth.
(e) Political Science: the forms and principles of civil government and how it affects public and private affairs.
(f) Geography: the study of the surface of the earth (its climate, animals, land, agricultural products, etc.)
(h) History: a systematic record of past events in which humans have taken part.

Geometry and Algebra: The Betrothed

The Greeks were the first to study geometry as a purely deductive system of mathematics (the form in which we know it today), although it was used for practical applications much earlier. Starting with certain definitions and axioms that were considered "self-evident," the Greeks constructed geometric figures, calculated areas and lengths, and deduced all sorts of relationships between geometric figures. They stated, in geometric terms, many relationships that would later be expressed much more concisely in the language of algebra. For example, Proposition 1 of Euclid's Book II stated that

> If there be two straight lines and one of them be cut into any number of segments whatever, the rectangle contained by the two straight lines is equal to the rectangles contained by the uncut straight line and each of the segments.[*]

The distributive law of algebra

$$a(b + c + d + \cdots + n) = ab + ac + ad + \cdots + an$$

is much clearer and says essentially the same thing.

Similarly, Proposition 4 of the same book stated that

> If a straight line be cut at random, the square on the whole is equal to the squares on the segments and twice the rectangle contained by the segments.[*]

Its algebraic equivalent

$$a^2 + b^2 + 2ab = (a + b)^2$$

is simpler, and more convenient for calculations.

Algebra, which was developed after geometry, was originally devised to solve arithmetic problems. The following two problems appeared in a collection of questions compiled about 310 A.D.:

> Four pipes discharge into a cistern: one fills it in one day; another in two days; the third in three days; the fourth in four days: if all run together how soon will they fill the cistern?

> Demochares has lived a fourth of his life as a boy; a fifth as a youth; a third as a man; and has spent thirteen years in his dotage: how old is he?[**]

This kind of problem has been around for a long time. In fact, one such problem ("A quantity, its $\frac{2}{3}$, its $\frac{1}{2}$, its $\frac{1}{7}$, its whole, amount to 33") was found in the Ahmes papyrus in Egypt, the oldest known mathematical handbook.

---

[*] As quoted in Morris Kline, *Mathematical Thought from Ancient to Modern Times* (New York: Oxford University Press, 1972), p. 65.

[**] As quoted in W. W. Rouse Ball, *A Short Account of the History of Mathematics* (*New York*: Dover Publications, 1908), p. 102.

"Demochares has lived a fourth of his life as a boy; a fifth as a youth; a third as a man; and has spent thirteen years in his dotage."

Basically, algebra is an abstract form of arithmetic in which specific numbers are replaced by symbols that may represent any number. These symbols, which are sometimes called *variables* or *unknowns,* may be manipulated according to certain rules. These rules of algebra reflect the rules of arithmetic—the relationships that exist between the numbers the symbols represent. For example, the distributive law of algebra, states that

$$a(b + c) = ab + ac$$

This is just a shorthand way of expressing the following arithmetic law:

"If you multiply one number by the sum of two others, you get the same answer that you would get if you multiplied the original number by each of the numbers that made up the sum and then added the products." It is easy to see that the algebraic statement is clearer and easier to work with than its verbal equivalent.

The substitution of symbols for numbers can serve several different purposes. Suppose that in the problem we posed earlier we represent the age of Demochares—a definite but unknown quantity—by the letter x. Then the number of years that he spent as a boy, as a youth, and as a man would be x/4, x/5, and x/3, respectively, and 13 years would have been spent in his dotage. The total time spent during all of these periods, x/4 + x/5 + x/3 + 13, must be the same as his age now, x. So we may write the equation

$$\frac{x}{4} + \frac{x}{5} + \frac{x}{3} + 13 = x$$

An equation like this one is really a question: What number or numbers can you substitute for x that will make the equation a true statement? By algebra you can deduce that x = 60, but in any case once you're told that x = 60 it is easy to confirm that

$$\frac{60}{5} + \frac{60}{4} + \frac{60}{3} + 13 = 60$$

We say that x = 60 is a **solution** to the equation because we obtain a true statement when we substitute the number 60, for the symbol x. In contrast, x = 2 is *not* a solution to the equation because $\frac{2}{4} + \frac{2}{5} + \frac{2}{3} + 13$ is *not* equal to 60.

In this example we found a solution, x = 60, and as it happens it is the only one. But there are also equations with no solutions, equations with an infinite number of solutions, and still others with something in between. The equation $x^2 + 1 = 0$ has no (real) solution (why?), but *every* number is a solution to the equation

$$x^2 - 1 = (x + 1)(x - 1)$$

And between these two extremes there is the equation

$$x^3 - 6x^2 + 11x - 6 = 0$$

which has three solutions: x = 1, 2, and 3.

We will be interested in finding solutions to equations that may be more complicated in two respects. Generally speaking, the equations we will consider will have more than one unknown (or symbol), each representing a different number. A solution is still defined in the same way, however: when you substitute the proposed numerical values for the symbols, the resulting statement must be true. So x = 1, y = 2 is a solution to 3x − y = 1, because 3(1) − 2 = 1; and x = 2, y = 1 is not a solution, because 3(2) − 1 ≠ 1. And for the same reason x = 1, y = 2, z = 3 is a solution to x − y + z = 2.

Finally, there may be several different equations each containing several different symbols. A solution to such a **system of equations** is a set of values which, when substituted into the equations, makes *all* of them true. Thus the system of equations

$$x + y = 2$$
$$x - y = 6$$

has the solution x = 4 and y = −2, while

$$x + y = 2$$
$$2x + 2y = 3$$

has no solution at all. (Can you see why?)

In the rest of this chapter we will discuss equations involving at least two unknowns. Such equations often express a relationship between two physical variables. If a ball is thrown upward in a vacuum with a velocity of 100 feet per second from a point 224 feet high, then its height, h, t seconds later may be expressed by the formula:

$$h = -16t^2 + 100t + 224$$

Just as in the equation we derived in calculating Demochares' age, some pairs of values for h and t make the equation true and others do not. If $t = 5$ and $h = 324$, for example, the equation holds, since at the end of 5 seconds the ball is 324 feet in the air.

An equation may express a relationship between three physical variables as well. If V, T, and P are the volume, temperature, and pressure of a gas expressed in appropriate units, then $V = T/P$.

When we are given a system of equations, we will be concerned about the number and nature of the solutions. Although this may be viewed as a purely algebraic problem, you can get a great deal of insight into what is happening by looking at the problem geometrically, as we will see in Section 4.

1 For each equation or set of equations, state whether each of the values given for the variables are solutions.

(a) $3x - 6 = 0$ (i) $x = 2$,   (ii) $x = 3$,   (iii) $x = 0$

(b) $x^2 - 5x + 6 = 0$ (i) $x = 1$,   (ii) $x = 2$,   (iii) $x = 3$

(c) $2x - 2y = 2$ (i) $x = 0$, $y = -1$;   (ii) $x = 6$, $y = -2$;
      $3x + 4y = 10$ (iii) $x = 2$, $y = 1$

(d) $x^2 - 9 = (x - 3)(x + 3)$ (i) $x = -43$,   (ii) $x = 3\frac{1}{2}$,
                         (iii) $x = 0$

2 How many solutions are there to the equation $x^2 + y^2 = 0$? Why?

3 Give three different solutions to the equation $x + 2y = 2$. How many different solutions to this equation are there?

4 Verify that $(3, 4)$, $(-3, -4)$, $(4, 3)$ and $(-4, -3)$ are all solutions to the pair of equations $xy = 12$ and $x^2 + y^2 = 25$.

5 Show that any solution to one of the following two equations is also a solution to the other: $x + y = 1$; $3x + 3y = 3$.

6 By subtracting the equation $x - y - z = 3$ from the equation $x + y + z = 1$, and comparing the result with the equation $2y + 2z = 0$, show that there is no pair of values (x, y) which satisfies all 3 equations.

1 How many different solutions are there to each of the following equations or sets of equations? (You may find some algebraic manipulation helpful.)

(a) $x + y = 1$

(b) $x + y = 1$
     $x - y = 9$

(c) $x + y = 1$
     $x - y = 9$
     $2x + y = 6$

(d) $x + y = 1$
     $x - y = 9$
     $3x + y = 4$

(e) $x + y - z = 2$
     $-x - y + z = -3$

2 How many solutions are there to the following set of equations?
$$x + y - z = 1$$
$$x - y + z = 3$$
Can you determine a specific numerical value for x, y, or z?

3 For what value of b does the equation $x^2 = b$ have
(a) no (real) solution?
(b) exactly one solution?
(c) exactly two solutions?

Analytic Geometry:
A Union of Algebra and Geometry

If you wanted to describe the location of every restaurant in a certain city, you might go about it in the following way: First pick a location and use it as a reference point. (Fifth Avenue and Forty-second Street would do nicely for New York City.) Then describe the location of a restaurant by using two numbers: The first number, which we call x, tells you how far east or west of the reference point the restaurant is, and the second number, y, indicates how far north or south of the reference point the restaurant is. We will agree in advance that a positive value for x means the restaurant is east of the reference point and a negative value for x means that it is to the west, while a positive y means the restaurant is north of the reference point and a negative y that it is to the south.

You can describe the location of a point in the plane in the same way. Draw a vertical and a horizontal line and let their point of intersection be the reference point, which is known as the **origin**. With each point associate two numbers: x indicates how far to the right or left of the vertical line the point is (x would be positive if the point is to the right of the vertical line and negative if it is to the left), and y indicates how far above the horizontal line (or below, if y is negative.) The vertical line through the origin, which consists of those points for which x = 0, is called the **y-axis**, and the horizontal line through the reference point, which consists of those points for which y = 0, is called the **x-axis**. The numbers x and y are called the **coordinates** of the point and are written (x, y.) Figure 1 shows a plane along with certain points and their coordinates. It should be clear that every point in the plane can be described by exactly one pair of coordinates and every pair of coordinates corresponds to exactly one point in the plane.

Now suppose we have an equation that relates two physical variables, for example, the height of a ball and time, which we mentioned earlier. In that case the height was given by the formula $h = -16t^2 + 100t + 224$. There are certain pairs of values for t and h that make the equation true, such as $t = 0$ and $h = 224$ [which we write as (0, 224)] and (1, 308), and other pairs that make it false, such as (3, 5). Let us take all pairs of values for t and h that make the equation true and then plot all such points (t, h) on the graph as we located the restaurants earlier (but using t rather than x to indicate how far to the right or left a point is and h rather than y to indicate how far up or down it is). We get a "picture" of the equation, in effect. This picture is called the **graph** of the equation.

Usually there will be an infinite number of pairs of numbers that satisfy an equation, and it is clearly impossible to plot them all. What we actually do is plot a few points and infer the rest of the graph from these.

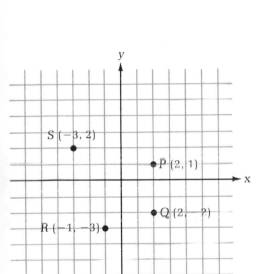

Figure 1

EXAMPLE 1

Plot the graph of the equation $-16t^2 + 100t + 224 = h$.

Solution

The first thing to do is to compile a table of pairs of numbers that make the equation true. In this case it is easiest to assign a value to t and calculate what h must be. One possible table of values is the following:

t	0	1	2	3	4	5	6	7	8
h	224	308	360	380	368	324	248	140	0

To each of these pairs of numbers there corresponds a point; each of these points is plotted on the graph shown in Figure 2. We estimate the rest of the points on the graph by smoothly connecting the plotted points.

Figure 2

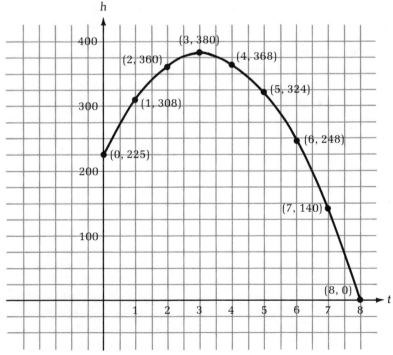

The graph of the curve $h = -16t^2 + 100t + 224$

EXAMPLE 2

An airplane is spotted overhead traveling north at 250 miles per hour. The time passed, T, and the distance traveled, D, are related by the equation

$$D = 250T$$

Draw the graph corresponding to this equation.

Solution

As before, we start by calculating which pairs of numbers satisfy the equation. We obtain the following table:

T	0	1	2	3	4	-1	-2
D	0	250	500	750	1000	-250	-500

In this case $T = -1$ should be interpreted as one hour ago and $D = -250$ should be interpreted as 250 miles south. The graph is shown in Figure 3.

Figure 3

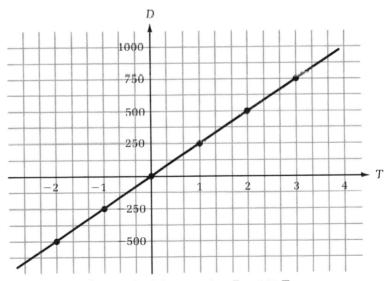

The graph of the equation $D = 250\,T$

(EXAMPLE 3)

When the temperature of a gas is held constant, the pressure, P, and the volume, V, are related by the equation $PV = 60$ (if expressed in appropriate units). Draw the graph of the equation.

Solution

First, we express one of the variables (either of them) in terms of the other:

$$V = \frac{60}{P}$$

Then we construct a table:

P	1	2	3	4	5	6	10	12	15	20	30	60
V	60	30	20	15	12	10	6	5	4	3	2	1

Figure 4

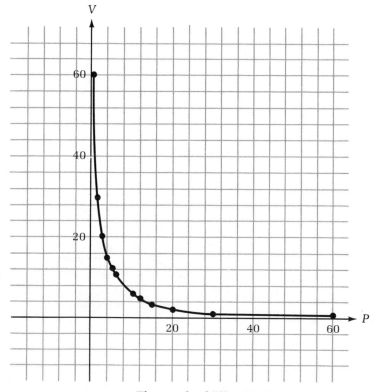

The graph of $PV = 60$

If we examine only the equation, it seems that negative numbers such as $P = -2$ and $V = -30$ should be on the graph, since these satisfy the equation. But physically these would be meaningless, because you cannot have a negative volume or pressure. The graph is shown in Figure 4.

You can graph an equation with three unknowns in the same way you graph one with two, except that it is done in three-dimensional rather than two-dimensional space. You find a trio of coordinates that satisfies the equation and use the first, second, and third coordinate to determine how far the point on the graph is to the east or west, north or south, and above or below the origin (the reference point).

1   What are the coordinates of the points A, B, C, and D shown?

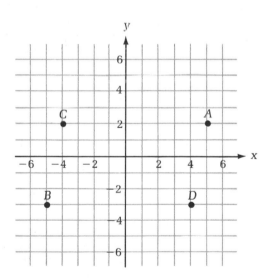

2   Graph the points with the following coordinates:
(a) (7, 2)   (b) (3, 6)   (c) (−2, −4)
(d) (−3, 5)   (e) (0, 0)

3 Suppose P has coordinates $(0, 0)$, Q has coordinates $(0, 2)$, R has coordinates $(2, 0)$, and S has coordinates $(2, 2)$. Plot these four points on a single graph. Then draw

(a) the line segment from P to Q
(b) the line segment from P to R
(c) the line segment from P to S
(d) the line segment from Q to R
(e) the line segment from Q to S
(f) the line segment from R to S.

For each of the line segments you have drawn, state what properties are common to the coordinates of all points on that segment.

4 How far and in what direction would you be traveling if you went
(a) from $(1, 4)$ to $(1, 8)$?
(b) from $(4, 2)$ to $(7, 2)$?
(c) from $(-3, 6)$ to $(-3, 2)$?

5 In each of the following equations make a table of values, plot the points, and try to guess what the whole graph looks like:
(a) $x - y = 4$ (b) $x^2 + y^2 = 9$
(c) $x = 3$ (d) $y = 1$
(e) $y = x^2 - 16$ (f) $x^2y = 40$

6 Graph each of the following equations. Determine whether the graph has a highest or lowest point and state what it is (they are). Estimate the values of x that make $y = 0$ and the values of y that make $x = 0$.
(a) $4x + 6y = 10$ (b) $y = 2x + 4$
(c) $x = y^2 - 4y - 8$ (d) $2x + 3y = 5$
(e) $y = x^2 - 6x + 4$ (f) $x = 3y - 7$
(g) $xy = 6$ (h) $4x^2 + 9y^2 = 36$

1 Imagine that the x-axis is a mirror and that the part of a curve above this line is the mirror image of the part of the curve below it. Then the curve is said to be *symmetric* with respect to the x-axis. We define symmetry with respect to the y-axis similarly. (See Figure 5.)

(a) What is the relationship between the coordinates of points P and Q if P and Q are mirror images with respect to (i) the x-axis? (ii) the y-axis?

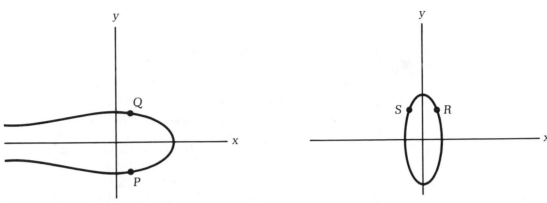

Symmetry with respect to the x-axis
(Points P and Q are mirror images)

Symmetry with respect to the y-axis
(Points R and S are mirror images)

Figure 5

(b) Consider the equation $y + x^2 = 25$. Observe that the points whose coordinates are $(5, 0)$ and $(-5, 0)$ are both on its graph, as are the points $(3, 16)$ and $(-3, 16)$ and the points $(4, 9)$ and $(-4, 9)$. In general, if you know that a point whose coordinates are (a, b) is on the graph, what conclusion can you draw?

(c) Can you deduce anything about the symmetry of the graph of the equation in part (b)?

2 The graph of the equation $y = (x - 2)^2$ is symmetric with respect to a line other than the axes. Plot the graph and determine that line. Can you see algebraically why this line is an axis of symmetry?

3 Graph the equations $x + 2y = 3$ and $2x - y = -4$ on the same sheet. Find the coordinates of their point of intersection. What is the algebraic significance of this point?

4 If points P and Q have the respective coordinates indicated below, what is the relationship between their positions in the plane?
(a) (a, b) and (b, a) (b) (a, b) and $(a, -b)$
(c) (a, b) and $(-a, -b)$ (d) (a, b) and $(2a, 2b)$
(e) (a, b) and $(0, b)$

5 Find the location of all points with coordinates (x, y) having the property that
(a) $-1 \leq x \leq 1$ and $-1 \leq y \leq 1$ (b) $y \geq x$
(c) $y^2 \geq x^2$ (d) $1 \leq x + y \leq 2$
(e) $2x + 2y = 2(x + y)$ (f) $2x^2 + 3y^2 = -1$

Linear Equations

In the last section we established a correspondence between geometric curves and algebraic equations: The value pairs that satisfy the equations are the coordinates of the points on the curve. In this section we will discuss equations that have a particularly simple graph, a straight line. Recall that the y-coordinate of a point indicates how far above or below the x-axis the point is and the x-coordinate indicates how far it is to the left or right of the y-axis.

Horizontal and vertical lines have the simplest equations. All points on a vertical line are the same distance to the right or left of the y-axis. For example, the vertical line 3 units to the right of the y-axis consists of all points whose x-coordinates are 3; thus this line is the graph of the equation $x = 3$. Similarly, the graph of the equation $y = -2$ consists of all points two units below the x-axis.

EXAMPLE 1

Plot the graphs of the following equations:
(a) $x = 3$ (b) $y = -1$ (c) $y = 3$

Solution   See Figure 6.

Figure 6

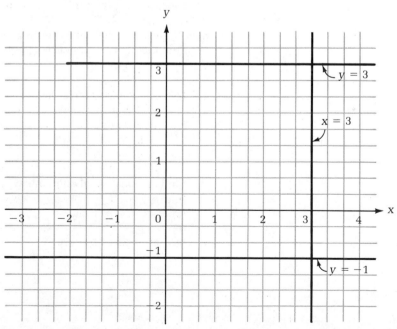

ELEVATION
1300ft.

ELEVATION
1200ft.

|——————— 1000 ft. ———————|

*After moving forward 1000 feet you have
risen 100 feet.*

To derive an equation of an *oblique* straight line (a line that is neither vertical nor horizontal), we use one of the line's basic properties: The direction of a straight line never changes. If you're climbing up a hill in a straight line and find that after moving forward 1000 feet you have risen 100 feet, then as long as you continue on that straight-line path you will always rise one tenth the distance you move forward. If you choose any pair of points on a non-vertical line and divide the difference in their y-coordinates by the difference in their x-coordinates, you will always get the same number, which is a characteristic of the line and independent of your initial choice of points. This ratio is called the slope of the nonvertical line.

The **slope** of a nonvertical line can be determined by taking any two different points on the line, say (a, b) and (c, d), and dividing the difference in their y-coordinates by the difference in their x-coordinates.* That is,

$$\text{Slope} = \frac{b - d}{a - c} \qquad \text{if } a \neq c$$

The slope of a line tells you its direction. If two roads both go one mile east for every five miles they go north, then both roads are going in the same direction and have the same slope, that is, they are parallel. So two different, nonvertical lines are parallel if and only if they have the same slope.

Euclid's Fifth Postulate states that for every line L and every point not on L there is exactly one line through P parallel to L. How can you find an equation corresponding to this line?

* The order of the points does not matter; that is, the slope may also be written as $(d - b)/(c - a)$. However, the point that has its y-coordinate subtracted in the numerator must be the point that has its x-coordinate subtracted in the denominator.

EXAMPLE 2

Find an equation of the line that passes through the point (2, 3) and has a slope of 2.

Solution

We know that (2, 3) is a point on the line. If the point (x, y) is also on the line, then the ratio of the differences in the y- and x-coordinates is $(y - 3)/(x - 2) = 2$ or $y = 2x - 1$. (See Figure 7.)

Figure 7

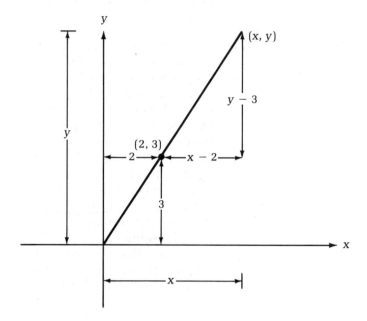

The last example is a special case of a more general rule:

A nonvertical line with slope m passing through the point (a, b) has the equation

$$y = m(x - a) + b$$

The statement "Two points determine a line" means that for every two points there is one and only one straight line passing through both.

> ## EXAMPLE 3

Find an equation of the line passing through the points (2, 4) and (6, 2.)

Solution

The first step is to calculate the slope:

$$\frac{4-2}{2-6} = -\frac{1}{2}$$

Now we can use our formula $y = m(x - a) + b$. We set the slope equal to $-\frac{1}{2}$ and the point (a, b) equal either to $(2, 4)$ to obtain the equation $y = -\frac{1}{2}(x - 2) + 4$ or to the point $(6, 2)$ to obtain the equation $y = -\frac{1}{2}(x - 6) + 2$. In either case the equation becomes, after simplification,

$$y = -\frac{1}{2}x + 5$$

Every straight line is the graph of some equation that has the form $Ax + By = C$ where A, B, and C are appropriate constants. Moreover, any equation in this form has a straight line as its graph unless A and B are both zero. (When A and B are both zero, there are two possibilities: If C is also 0, the graph contains every point in the plane; and if C is not 0, the graph contains no points in the plane. Explain why.) If $B = 0$ and $A \neq 0$, the graph will be a vertical line; if $A = 0$ and $B \neq 0$, it will be a horizontal line; and if $C = 0$, the line will go through the origin, $(0, 0)$. Not surprisingly, equations in the form $Ax + By = C$ are called **linear equations**.

Although every equation has exactly one graph, a single graph will correspond to infinitely many equations. The graph of the equation $x - 2y = 4$ is also the graph of the equation $4x - 8y = 16$ and of any equation obtained by multiplying both sides of $x - 2y = 4$ by a constant. If the coordinates of a point satisfy one equation, they will satisfy all equations obtained by such multiplications.

Now suppose you are given a system of two equations and want to find all the values that satisfy both. The coordinates of a point on an equation's graph must satisfy the equation, so a point on each of two graphs must have coordinates satisfying both of them. If you want to know the solutions of a set of equations, you can determine the points their graphs have in common; the coordinates of those points will be the solutions.

In the special case in which there are two linear graphs, there are three possible kinds of solutions. One possibility is that one equation is just the other one in disguise. This would occur if the two equations were those we just mentioned, $x - 2y = 4$ and $4x - 8y = 16$. Since the solutions to one equation are precisely the solutions to the other, there would be an infinity of common solutions, corresponding to the entire straight line.

If the graphs of the two straight lines are really different, then there are two more possibilities: The lines may be parallel or nonparallel, and this can be determined by checking whether they have the same slopes or not. If the two equations are $2x - y = 6$ and $x - 3y = -2$, for example, we may rewrite them as $y = 2x - 6$ and $y = \frac{1}{3}x + \frac{2}{3}$. Thus we see that they have different slopes, 2 and $\frac{1}{3}$. Geometrically we may deduce that the lines meet at a point, and indeed they do—the point with coordinates $x = 4$ and $y = 2$.

On the other hand, the pair of equations $2x - y = 6$ and $-6x + 3y = 10$ may be rewritten as $y = 2x - 6$ and $y = 2x + \frac{10}{3}$. We can see that both have a slope of 2. Since parallel lines never meet, we may deduce that there will be no solution. This can be confirmed algebraically by looking at the pair of equations as rewritten and observing that $2x - 6$ cannot be equal to $2x + \frac{10}{3}$ for any value of x, so y cannot be equal to both at the same time.

1 Graph each of the following equations and indicate whether the graph is a horizontal, vertical, or oblique line:

(a) $x = -1$ (b) $x = 2y - 6$
(c) $x = y - 6$ (d) $3y - 6x = 4$
(e) $27 = 4y$ (f) $x + y = 2$

2 Find the slope of each of the following lines:

(a) $y = 2x - 6$ (b) $x + y = 2$
(c) $x - y = 7$ (d) $x = 4y - 7$

3 Find an equation of the line through the point P with slope m for each of the following values of P and m:

(a) $P = (1, 2)$ and $m = 1$ (b) $P = (-1, 2)$ and $m = -3$
(c) $P = (0, 0)$ and $m = -3$ (d) $P = (2, 1)$ and $m = -7$

4   Find an equation whose graph is a line with slope 2 passing through the point
(a) (1, 1)   (b) $(-1, 1)$ (c) $(1, -2)$

5   Graph the equations you found in Exercise 4 and confirm that all the lines obtained are parallel.

6   Choose three different *pairs* of points that lie on the line $x - 2y = 6$. Calculate the slope three times, using a different pair of points each time, and confirm that the answer is always the same.

7   Find an equation of the line passing through the points
(a) (2, 1) and $(-2, -1)$
(b) (1, 1) and (4, 3)
(c) $(-2, 1)$ and (2, 3)

8   Suppose you are given the line $Ax + By = C$ with $B \neq 0$ and are told that the points (x', y') and $(x' + h, y' + k)$ are on its graph.
(a) Why must it be true that $Ax' + By' = C$ and $A(x' + h) + B(y' + k) = C$?
(b) Using the equations in part (a), show that $Ah + Bk = 0$.
(c) Use the definition of slope to show that the slope of the line is k/h, which is also equal to $-A/B$.
(d) Rewrite the original equation in the form $y = -Ax/B + C/B$ and confirm that $-A/B$ is the slope.

9   Why do we assume that $B \neq 0$ when we calculate the slope in Exercise 8?

10 Suppose you are given more than two linear equations in two unknowns (x and y) and you want to find values of x and y that satisfy all of the equations. Use your geometric intuition to explain why the solution must be either (i) every point on a straight line, (ii) a single point, or (iii) no solution at all.
    In each of the following cases there are three equations in two unknowns. State into which of the three categories these equations fall. (Do this algebraically if you can, and then confirm your answer graphically.)

(a) $x + y = 6$ (b) $x + y = 6$ (c) $x + y = 6$
      $x - y = 2$ $x - y = 2$ $-2x - 2y = -12$
      $5x - y = 18$ $3x + y = -2$ $3x + 3y = 18$

11 For what value of k will there exist a solution to the following system of three equations?
$$x - y = 1$$
$$2x + 2y = 22$$
$$-x + 3y - k$$
What is the solution?

12 We defined the slope of the line passing through the points (a, b) and (c, d) to be the ratio of the difference in the y- and x-coordinates: $(d - b)/(c - a)$. What would happen if we reversed the points, that is, formed the ratio $(b - d)/(a - c)$? Why do we insist that $a \neq c$? What would the line through the two points be if a were equal to c?

13 What is the slope of a horizontal line? Why is it meaningless to speak of the slope of a vertical line?

1 In general, an equation involving three variables has as its graph a surface in three-dimensional space. In particular, the equation $Ax + By + Cz = D$, where A, B, C, and D are constants, is a plane.
   (a) If you have a system of two such equations, there are three possible solution sets. What are they?
   (b) Answer part (a) if there are three equations.
   (c) Suppose there are n equations in three variables and the graph of each is a plane. If they have exactly one common solution, what are the possible values of n?
   (*Hint:* Answer Question 1(c) by thinking geometrically.)

2 Consider the two sets of equations shown below and check the solutions given:

(a) .9x + y = 11   (b) x + y = 11
       x + y = 10.9   x + .9y = 10.9
     solutions: x = −1 and y = 11.9   solutions: x = 10, y = 1

Can you guess (using your knowledge of geometry) why a slight change in the equations leads to such a large change in the solution?

3 An elevator is climbing at 3 feet per second. At time $t = 0$ it is 100 feet up. The altitude, A, of this elevator at time t is given by the formula $A = 3t + 100$.
   (a) Graph the equation. When will the elevator be 130 feet high (assuming it maintains a constant speed)? When was it 85 feet high?
   (b) Another elevator, also traveling at a constant speed, has its altitude expressed by the formula $A = 5t + 60$. When will the elevators be at the same height? At what height will this occur?

4 (a) Write the equation of a line that passes through the point $(0, a)$ with slope 2.
   (b) Write the equation of a line that passes through the point $(0, b)$ with slope 2.

5 The lines described in Exercise 4 have the same slope; thus, if $a \neq b$, they are parallel (that is, they never meet.) Show algebraically that the equations have no common solution when $a \neq b$.

The Distance Between Two Points

The concept of *length* is fundamental in geometry. The area of geometric figures is expressed in terms of lengths, geometric figures are similar if certain lengths are proportional, and triangles are congruent if the lengths of corresponding sides are equal. Since the length of a line segment is simply the distance between its endpoints, it will be useful to derive an algebraic expression for distance.

It is easy enough to find the distance between two points when they are on the same horizontal or vertical line. If one point is due east of the other, both points have the same y-coordinate, so they might be represented as (a, c) and (b, c) (see Figure 8.) Clearly, the **distance** between the points is $b - a$ if b is larger than a, and it is $a - b$ otherwise.

Figure 8

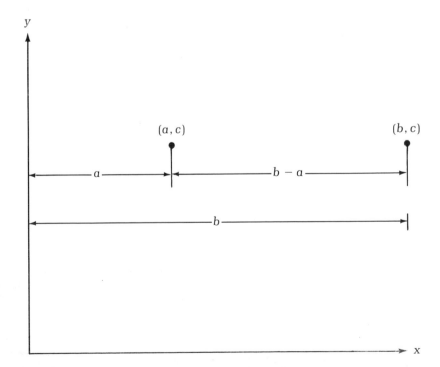

This is a satisfactory answer if we know which x-coordinate is larger, but it will be convenient to have a formula that doesn't require this information. To give such a formula we need to know what is meant by the *absolute value* of a number.

The **absolute value** of a number x is a positive number with the numerical value (but not necessarily the sign) of x. So the absolute value of -3 is 3, of 4 is 4, of 6 is 6, and of -10 is 10. The absolute value of x is denoted by $|x|$. In our original problem, then, the distance between the points (a, c) and (b, c) is $|a - b|$.

For any number x (positive, negative, or zero,) we know that x^2 cannot be negative. Thus $x^2 = |x|^2$ for any number x. This fact will be useful in finding the distance between two points that are not on the same vertical or horizontal line.

To see why this is reasonable, let us recall that a distance is always a nonnegative number and the distance from a point P to a point Q is always the same as the distance from Q to P. Now $|a - b|$ is never negative whatever the magnitudes of a and b. Moreover, since $|a - b| = |b - a|$, it doesn't matter whether you interpret the formula as giving the distance from (a, c) to (b, c) or the other way around; the distance is the same in either direction, as it should be.

When two points, say (e, f) and (e, g), are on the same vertical line, the distance between them is $|f - g|$.

EXAMPLE 1

Find the distance between each of the following pairs of points:
(a) $(4, 2)$ and $(1, 2)$
(b) $(1, 2)$ and $(4, 2)$
(c) $(0, 4)$ and $(0, -2)$

Solution

(a) $|4 - 1| = |1 - 4| = |3| = |-3| = 3$
(b) The same answer as to part (a)
(c) $|4 - (-2)| = |4 + 2| = |6| = 6$ or $|-2 - 4| = |-6| = 6$

Now suppose we take any two points with coordinates (a, b) and (c, d). (See Figure 9.) The **distance** between these points is the length of the line joining them, and that line is the hypotenuse of

Figure 9

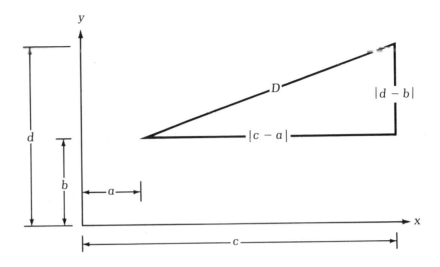

a right triangle. The length of the hypotenuse, D, may be expressed in terms of the other sides of the triangle by means of the Pythagorean theorem:

$$D^2 = (|c - a|)^2 + (|d - b|)^2$$
$$= (c - a)^2 + (d - b)^2$$

(because $x^2 = |x|^2$ for any number x), or

$$D = \sqrt{(c - a)^2 + (d - b)^2}$$

EXAMPLE 2

The points $A\,(1, 2)$, $B\,(4, -2)$, and $C\,(2, 5)$ are the vertices of a triangle. Calculate the length of each of the three sides.

Solution

See Figure 10 (on the next page). The lengths of the sides are

$$\overline{AB} = \sqrt{(1 - 4)^2 + (2 - (-2))^2} = 5$$

$$\overline{BC} = \sqrt{(4 - 2)^2 + (-2 - 5)^2} = \sqrt{53}$$

and

$$\overline{AC} = \sqrt{(1 - 2)^2 + (2 - 5)^2} = \sqrt{10}.$$

Figure 10

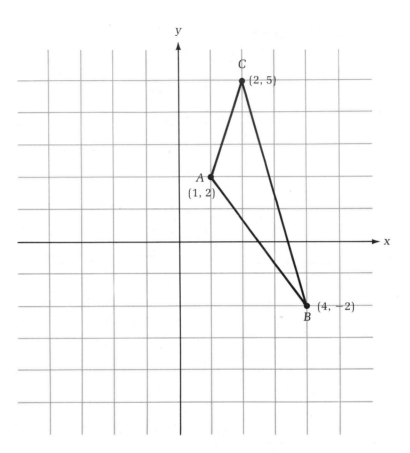

1   Find the distance between the pairs of points that have the following coordinates:
(a)   (4, 0) and (4, 2)
(b)   (−1, 2) and (3, 2)
(c)   (2, 3) and (5, 7)
(d)   (2, 3) and (−3, 9)

2   P (2, −2), Q (3, 2) and R (−3, −4) are the vertices of a triangle. Find the length of each of the three sides.

3   Calculate the distance from (0, 0) to (a, 0) and add it to the distance from (a, 0) to (a, b). Show that this sum is greater than the distance from (0, 0) to (a, b) if a ≠ 0 and b ≠ 0. What does this mean geometrically?

4   Evaluate:
     (a) $|4|$ (b) $|-6|$ (c) $|0|$ (d) $|-17|$

5   Show that $|x| = x$ if $x \geq 0$
                      $= -x$ if $x \leq 0$

6   Show that $|a - b|^2 = |b - a|^2 = (a - b)^2$.

7   Show that $|x| = |-x|$.

8   What can you say about the value of x if you know that
     (a) $|x|^3 = x^3$?
     (b) $|x|^2 = x^2$?
     (c) $|x|^2 = -x^2$?

9   If you wanted to walk from the point $(-4, 0)$ to the point $(6, 0)$, you would first walk 4 units to get to $(0, 0)$ and then 6 units beyond that to $(6, 0)$, a total of 10 units in all. Show that the two points are 10 units apart by using the distance formula.

1   Suppose a point (x, y) is as far from the point $(1, 3)$ as it is from the point $(5, 1)$. Expressed algebraically, this means that

$$\sqrt{(x - 1)^2 + (y - 3)^2} = \sqrt{(x - 5)^2 + (y - 1)^2}$$

     (a)   By squaring both sides of the equation above and simplifying, show that $2x - y = 4$.
     (b)   Show that the line passing through $(1, 3)$ and $(5, 1)$ is the graph of the equation $y = -\frac{1}{2}x + \frac{7}{2}$.
     (c)   Show that the point common to both lines is $(3, 2)$; confirm that this point is midway between $(1, 3)$ and $(5, 1)$.
     (d)   From geometry it is known that $2x - y = 4$ is the perpendicular bisector of the line segment joining $(1, 3)$ and $(5, 1)$. Verify that the product of the slope of this line and the slope of the line through $(1, 3)$ and $(5, 1)$ is -1.

2   Show that the line joining two points (a, b) and (c, d) and the line perpendicular to it (constructed as in Question 1) have slopes whose product is -1. (Assume that neither of the lines is vertical.)

3   Find the distance between a point (x, y) and a point $(2, -1)$. Show that all points five units away from $(2, -1)$ satisfy the equation $x^2 - 4x + y^2 + 2y = 20$.

4   Plot and describe each of the following graphs:

(a) $y = |x|$ (b) $y = -|x|$

(c) $y^2 = |x|^2$ (d) $|y| = |x|$

(e) $|x| + |y| = 0$ (f) $|x + y| = 0$

5   Show that all points that are twice as far from (4, 0) as they are from (0, 0) satisfy the equation $3x^2 + 8x + 3y^2 = 16$.

Conic Sections and Other Graphs

More than 2000 years ago the Greeks knew that the circle, ellipse, hyperbola, and parabola are related, that all these curves are conic sections. A **conic section** is the curve in which a plane intersects a right circular cone. By a *right circular cone* we mean the surface formed when two surfaces in the shape of ice cream cones (without ice cream) are placed end to end so that they have a common axis. If you take a circle and point *P* directly above the center and then take all the lines that join *P* to points on the circumference of the circle, you will form a right circular cone. (See Figure 11.)

     Conic sections arise frequently, both in the works of humans (the manufacture of headlights and antennas, navigation, etc.) and in nature (the orbits of planets, trajectories in a vacuum, etc.). If

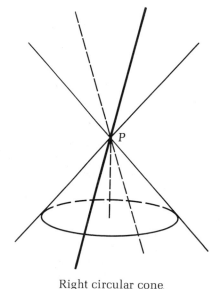

Right circular cone

Figure 11

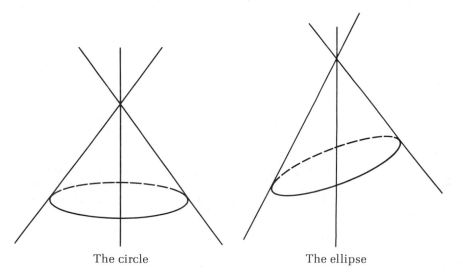

The circle

The ellipse

Figure 12   **Figure 13**

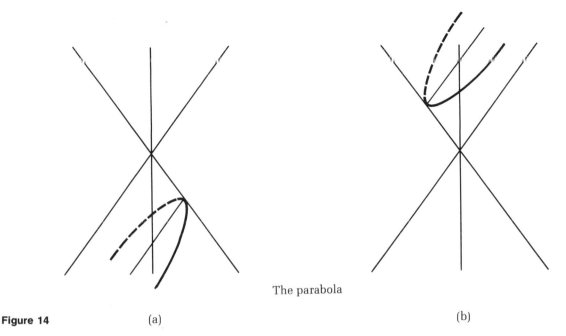

The parabola

Figure 14   (a)   (b)

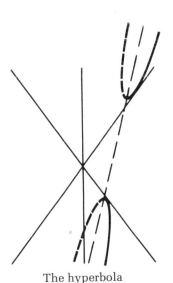

The hyperbola

Figure 15

you want to analyze these curves (perhaps because you are manufacturing antennas or calculating orbits), it is useful to have an
algebraic expression for them. Analytic geometry provides the
means of obtaining such expressions.

When a plane intersects a right circular cone, there are four
types of curve that may be formed. If we assume that the axis of
the cone is vertical, the simplest curve is obtained when the
intersecting plane is horizontal (see Figure 12). In this case we have
a **circle**. If we tilt the plane slightly from the vertical, the circle is
stretched in one direction but keeps the same dimensions in the
other, so that we have an **ellipse** (see Figure 13). A cross-section of
an egg or a football (taken with a plane parallel to the longest
dimension) has approximately the shape of an ellipse.

If we tilt the plane somewhat more, until it is parallel to one of
the lines that forms the cone, we obtain a **parabola**. The parabola
shown in Figure 14 (a) has a finite upper point but extends infinitely downward. Other parabolas have lower points that are
formed when planes intersect the upper half of the cone (Figure 14
(b). When the intersecting plane is tilted still more, it intersects
both halves of the cone and the resulting curve extends infinitely
in both directions (see Figure 15). This curve is called a **hyperbola**.

The circle, ellipse, parabola, and hyperbola are the four basic conic sections, but one special case, referred to as the *degenerate case,* can arise in which the curve takes an extreme form. This is the case when the plane passes through the point connecting the two half-cones. In such a case the circles and ellipses shrink to a point, the parabola becomes a straight line, and the hyperbola becomes two intersecting straight lines.

For each of the four basic conic sections we will state one of its basic properties and use this property to derive an algebraic expression for the curve. We will also mention a few places where one may encounter the curve, either in nature or in our everyday life.

The Circle

A circle may be defined as the set of all points in a plane that are at some given distance, called the *radius* of the circle, from a fixed point called the *center.* We will derive the equation of a particular circle by using this property.

EXAMPLE 1

Find an equation of the circle consisting of all points (x, y) that are 2 units from the point $(3, 1.)$

Solution

See Figure 16. The distance between $(3, 1)$ and the point (x, y) is $\sqrt{(3 - x)^2 + (1 - y)^2}$, so any point (x, y) at a distance 2 from the

Figure 16

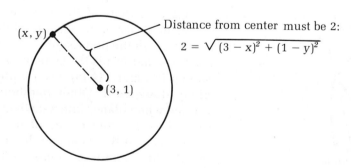

center of the circle must satisfy the equation

$$2 = \sqrt{(3-x)^2 + (1-y)^2}$$

When we simplify this, we find that the equation of the circle is

$$4 = (3-x)^2 + (1-y)^2$$

The circle is a very common geometric figure; it is the shape of wedding rings, coins, and dinner plates. It also has the property of having the most area for a given length of boundary.

The Ellipse

The ellipse, which is shaped something like the cross-section of an egg, may be defined in the following way: If you take two points in the plane, called the *foci* of the ellipse, and then draw a curve consisting of all points whose distances from these foci add up to some fixed number, the curve will be an ellipse.

(EXAMPLE 2)

Find an equation of the ellipse consisting of all points of the plane whose distances from the points $(3, 0)$ and $(-3, 0)$ add up to 10.

Solution

The distances from the point (x, y) to the points $(3, 0)$ and $(-3, 0)$ are

$$\sqrt{(x-3)^2 + y^2} \quad \text{and} \quad \sqrt{(x+3)^2 + y^2}$$

So any point of the ellipse must satisfy the equation

$$\sqrt{(x-3)^2 + y^2} + \sqrt{(x+3)^2 + y^2} = 10$$

If we rewrite this equation as

$$\sqrt{(x-3)^2 + y^2} = 10 - \sqrt{(x+3)^2 + y^2}$$

then after squaring and simplifying we have

$$5\sqrt{(x+3)^2 + y^2} = 25 + 3x$$

Squaring and simplifying again, we obtain finally $16x^2 + 25y^2 = 400$. The graph of the ellipse is shown in Figure 17.

Figure 17

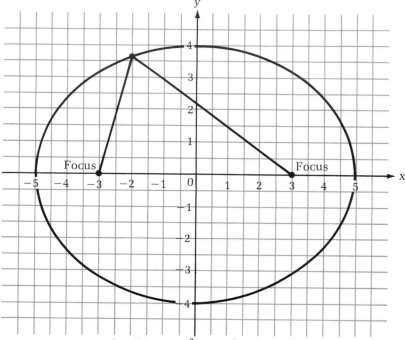

The ellipse $16x^2 + 25y^2 = 400$

It is easy to construct a homemade ellipse using our definition. Just put two nails at the foci, attach the ends of a piece of string 10 units in length to them, and then extend the string in all directions (in the horizontal plane) with your finger; the path of your fingertip will be an ellipse.

Perhaps the best-known application of the ellipse is found in the heavens. The orbits of the earth and the other planets around the sun are ellipses with the sun at one of their foci. The ellipse has another interesting property as well: If a source of light or sound is at one focus of the ellipse, the light or sound waves will emanate from the source in all directions, strike the side of the ellipse, and then converge at the other focus. This principle is often exploited in science museums: If a portion of the wall is built in the form of an ellipse, visitors speaking at one end of a hall (at a focus of the ellipse) can be heard by other visitors standing at the other focus (see Figure 18). A room with this property is called a *whispering gallery*.

Figure 18

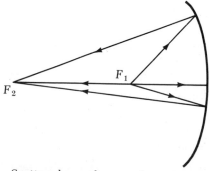

Scattered sound waves from F_1 hit elliptical wall and converge at F_2.

The Parabola

The curve consisting of all points that are as far from a given line (called the *directrix*) as from a given point (called the *focus*) is a parabola. The equation of the parabola may be derived from this property. The parabola, unlike the circle and ellipse, is unbounded.

EXAMPLE 3

Find an equation of the parabola consisting of all points that are as far from the point $(0, 2)$ as they are from the line $y = -2$.

Solution

The distance from a point (x, y) to the line $y = -2$ (that is, the shortest distance) is the distance from (x, y) to the point $(x, -2)$, which is $|y + 2|$. The distance from the point (x, y) to the point $(0, 2)$ is $\sqrt{x^2 + (y - 2)^2}$, so any point of the parabola must satisfy the equation

$$|(y + 2)| = \sqrt{x^2 + (y - 2)^2}$$

Squaring both sides and simplifying, we obtain the equation of the parabola: $8y = x^2$. The graph of the parabola is shown in Figure 10.

Figure 19

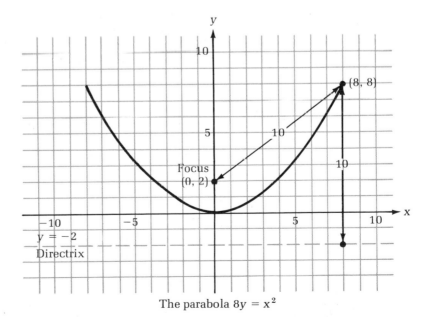

The parabola $8y = x^2$

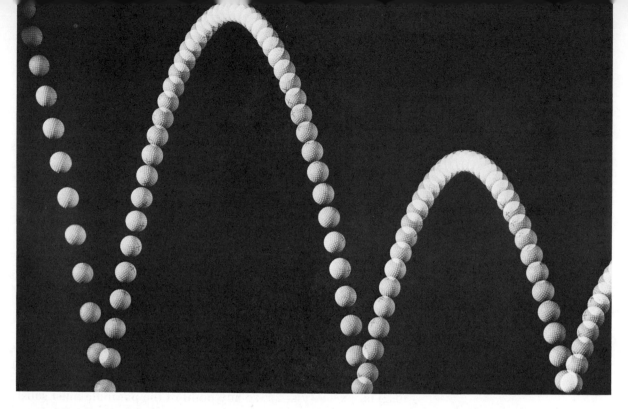

A bouncing ball also traces out a parabolic path.

The parabola, like the ellipse, is a fairly common curve. A ball thrown into the air in a vacuum will take a parabolic path. A thin cable suspended freely will approximate a parabolic arc. A particularly useful property of parabolas is this: If a source of light or sound is placed at the focus of a parabola, the light or sound waves will emanate from the source in all directions, but when they hit the parabola they will be reflected in parallel rays. And conversely, if parallel rays strike the surface of a parabola, they will be reflected in such a way that they will converge at a single point, the focus. (See Figure 20).

Figure 20

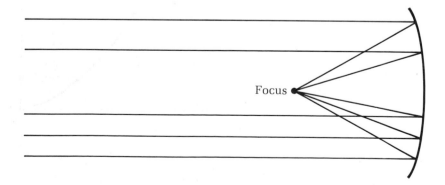

This last property of the parabola may be exploited in several ways. By putting a bulb at the focus of a parabolic reflector you will send out parallel rays straight ahead instead of in all directions; this fact is useful in designing automobile headlights, flashlights, etc. An antenna receiving distant signals that are almost parallel reverses the process, it takes all these parallel rays and concentrates them at the focus, where the signal is strongest.

The effect of having sound projected across a room may be obtained by using parabolas as well as ellipses. If a parabolic surface is placed at each end of the room, both parabolas having the same axis, and a speaker and listener are placed at the foci, the sound waves will go out from the speaker, be reflected by the first parabola in parallel rays, hit the second parabola, and be reflected by it to converge at the other focus (Figure 21.) It was observed many years ago that people standing in one location in the U.S. Congress could overhear whispering a considerable distance away; this effect was no doubt caused by the shape of the ceilings.

Figure 21

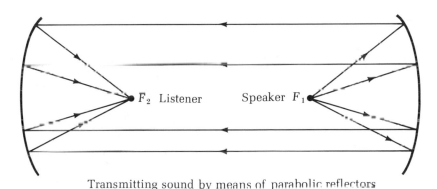

Transmitting sound by means of parabolic reflectors

The Hyperbola

At first glance the hyperbola looks like two parabolas with a common axis and with their closed ends facing each other. (In fact, the two halves of a hyperbola are not parabolas, although they are somewhat like parabolas in shape.) The following property of the hyperbola can be used to derive its equation: Suppose we select any two points in the plane and call them foci. If we take all points of the plane the difference of whose distances from the foci is some fixed constant, then these points will form a hyperbola.

EXAMPLE 4

Find an equation of the hyperbola consisting of all points in the plane such that the difference between their distances from the points $(5, 0)$ and $(-5, 0)$ is 6.

Solution

Any point (x, y) on the hyperbola must be 6 units closer to one of the foci than to the other. So it must satisfy the equation

$$\sqrt{(x + 5)^2 + y^2} - \sqrt{(x - 5)^2 + y^2} = \pm 6$$

or

$$\sqrt{(x + 5)^2 + y^2} = \pm 6 + \sqrt{(x - 5)^2 + y^2}$$

Squaring both sides of the equation and simplifying, we obtain $5x - 9 = \pm 3\sqrt{(x - 5)^2 + y^2}$. Squaring and simplifying once again, we obtain $16x^2 - 9y^2 = 144$. The graph of the hyperbola is shown in Figure 22.

Figure 22

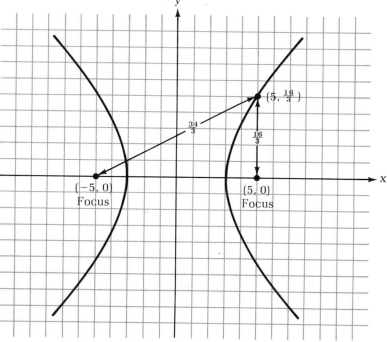

The hyperbola $16x^2 - 9y^2 = 144$

One of the interesting applications of the hyperbola arises in navigation. Two transmitters located at different locations send signals simultaneously at regular intervals. A ship receiving the two signals can observe the difference in arrival time and can easily calculate the difference in the transmitter distances from the known signal speed. From this information the ship can deduce that its position lies along a certain hyperbola. (See Figure 23.) By gathering information from other sources, such as the position of certain stars, it can determine its position.

Figure 23

The number on each part of a given hyperbola indicates how much earlier the signal would arrive from focus I than from focus II if the ship is situated on that hyperbola. If the two signals are received at the same time, the ship is on the perpendicular bisector of the line segment connecting the transmitters.

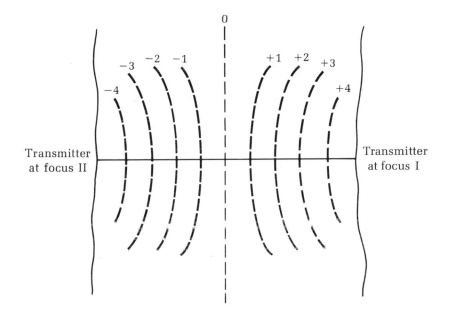

1 Find an equation of the circle with center at $(0, 0)$ and radius
   (a) 3
   (b) 4
   (c) 5

2 Show that the point $(4, 1 - \sqrt{3})$ is on the circle $4 = (3 - x)^2 + (1 - y)^2$ (shown in Figure 16). Then confirm that the distance from that point to the center is 2.

3   Show that the point $(3, \frac{16}{5})$ is on the ellipse $16x^2 + 25y^2 = 400$ (shown in Figure 17). Then confirm that the distances from that point to the foci, $(3, 0)$ and $(-3, 0)$, add up to 10.

4   Show that the point $(12, 18)$ is on the parabola $8y = x^2$ (shown in Figure 19). Then confirm that the distances from the point $(0, 2)$ and the line $y = -2$ are the same. What is this common distance?

5   Show that the point $(4, \frac{4}{3}\sqrt{7})$ is on the hyperbola $16x^2 - 9y^2 = 144$ (shown in Figure 22). Then confirm that the difference between the distances from the point to the two foci, $(5, 0)$ and $(-5, 0)$, is 6.

Dorothea Rockburne,
Velar—Combination Series,
1978.

6   Find the circle formed by all points at a distance 10 from $(-1, 4)$.

7   If the sum of the distances from (x, y) to $(0, 1)$ and $(0, -1)$ is 2, what must be the value of y? The values of x? (Look at this problem both algebraically and geometrically.)

8   Find the set of all points P with coordinates (x, y) such that the sum of the distances from P to $(0, 12)$ and P to $(0, -12)$ is 26.

9   Find the points which are as far from the line $x = -2$ as they are from the point $(4, 0)$.

10   If the distance from (x, y) to $(0, 6)$ is subtracted from the distance from (x, y) to $(0, -6)$ the absolute value of the result is 8. Find an equation satisfied by all such (x, y).

11   (a)   The graphs of equations in the form $Ax^2 + By^2 = C$, where A, B, and C are positive constants, are ellipses. What is the greatest possible value of x? of y?
　　 (b)   We mentioned earlier that an ellipse may degenerate into a point. For what value(s) of the constant(s) will the graph of the equation be a point?

12   The graph of $x^2 + y^2 = C^2$ is a circle. For what value of C will the graph become a point?

13   The graph of $Ax^2 + By = C$ is generally a parabola. For what value of the constants A, B, and C will the curve degenerate into a line? What line?

14   The graph of $Ax^2 - By^2 = C$ is a hyperbola if A and B are positive. For what value of C will they become two straight lines? What lines?

1   Show that all points (x, y) which are three times as far from the point $(-4, 0)$ as they are from the point $(4, 0)$ have coordinates satisfying the equation

$$(x - 5)^2 + y^2 = 9$$

2   Show that all points (x, y) which are twice as far from the line $x = -4$ as they are from the point $(2, 0)$ have coordinates satisfying the equation

$$3(x - 4)^2 + 4y^2 = 48$$

3 Show that all points which are twice as far from the point (4, 0) as they are from the line x = −2 have coordinates satisfying the equation

$$3(x + 4)^2 - y^2 = 48$$

4 Plot the equation
(a)   y − x + 1 = 0
(b)   2y + x − 4 = 0

5 What would the graph of (y − x + 1) (2y + x − 4) = 0 look like?

6 What would the graph of (y − x + 1)² + (2y + x − 4)² = 0 look like? How would you relate the answers to Questions 5 and 6 to the answer to Question 4?

Summary

In this chapter we described a mathematical model (analytic geometry) that was constructed by combining two other mathematical models (algebra and geometry). We saw that problems in geometry may be solved using algebraic methods and algebraic problems may be attacked geometrically. As a simple illustration we considered the problem of how many different kinds of solutions there can be to two first-degree equations in two unknowns. When this question is viewed algebraically, the answer is far from clear; but when the problem is translated into the language of geometry (how many points can two straight lines have in common?), the answer is immediate: There may be no points in common (if the lines are parallel), there may be one point in common (the usual case when the lines intersect), or there may be an infinite number of points in common (if the "two" lines are really two equations of the same line). You can be sure that there will not be exactly two solutions—a result that would be hard to derive by purely algebraic methods.

Our goal in this chapter has been to get an idea of what analytic geometry is about without using an excessive amount of computation. We might pursue the uses of analytic geometry further (deriving algebraic proofs for geometric theorems, for example), but this would involve using relatively complicated calculations, which is not our purpose. As an entertaining application of analytic geometry, we will give an equation whose graph is a rough approximation to a human face. The equation is

$$(9y^2 + 16x^2 - 14{,}400)[y + (\sqrt{8 - \tfrac{1}{2}x^2})^2][(y - 20)^2 + (x + 8)^2] \cdot$$

$$[(y - 20)^2 + (x - 8)^2](4x^2 + 9\sqrt{-y - 12^4} - 576) \cdot$$

$$[(x - 8)^2 + (y - 20)^2 - 16][(x + 8)^2 + (y - 20)^2 - 16] = 0$$

(Note that the left side of this equation is a product of seven factors. The product will be zero whenever any one of the factors is zero.) The graph is shown in Figure 24.

Figure 24

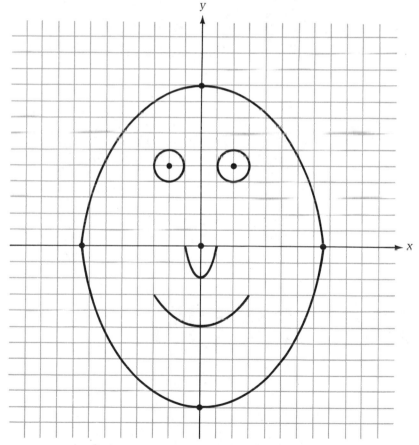

Analytic geometry sometimes wears a happy face.

MATHEMATICS

OF CHANGE

CALCULUS

Introduction

When a cube of sugar is dipped into a cup of tea, the rate at which the sugar dissolves is proportional to the surface area exposed (the part of the sugar that is in contact with the tea). If it takes ten minutes for the sugar to shrink to half its original size, when will it be one quarter of its original size?

Gossip spreads rapidly in a certain city. The number of new people who hear a rumor in a given hour is proportional to the number of people who have already heard it. One person starts the rumor initially, and in ten hours 100 people have heard it. How many people will have heard the rumor after twenty hours?

An automobile company advertises that when you step on the gas in their new model, your speed increases steadily from 0 to 60 miles an hour in 40 seconds. How far does the car travel in that 40-second period?

You have 1000 feet of fence to enclose three sides of a rectangular area (there is a river that serves as the fourth side). What is the largest area that you can enclose?

Halley's Comet was last seen in this part of the world Y years ago. When will it be seen again?

A football is thrown into the air with a velocity of 44 feet per second and at an angle of 45° to the horizontal. How high will it go, how far away will it land, and how long will it take to reach its landing place?

The element that is common to all these problems is change—changes in volume, information, area, distance, population size, or profit. Some other problems that involve change are finding the half-life of radioactive material or the speed of a chemical reaction, calculating the rate of growth of a colony of bacteria, and figuring out how many hot dogs to buy and what price to charge in order to maximize your profits when you know how many hot dogs will be sold at each possible price. And it was precisely to handle problems involving change that the calculus was invented.

It is easy to see from this small sample that the calculus is a tool for solving practical problems. One of its earliest applications was to the computation of the orbits of heavenly bodies, and it has since been used to calculate the paths of rockets, to analyze electrical circuits, and to describe the flow of heat and fluids.

But quite aside from its practical value, the calculus is a great intellectual achievement. Its basic ideas are quite deep; they were not developed quickly, nor were they mastered easily. Isaac Newton and Gottfried von Leibniz, who independently developed most of the rules of manipulation, were themselves somewhat shaky about the theory. So if you find the ideas presented in this chapter difficult, you are in distinguished company.

Students in various academic disciplines such as engineering, chemistry, and physics take a course in calculus to master manipulative skills. This is *not* our purpose here. We are interested in ideas, and we will do calculations and develop technical resources only as required for that purpose. The concepts behind the calculus are not easy, but even if you do not acquire a deep understanding of them, you should get a glimpse into the inner workings of this branch of mathematics. We will try to describe the basic problem of the calculus in terms of a very simple context: in terms of two passengers taking a trip in a car.

The Description of a History: Distance and Velocity

Imagine that a car starts on a trip at some particular time, which we designate as $t = 0$, along a straight road that runs in a north–south direction. After driving awhile—sometimes north, sometimes south, sometimes not moving at all—the car stops. (To change direction the car simply goes into reverse, not turning around.) Suppose you want to write a "history" of this trip in enough detail to give a complete description of the car's motion during the entire period. There are two ways of particular interest in which this can be done:

1. You can give a **distance history**, which indicates how far the car is from the starting point at every moment of the trip.
2. You can give a **velocity history**, which describes how fast the car is going at every moment of the trip and whether it is going north or south.

Both distances and velocities are assumed to be *signed quantities*; that is, they may take on either positive or negative values.

When the car is north of the starting point, the distance traveled will be considered positive, and when it is south of the starting point, the distance traveled will be considered negative. Similarly, when the car is traveling north, it will be considered to have a positive velocity, and when traveling south, a negative one. A car traveling rapidly south has a high speed but a low (negative) velocity.

Since either the distance or the velocity history taken alone provides a complete description of the car's motion, it should be possible to start with either description and derive the other from it. And with one minor exception, which we will discuss later, you can do just that; and, at least for a few very simple examples, we will see how.

Suppose, then, that Smith and Jones are passengers in a car in which there are two instruments: an odometer and a speedometer. These particular instruments are manufactured a little differently than the ones you generally see in cars, however. This speedometer indicates a positive velocity when the car is traveling north and a negative velocity when it is traveling south. (Ordinary speedometers simply indicate the speed and ignore the direction.) The

odometer interprets any distance that the car has traveled north as positive and any distance south as negative. So if you start a trip with an odometer reading of 10,000 miles and go 1000 miles south, the odometer reading will be 9000 miles at the end of the trip. At any time the odometer reading reflects all of the north–south traveling that has been done by the car since it was bought.

On the trip Smith will only be allowed to see the odometer and Jones will only be allowed to see the speedometer. By looking only at his own instrument, each passenger must deduce what he can about the readings of the other passenger's instrument. This may seem to be a fairly straightforward problem, and yet it is the fundamental one of the calculus.

Smith can obtain information about the car's motion only *from the odometer. Jones can obtain information only from the speedometer.*

By looking at the odometer during the trip, Smith will obtain a distance history, and Jones, using the speedometer, will obtain a velocity history. But how are Smith and Jones to describe what they observe? Jones might say something like, "The trip took 6 hours. The car went at a steady 20 miles per hour for the first 2 hours, then accelerated steadily to 40 miles per hour during the next 30 seconds, and finished the last 3 hours and $59\frac{1}{2}$ minutes at 40 miles per hour." Or Smith might say, "The 40-mile trip took 5 hours. During the first hour 6 miles were covered. During the next hour 7 miles were covered," and so on. But a verbal description like this is clearly inadequate. Smith's description is incomplete, since it doesn't say how the distance was covered within each hour; and Jones's description, while it might be adequate here, wouldn't serve to describe a much more complicated travel history. What is needed is a formula that translates each instant in time into the distance traveled up to that instant (for Smith) or the velocity at which the car was traveling at that instant (for Jones). And it will be helpful if these distance and velocity formulas are illustrated by graphs.

Suppose the graph shown in Figure 1 reflects the distance history of the car. If you choose any point on the time axis, find the point on the graph directly above that point, and then look directly to the left, you will see the odometer reading that corresponds to that time. For example, if you start with $t = 3$ (hours) and move up to point A on the graph and across to the left, you will have a reading of 5100 (miles). This means that the odometer reading was 5100 when $t = 3$. In the same way you can deduce that D (the odometer reading) was 5050 when $t = 2$ and again when $t = 4$. Apparently the car was traveling north until the third hour, turned south during the fourth hour, and then turned north again.

The odometer, or distance, graph can also be used to calculate the average velocity during any period of time. If it takes 18 hours to travel from Raleigh, N.C. to Miami, Fla., a distance of 900 miles, then the average velocity for the trip would be

$$\frac{900 \text{ miles}}{18 \text{ hours}}$$

or 50 miles per hour. In general, the **average velocity** during a period of time is defined to be

$$\text{Average velocity} = \frac{\text{Distance traveled}^*}{\text{Time elapsed}}$$

---

*By "distance traveled" we mean the change in the odometer reading. If you go halfway around the world and then come back to the same place, you may need a change of oil and new tires, but your "distance traveled" would be zero for the present purpose.

Figure 1

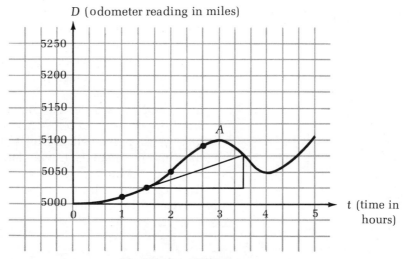

D (odometer reading in miles)

t (time in hours)

The Distance History

Looking again at Figure 1, we can calculate the average velocity between $t = 1\frac{1}{2}$ and $t = 3\frac{1}{2}$. When $t = 1\frac{1}{2}$ the odometer reads 5025, and when $t = 3\frac{1}{2}$ the odometer reads 5075. The distance traveled in this period is the difference between these odometer readings (50 miles) and is represented on the graph by the vertical side of the triangle. The time elapsed in this interval is $3\frac{1}{2} - 1\frac{1}{2} = 2$ hours, which is represented by the horizontal side of the triangle. The ratio of the vertical side to the horizontal side, which is just the slope of the line segment joining the two points on the graph, is the average velocity during this interval: 25 miles per hour.

From the diagram you can confirm that the car also averaged 25 miles per hour during the first two hours of the trip and -50 miles per hour between $t = 3$ and $t = 4$ (the car was traveling south between the end of the third hour and the end of the fourth hour).

Jones's speedometer history of the trip is expressed in much the same way as Smith's odometer history, by graphing the speedometer reading (rather than the odometer reading) against time. An example of a speedometer history graph is shown in Figure 2.

This graph shows the velocity history of the same trip that was described by the distance history graph in Figure 1. From this graph we can draw the conclusions listed on the next page.

Figure 2

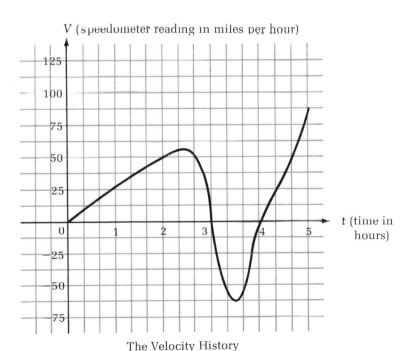

The Velocity History

1. The car started from a position of rest and moved north at a steadily increasing rate until, at the end of the second hour, a velocity of 50 miles per hour was reached.
2. The velocity increase tapered off until a maximum velocity of about 60 miles per hour was reached, at $t = 2\frac{1}{2}$.
3. The velocity decreased sharply until, at $t = 3$, the car was not moving at all (since the speedometer read 0 miles per hour).
4. The car started moving in a southerly direction at greater and greater speeds until a maximum negative velocity of about -62 miles per hour was reached at $t = 3\frac{1}{2}$.
5. Once again the speed diminished until, at $t = 4$, the car was at rest.
6. Finally, the car reversed direction once again, going north at an increasing velocity until $t = 5$, when the velocity reached 100 miles per hour.

Use the distance graph in Figure 1 to answer Exercises 1–5.

1 Calculate how far the car traveled in each of the first five hours of the trip.

2 During which of the first five hour periods did the car travel furthest north? In which period did it travel least in either direction?

3 Calculate the average velocity of the car in the period between the end of the second hour and the end of the fifth hour.

4 In which one-hour interval(s) did the car average less than 50 miles per hour in magnitude (going either north or south)?

5 (a) What is the average velocity over each of the five one-hour intervals? What is the average velocity over the entire five-hour interval? Could the average velocity over the five-hour interval exceed the largest of the average velocities over the one-hour intervals? Could it be exceeded by the smallest of the average velocities?

   (b) Suppose the average velocity over the five-hour interval was exactly equal to the largest of the average velocities of the one-hour intervals. What could you conclude? (Use common sense rather than a formal calculation.)

6   Using the velocity graph in Figure 2, find (approximately) the three times at which the car was traveling at an instantaneous speed of 25 miles per hour.

7   For the velocity history graphed in Figure 2, what were the greatest velocity and the least velocity attained in each of the following time intervals?
   (a) $t = 2\frac{1}{2}$ to $t = 3$
   (b) $t = 1$ to $t = 1\frac{1}{2}$
   (c) $t = 2\frac{3}{4}$ to $t = 3\frac{1}{4}$
   (d)   What is the greatest distance you could have traveled in each of the periods listed above? The smallest distance? If you calculated the average velocity during each of these periods by using the distance graph, what do you think the relationship would be between the average velocity during a period and the highest and lowest instantaneous velocities you just found?

8   Look at the velocity and distance history graphs (Figures 1 and 2) together. See at what times the speedometer reads 0, and then look at the distance graph to see what is happening there at these times. Can you find a relationship between the two graphs when this happens and give an informal, intuitive reason for it? If the distance graph is moving up (down), what must be true about the speedometer reading? Use this last answer to deduce what must happen to the speedometer reading when the car changes from a southward to a northward direction or vice versa.

The General Relationship Between the Histories

We said earlier that either the velocity or distance history graph, taken alone, is enough to describe the trip completely. Thus it should be possible to take either one of these graphs and derive the other from it. And in fact you can take Smith's distance graph and deduce Jones's velocity graph from it, but you cannot quite reverse the procedure. To see why, let's look at an example:

   Take the very simple case in which the velocity of a car is a constant 40 miles per hour for two hours. We know exactly how the car traveled during the entire two-hour period, and we know how much the odometer reading *increased* as well, but we don't know the *actual* odometer reading because we don't know what the odometer read at the beginning of the trip. All we know is this: If the odometer reading at the beginning of the trip was S miles, the odometer reading at the end of the trip was $S + 80$ miles.

To get some idea of the relationship between distance and velocity graphs, we will use common sense to solve a few simple problems.

EXAMPLE 1

The first graph shown in Figure 3 is the distance history of a certain trip. Which of the four speedometer graphs shown just below it could correspond to the same trip?

Figure 3

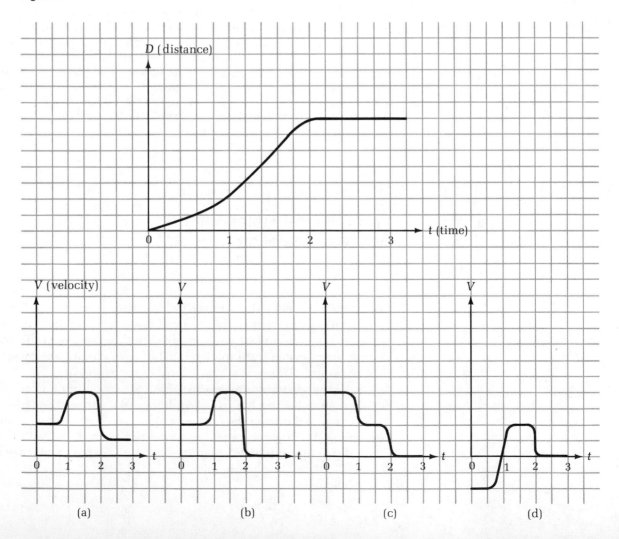

Solution

By looking at the distance graph we see that in the first hour the distance traveled increases at a constant rate (since the distance graph is a straight line). After a brief transition period the distance is again increasing at a (greater) constant rate during the second hour. In the last hour the odometer reading does not change.

If the odometer remains unchanged throughout a period, the speedometer reading must be constantly zero during that period. Since the velocity graph (a) is not zero during the third hour, this case can be eliminated.

Since the distance is increasing at a steady rate in each of the first two hours, the speedometer should have a constant reading for each of those periods; moreover, the rate is higher in the second hour, so the speedometer should have a higher reading for that period. Graph (c) does not satisfy the last requirement, so that case can be eliminated. Finally, in case (d) we have a negative speedometer reading during the first hour, which indicates that the car is moving south; but an increasing distance indicates that the car is moving north, so that cannot be right. That leaves (b) as the only plausible case.

The distance and velocity histories of a car in a race will be more complex than those we are analyzing in this chapter. However, the same general relationship between the histories holds for any trip.

EXAMPLE 2

The speedometer graph shown in Figure 4 is the speed history of a certain trip. Which of the four odometer (distance) graphs shown below it might have been taken from the same trip?

Figure 4

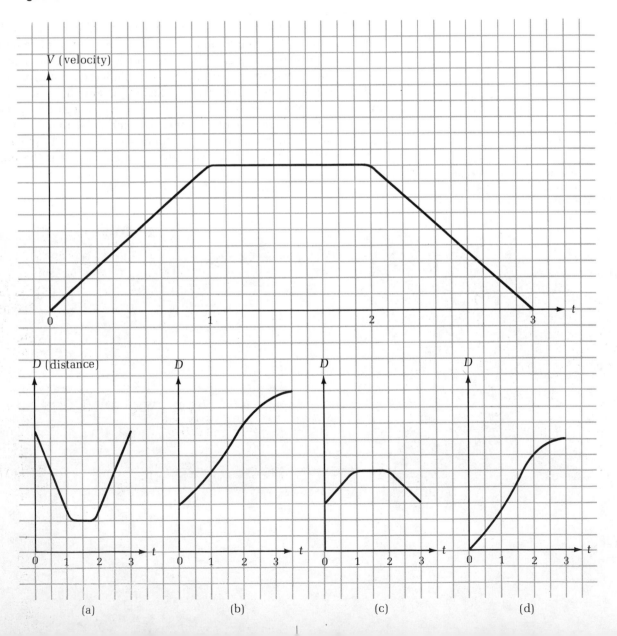

Solution

We can see readily that (a) cannot be the correct graph. During the first hour the velocity graph is always positive, which means that the car is going north; but the value of the distance is decreasing in (a), which means that the car is going south. The same objection can be made about graph (c): In the third hour it would appear that the car is moving south, but the velocity graph is positive for that period, and this means that the car is moving north.

Of the remaining graphs, (b) and (d), *either* might be correct. Graphs (b) and (d) are the same except that the distance in (b) is always greater by some constant value. The physical interpretation of this observation is that in graph (d) the odometer read zero at the beginning of the trip while in (b) the odometer had a positive reading at first.

The twin problems of translating distance and velocity graphs into one another are basic to the calculus. The velocity graph is called the **derivative** of the distance graph corresponding to it, and the distance graph is called the **integral** of the velocity graph. Much of the usual freshman calculus course is spent differentiating and integrating, that is, finding the derivatives and integrals of various curves. Although it is not our purpose to master the technique, it will be worthwhile to see how it is done in a few simple cases.

A Car at Rest

Each of the two graphs shown in Figure 5 reflects the fact that a car has been at rest during the interval from $t = 0$ to $t = 2$. On the distance graph the odometer reading never varies from S, and there is no difficulty concluding that the speedometer graph is always at zero. Similarly, one can conclude that the odometer reading is fixed from the fact that the speedometer stays at zero, but the exact reading cannot be determined. So

> *The derivative of a graph that is constant is a graph that is zero. The integral of a graph that is always zero is a graph that is a constant (which can be determined only if there is more information).*

As we have said, it is Smith's job to deduce the velocity graph from the distance graph and Jones's job to deduce the distance

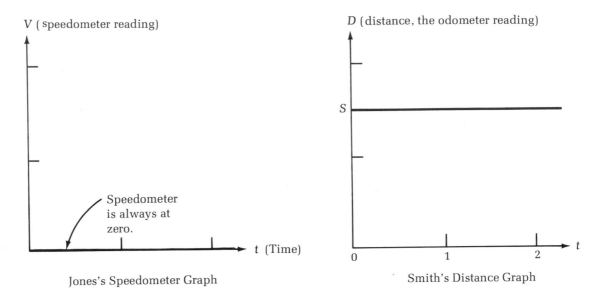

Jones's Speedometer Graph

Smith's Distance Graph

Figure 5

graph from the velocity graph. To put this another way, Smith's question is Jones's answer, and Jones's question is Smith's answer. When Jones is given a velocity graph, he can either work out the corresponding distance graph directly or ask himself, "What distance graph of Smith will produce this velocity graph I was just given?" The answer to this question is the solution to his problem. And Smith can also solve his problem by working backwards in this way.

An analogous situation would be a bilingual class with French and English speaking students. Imagine that the teacher asked each student to translate a word from his own language into the other language with the aid of a dictionary. If Pierre were asked to translate the word "fromage," he would check the dictionary and say "cheese." And if Edna, an English-speaking student, were given the word "cheese" to translate, she could do either of two things: She could look up the word in her own dictionary, or she could recall the answer Pierre gave when he was asked to translate "fromage."

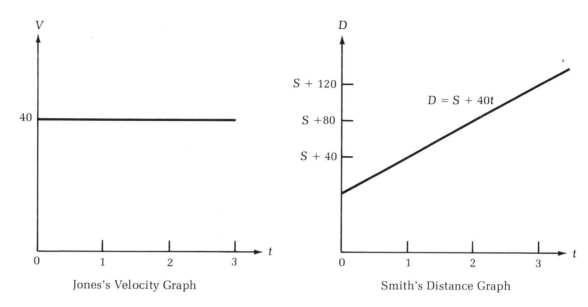

Jones's Velocity Graph

Smith's Distance Graph

Figure 6

A Car Traveling at Constant Velocity

The two graphs shown in Figure 6 each reflect the motion of a car traveling at a constant velocity. Looking at Jones's velocity graph first, we see that the velocity is always 40 miles per hour. Since rate × time is equal to distance, the odometer will read $D = S + 40t$ at a time t hours after the trip starts if S is the initial odometer reading. The value of S cannot be determined from the velocity graph.

Smith must answer the reverse problem: If the distance graph is the straight line $D = S + 40t$, what is the corresponding velocity graph? He can answer this directly or ask himself, "To what graph of Jones does my graph correspond?" If he can figure this out in any way (from memory, from looking over Jones's shoulder), he will have answered his own problem.

> The derivative of a straight line graph in the form $D = A + Bt$ is the graph $V = B$.

> The integral of a graph in the form $V = C$ is a straight line in the form $D = K + Ct$ where K is some undetermined constant.

1 Assume that when a car starts on a trip, both the odometer and speedometer read zero. State which of the following are necessarily true:

(a) If the odometer reading is always at least zero, the speedometer reading will also always be at least zero.

(b) If the speedometer reading is always at least zero, then the odometer reading will also always be at least zero.

(c) If the odometer reading never gets smaller, the speedometer reading won't either.

(d) If the speedometer reading never gets smaller, the odometer reading won't either.

(e) The odometer readings and speedometer readings can be going in different directions, that is, one can be increasing while the other is decreasing.

(f) If the speedometer shows a negative reading, the odometer will too.

(g) If the odometer shows a negative reading, the speedometer will too.

2 Listed below are a number of verbal descriptions of how the velocity of a car changed with time during the course of a trip. Following each description are three distance graphs. Which of them might correspond to that trip?

(a) "We started going north slowly, but as time went by we went faster and faster."

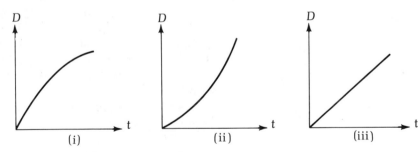

(b) "Starting north at 40 miles per hour, we braked, and the car slowed steadily until we came to rest at the end of 4 minutes (so that we were going 30 miles per hour after one minute, 20 miles per hour after two minutes, etc.)."

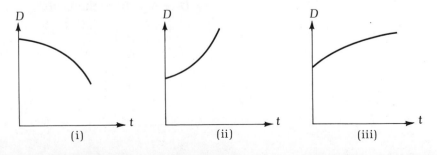

(c)   "We traveled south at a constant speed."

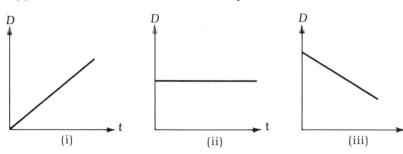

(i)   (ii)   (iii)

3 Figure 7 shows the velocity history of a trip. Which of the three graphs below might give the corresponding distance history?

Figure 7

Velocity Graph

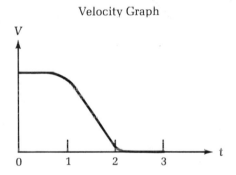

Distance Graphs

(a)   (b)   (c)

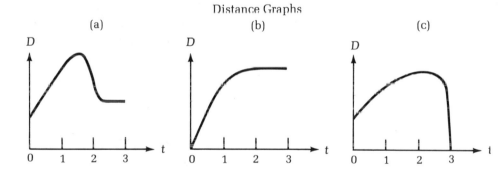

4 If the diagram in Figure 8 is the distance history of a trip, which of three graphs below might give the corresponding velocity history?

Figure 8

Distance Graph

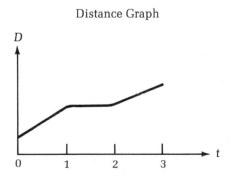

Velocity Graphs

(a)   (b)   (c)

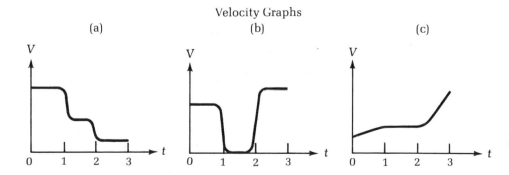

5   On the left of Figure 9 there are five velocity graphs, and on the right there are five distance graphs. Match them so that each pair might correspond to the same trip.

Velocity Graphs   Distance Graphs

Figure 9

(a)   (i)

(b)   (ii)

(c)   (iii)

(d)   (iv)

(e)   (v)

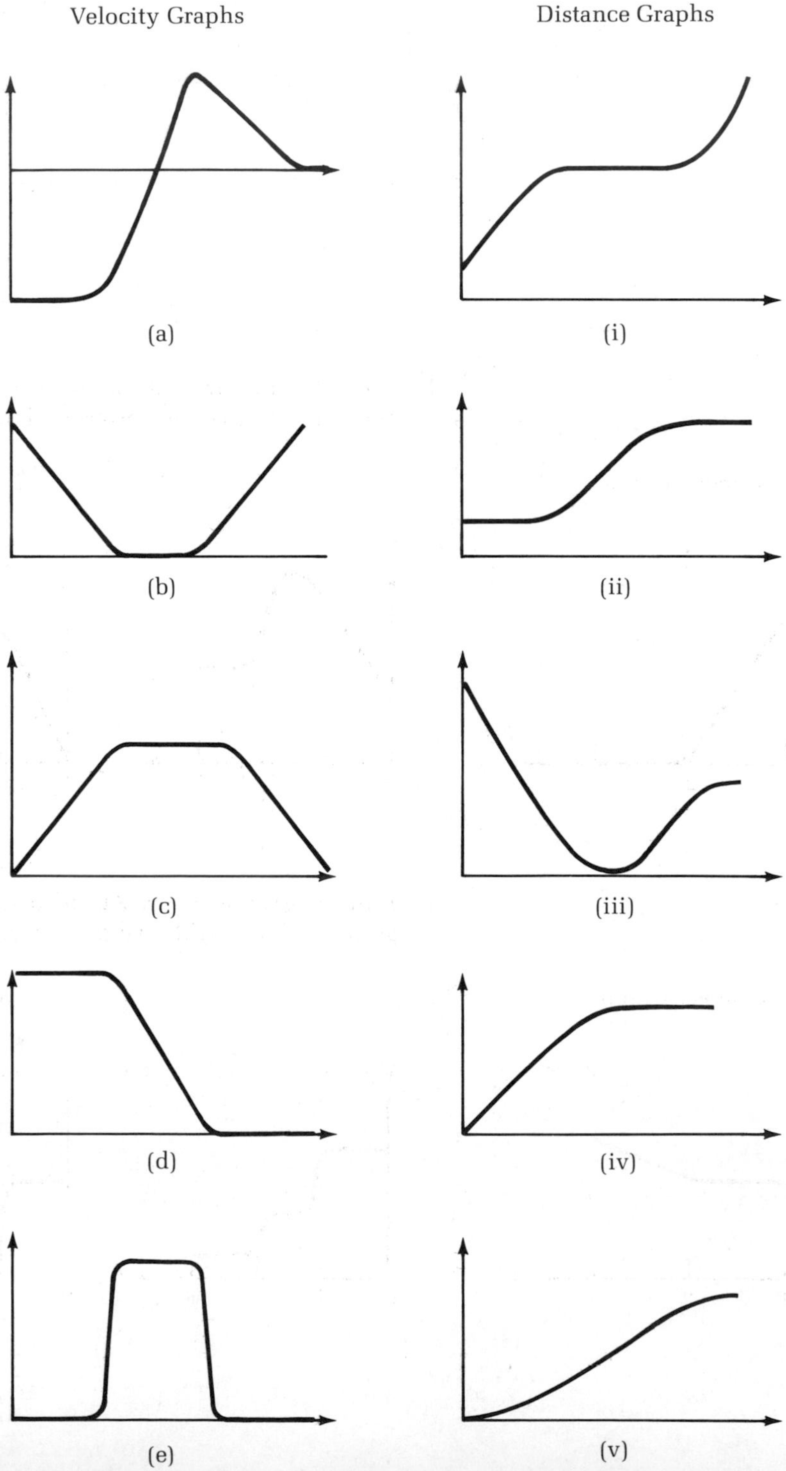

6 If Smith's distance graph is of the form $D = S + 40t$, then Smith can deduce that the velocity graph is $V = 40$ by putting himself in Jones's shoes and working backwards. But the velocity graph can also be worked out directly in the following way:

(a) Show that if the distance traveled is given by the formula $D = S + 40t$, then the distance traveled between the times $t = a$ and $t = b$ is $40(b - a)$.

(b) Show that the average velocity between any two moments $t = a$ and $t = b$ is always the same, whatever the values of a and b. What is the value of this average velocity?

(c) Since the average velocity is the same for *any* time interval, we may conclude that the speedometer reading never varies from that average velocity.

7 In each of the following cases the odometer readings are expressed in terms of the time t. Find the average velocity in the interval between $t = a$ and $t = b$.

(a) $D = 400 + 5t$

(b) $D = 500 + 5t$

(c) $D = 400 + 10t$

(d) $D = 500 + 10t$

Instantaneous Velocity

It is fairly easy to analyze the motion of a car traveling at a constant velocity, but when the car's velocity is always changing the problem is much more difficult. It isn't just that the calculations become harder; the very meaning of the term "velocity" turns out to be quite elusive when you try to pin it down. Let us see why.

If a traveler drives the 900 miles from New York to Chicago in 20 hours, the *average* velocity will be 45 miles per hour. The idea of "average velocity" is simple enough; the actual time that elapsed during the trip is the same as it would have been had the speedometer remained constantly at 45 miles per hour. And the calculation of the average velocity is also very easy; you divide the distance traveled during the trip by the time spent.

If the velocity of the car varies from moment to moment, it is still reasonable to ask what the "instantaneous velocity" is at a given moment. Intuitively, instantaneous velocity should correspond to the reading on the speedometer. But if we have only the

odometer readings, how are we to calculate instantaneous velocity? During a single instant, which is just a point in time, *no*
distance is covered and *no* time elapses. If we resort to the earlier
definition of velocity as distance/time and mechanically insert the
values of zero for both distance and time, we get 0/0, a meaningless expression.

It seems reasonable to expect that the notion of instantaneous
velocity will somehow be based on that of average velocity, but it
should be clear from what we have said that defining the precise
relationship between them will require some care. To explore this
relationship, let us take a new look at an old problem.

We have already mentioned that there is a connection between the distance and velocity graphs of a trip, and we spelled
out what that connection was in a few simple cases. Now we will
generalize these results, starting with a simple example that we
discussed earlier: a car traveling at a constant velocity.

In the distance graph shown in Figure 10, the odometer reads
40 miles at the start of the trip and the car travels at a steady 20
miles per hour during the trip. Thus the distance traveled, *D*, is
given by the formula

$$D = 40 + 20t$$

If you want to find the average velocity during any portion of
the trip, divide the distance traveled by the elapsed time. Geometrically, this is the vertical change divided by the horizontal
change. So the average velocity in the first 6 hours going from *P* to
Q on the graph is

$$\frac{120 \text{ miles}}{6 \text{ hours}} = 20 \text{ miles per hour}$$

Similarly, the average velocity during the first 9 hours (going from
P to *R* on the graph) is

$$\frac{180 \text{ miles}}{9 \text{ hours}} = 20 \text{ miles per hour}$$

as before. This ratio of the vertical change to the horizontal change
of a graph never changes if the graph is a straight line; as we
mentioned earlier, this ratio is the slope of the line. So in the
simple case when the distance graph is a straight line, the definition of instantaneous velocity is easy enough. Since the average
velocity in *any* interval is always the same constant (20 miles per
hour in our example), we may define the instantaneous velocity, at
any time, to be this constant as well.

Figure 10

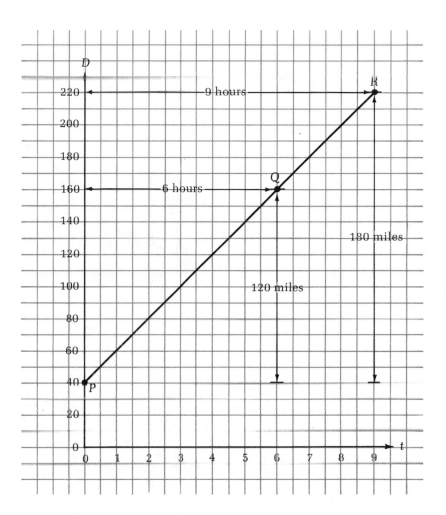

(**EXAMPLE 1**)

For the distance graph shown in Figure 10, find the average velocity between the sixth and ninth hours.

Solution

The odometer reads 160 miles at the end of the sixth hour and 220
miles at the end of the ninth hour, so the car traveled
$220 - 160 = 60$ miles in 3 hours, or 20 miles per hour on the
average.

Figure 11

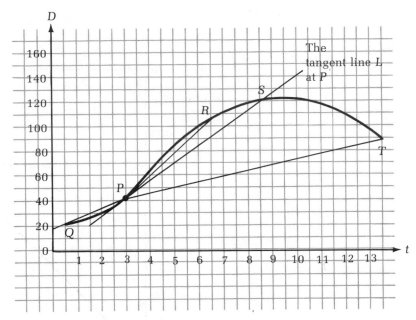

The Distance Graph for a Car Moving at Varying Speeds

When the car's speed is changing, the problem is more complex. The average velocity between two points is calculated in the same way (by dividing the vertical change by the horizontal change to obtain distance divided by time), but the average velocity differs from one interval to another.

Now suppose we want to define the "instantaneous velocity" at $t = 3$ in the distance graph shown in Figure 11. Although we don't know exactly what instantaneous velocity is, we should be able to approximate it by taking average speeds over intervals close to $t = 3$; and the closer the intervals are, the better the approximations should be. We would also expect that the average velocity from $t = 3$ to $t = 13\frac{1}{2}$ (the slope of the line segment $\overline{PT}$) won't be as good an approximation as the average velocity between $t = 3$ and $t = 6\frac{1}{2}$, and a smaller interval should be better yet.

Geometrically, the average velocity during a period of time is the slope of the secant line to the distance graph (the *secant line* is a line segment that connects two points on the graph). As you take average velocities over intervals which are closer and closer to the point $t = 3$ (or, equivalently, calculate the slopes of nearby secant lines $\overline{PT}$, $\overline{PS}$, and $\overline{PR}$ in succession), you come closer to the **slope of the tangent line** to the curve at the point P, that is, to the slope of the line L. (A *tangent* to a curve at a point is a line through the point going in the same direction as the curve at that point. If a car on a road suddenly hit an ice patch and lost all traction, it would go off in the direction of the tangent.) And if we take secant lines to the left of P, such as $\overline{OP}$, their slopes too will come close to the slope of the tangent line at P. In a nutshell:

> The **instantaneous speed** at a given time is defined to be the slope of the tangent line at the corresponding point on the distance graph if there exists such a tangent line.

EXAMPLE 2

For the distance graph in Figure 11:
(a) Estimate the time when the instantaneous velocity is zero.
(b) Find the average velocity between $t = 3$ and $t = 4$.

Solution

(a) When $t = 9$ the tangent line is approximately horizontal, so the slope and therefore the instantaneous velocity is zero.
(b) The average velocity is $(60 - 42)$ miles/1 hour $= 18$ miles per hour.

Now we will drop Smith's point of view and adopt Jones's. Suppose we are given a speedometer graph and want a geometric interpretation of the distance traveled. Once again we'll start with the simple case in which the velocity is constant. The speedometer graph for the case $V = k (k \geq 0)$ is shown in Figure 12.

> **To find the distance traveled in a certain period of time, find the area under the speedometer graph between the starting time and the stopping time of the period.**

Figure 12   Speedometer reading

If the graph shown in Figure 13 were the speedometer graph, then the shaded area would represent the distance traveled between $t = a$ and $t = b$. In this case, however, it would be more difficult to find a value for the area.

Figure 13

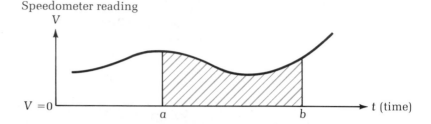

Speedometer reading

If the speedometer reading is never greater than zero (see Figure 14), the distance traveled is calculated in a similar way, as the area *above* the speedometer graph (and below the line where the velocity is zero) between the starting and stopping times. This area indicates that the distance is negative, that is, it indicates the distance you've gone south.

Figure 14

The area below the line $V = 0$ is the distance traveled south.

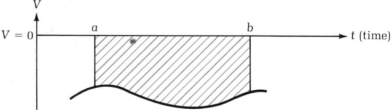

Speedometer reading

Finally, if the speedometer reading is sometimes positive and sometimes negative (see Figure 15), you calculate the areas of all the regions above the line $V = 0$ and those of all the regions below it. If the sum of the regions above the line exceeds the sum of the

Figure 15

Add all the areas labeled "plus" and then add all the areas labeled "minus." The difference between the sums is the net distance traveled *north* if the sum of the "plus" areas is greater and is the net distance traveled *south* if the sum of the "minus" areas is greater.

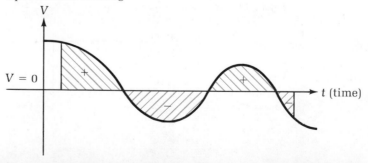

Speedometer reading

regions below the line, the difference in these two sums is how far north you've traveled; if there is more area below the line, the difference indicates how far south you've traveled.

The two geometric relationships between the distance and velocity graphs that we just observed are the tools for calculating many of the elementary formulas in calculus:

1. The slope of the tangent line at a point on the distance graph is the instantaneous velocity.
2. The area under or above the velocity graph between two different times is the distance traveled during that period.

As we said earlier, we will not build up a large inventory of such formulas for translating velocity graphs into distance graphs or distance graphs into velocity graphs; but it will be useful to see how these relationships can be used in at least one case where the speed is not constant.

Suppose the buyer of a new car starts his first ride by pushing the gas pedal down as far as it can go. As it happens, the distance he travels, D (measured in feet), is related to time, t (measured in seconds), by the formula $D = t^2$. How can we find the car's instantaneous velocity at some time t?

Before we look closely at the problem, let's see what we can deduce by common sense. From Table 1 it's clear that the distance traveled during successive seconds keeps increasing; the car travels 1 foot during the first second, 3 feet during the next second, 5 feet during the next second, and so on. It would seem that the instantaneous speeds (the speedometer readings) are rising as well.

We will try to find out precisely how the speedometer readings are changing with time. As a start, we will deduce the speedometer reading (that is, find the *instantaneous velocity*) at $t = 3$. We do this indirectly, by first finding the *average* velocities between $t = 3$ and $t = \bar{t}$ for various values of $\bar{t}$ and observing what happens to these average velocities as we take values of $\bar{t}$ closer and closer to (but not quite reaching) 3.

In Figure 16 we have drawn the secant lines between the point $P(t = 3)$ and the points $T (t = 6)$, $S (t = 5)$, $R (t = 4)$, and $Q (t = 3\frac{1}{2})$. The average velocities during the four periods are indicated in Table 2. Examination of the table suggests that the "real" instantaneous velocity at $t = 3$ is 6 feet per second. But we certainly have not proved this. We could keep repeating the process, taking intervals from $t = 3$ to $t = 3 + h$ for smaller and smaller positive numbers h or, alternatively, taking intervals from $t = 3 - h$ to $t = 3$. In either case each approximation is only that—an approximation.

Table 1

Time	Distance
0	0
1	1
2	4
3	9
4	16
5	25
6	36

Figure 16

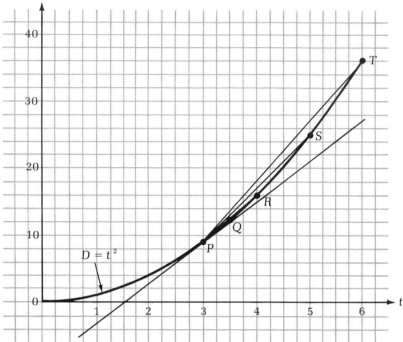

Distance
(odometer reading)

$D = t^2$

Table 2

Start of time interval	End of time interval	Distance traveled during time interval	Time elapsed (in secs.)	Average velocity (in feet per sec.)
3	6	$36 - 9 = 27$	3	9
3	5	$25 - 9 = 16$	2	8
3	4	$16 - 9 = 7$	1	7
3	$3\frac{1}{2}$	$12\frac{1}{4} - 9 = 3\frac{1}{4}$	$\frac{1}{2}$	$6\frac{1}{2}$

If you approximate the instantaneous velocity at $t = 3$ by the average velocity between $t = 3$ and some later time, you will inevitably obtain an overestimate; if you calculate the average velocity from some earlier time to $t = 3$, the result will be an underestimate. In either case it will be incorrect. The challenge is to move from this variety of wrong answers to the single right one.

Let us algebraically calculate the average velocity from $t = 3$ to $t = 3 + h$, where h is some positive number that is not yet determined. This average velocity should be somewhat greater than the instantaneous velocity at $t = 3$, because the car's velocity is steadily increasing. Since the distance traveled during this period is

$$(3 + h)^2 - 3^2 = (9 + 6h + h^2) - 9 = 6h + h^2$$

and the time elapsed is h, we conclude that the average velocity is $(6h + h^2)/h = 6 + h$. So *the instantaneous velocity at $t = 3$ is less than $6 + h$ for any positive value of h.*

In a similar way it can be shown that for *any* positive value of k, *the instantaneous velocity is greater than $6 - k$,* which is the average velocity between $t = 3 - k$ and $t = 3$. (This will be left as an exercise.)

So we know that the instantaneous velocity is a number greater than $6 - k$ and less than $6 + h$ for any positive value of k and h. We must conclude that the instantaneous velocity at $t = 3$ must be 6.

There is nothing special about $t = 3$, of course; the instantaneous velocity can be calculated in a similar way for any time t. So let's calculate the average velocity in the period from t to $(t + h)$ where both t and h are not yet fixed. The average velocity during this period is

$$\frac{(t^2 + 2ht + h^2) - t^2}{h} = 2t + h$$

In a similar way it can be shown that the average velocity in the period from $t - k$ to t is $2t - k$. (See the exercises.) When the approximating interval becomes smaller (when "h and k approach zero"), the average speed in the interval becomes arbitrarily close to $2t$. So the instantaneous velocity at time t is defined to be $2t$.

Our conclusion is this: If $D = t^2$ is the relationship between time elapsed and distance traveled on a trip, then $V = 2t$ is the relationship between the speedometer reading and time.

Having solved this problem, we have automatically solved another problem as well: "If the speedometer readings are given by the formula $V = 2t$, what is the formula that relates odometer readings to time?" The answer is, of course, $D = S + t^2$ where S is the initial odometer reading. In our example the initial odometer reading is 0, so $D = t^2$. Still, it is worthwhile, deducing the odometer graph that corresponds to the speedometer graph $V = 2t$, starting from scratch and ignoring the information that we ob-

Figure 17

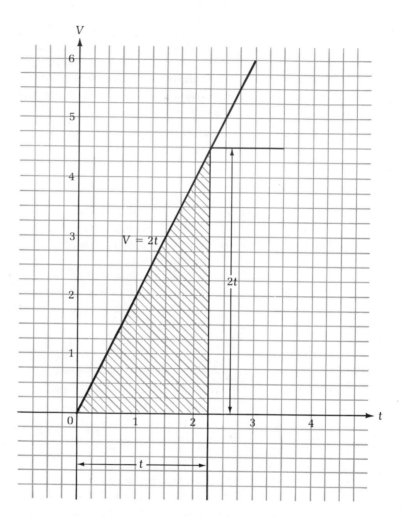

tained by solving the inverse problem. To see how this might be done let us first draw the speedometer graph, as in Figure 17.

To find the distance traveled from the beginning of the trip, $t = 0$, to some arbitrary time t we must calculate the area of the shaded region underneath the curve. In this case the calculation is quite simple. The area of a triangle is half the base times the height, which is $\left(\frac{1}{2}\right)(t)(2t)$ or t^2. The odometer reading at any time t will be the initial odometer reading at the start of the trip plus the distance traveled during the trip at time t, that is, $D = 0 + t^2 = t^2$.

In the usual language of the calculus, the derivative of the curve $D = t^2$ is $V = 2t$ and the integral of the curve $V = 2t$ is the curve $D = t^2 + S$ where S is the initial odometer reading.

exercises

1 From the distance graph in Figure 10:
    (a) Calculate the average velocity in the following periods:
        (i) $t = 0$ to $t = 5$
        (ii) $t = 2$ to $t = 8$
        (iii) $t = 1$ to $t = 4$
    (b) Confirm that the answers are all the same.
    (c) Use what you know about similar triangles to show that the answer will always be the same.

2 From the distance graph in Figure 11:
    (a) Calculate the average velocity in each of the following intervals:
        (i) $t = 3$ to $t = 13\frac{1}{2}$
        (ii) $t = 3$ to $t = 8\frac{1}{2}$
        (iii) $t = 3$ to $t = 4$
    (b) Estimate the slope of the tangent line L. Do the slopes of the line segments determined by P and a second point on the distance graph approach the slope of L as the second point gets close to P?
    (c) Show that it need not always be true that closer intervals yield better approximations to the instantaneous velocity. (Compare the slope of line L with the slope of the secant line passing through P and S.)

3 From the distance graph in Figure 16 (in which $D = t^2$):
    (a) Find the average velocity in the interval between $t = 3$ and
        (i) $t = 3\frac{1}{4}$
        (ii) $t = 3.1$
        (iii) $t = 3.01$
    (b) Find the average velocity between $t = 3$ and
        (i) $\bar{t} = 1$
        (ii) $\bar{t} = 2$
        (iii) $\bar{t} = 2.9$
    (c) Calculate the average velocity between 3 and $3 + h$ and the average velocity between $3 - k$ and 3.
    (d) Show that the average velocity between $t - k$ and t is $2t - k$.

These off-screen photos capture moments in the motion of dancers. When we speak of instantaneous velocity, we are thinking of velocity at such a "frozen moment."

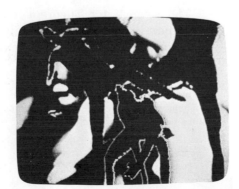

1 If $D = t^2$, we have shown that $V = 2t$; and if $D = 4t$, we have shown that $V = 4$. Show directly (by taking average velocities over "small" periods of time) that when $D = t^2 + 4t$, $V = 2t + 4$.

2 Show directly that if $D = 8t^2$, then $V = 16t$.

3 Suppose that $D = At^2 + Bt + C$ where A, B, and C are constants. Guess what V is and then verify your guess by taking average velocities over "small" periods of time.

Some Techniques and Theorems

Along with the calculating techniques that are usually taught in calculus, there are a few theorems with particularly simple interpretations in the automobile trip model. We will state several of them.

> **The Intermediate Value Theorem:** If a graph takes on two different values, it must take on all the values in between.*

For distance graphs this means that a car that had traveled 50 miles at the end of one hour and 90 miles at the end of two must have traveled exactly 80 miles at some instant during the second hour. For velocity graphs it means that a car moving at 40 miles per hour at $t = 1$ and 50 miles per hour at $t = 2$ must have been moving at an instantaneous speed of exactly 46.2 miles per hour at some time during the second hour.

> **The Mean Value Theorem** relates the velocity and distance graphs: If a car averages X miles per hour during some period of time, it must be moving at exactly X miles per hour at some time during the period.*

---

*We assume that the velocity and distance graphs are what mathematicians call *continuous*. That means the odometer and speedometer move gradually from one reading to another and don't change their readings in fits and starts.

For example, a car traveling the 900 miles from New York to Chicago in 18 hours must have actually been going 50 miles per hour at some time during the trip. It is easy to show that the theorem is plausible: If the car always traveled faster (or always slower) than 50 miles per hour, it could not possibly average 50 miles per hour on the trip. And if the instantaneous speed was sometimes faster and sometimes slower than 50 miles per hour, the Intermediate Value Theorem assures us that the car must have been moving at exactly 50 miles per hour at some time during the trip.

Maxima and Minima

There is an extremely important relationship between the distance and velocity graphs that is the basis of a great many applications.

Recall that when the car is moving north the distance graph is rising and the speedometer reading is positive, and when the car is moving south the distance graph is falling and the speedometer reading is negative. Suppose a car starts driving north and then reverses its direction. Just before it reverses, the speedometer reading is positive, and afterward it is negative. By the Intermediate Value Theorem there must be some instant when the speedometer reads zero (when the car is instantaneously at rest), and that is the instant when the car reaches the northernmost point. By similar reasoning we conclude that the speedometer will read zero when the car reaches its southernmost point as well. This suggests the following theorem:

> Suppose a car is moving in a north–south direction during the period from $t = a$ to $t = b$. When the car reaches its northernmost and southernmost positions, the speedometer will read zero (unless these positions are reached at the start or end of the period, when $t = a$ or $t = b$).

It should be stressed that the argument we just gave holds only if the car reaches its northernmost or southernmost point within the period and then reverses its direction. If the car is at an extreme point at the beginning or end of the period, it may not be reversing and there is no way to tell what the speedometer reading will be.

In the automobile-trip example you can find how far north and how far south a car has traveled by checking the odometer when the speedometer reads zero, and you can use the same technique to solve a great many other problems as well. Because there are many practical problems in which you want to maximize something (such as profits, area, or reliability) or minimize something (such as waste, death, or delay), this theorem, which we will now express in a more general form, is very useful.

> **Theorem:** If a graph that is smooth enough to have a tangent at every point has maximum or minimum points inside an interval, then the graph that is its derivative will have the value zero at these points.

Although we have not developed the techniques of translating distance and velocity graphs into one another, we can describe some typical applications of this theorem. Some of these applications will require us to use a few elementary theorems about derivatives, which we will mention and use as they are needed.

In the first part of this chapter we described the relationship between distance graphs and their corresponding velocity graphs and mentioned a number of simple examples, specifically:

Distance graph	*Corresponding* *velocity graph*
$D = 0$	$V = 0$
$D = $ any constant	$V = 0$
$D = t$	$V = 1$
$D = t^2$	$V = 2t$

We also mentioned that the distance graph is sometimes referred to as the integral of the velocity graph and the velocity graph is sometimes referred to as the derivative of the distance graph. In fact, while graphs are helpful in visualization, they are not essential to the concept of integral or derivative. So that $D = t^2$ is often referred to as the integral of $V = 2t$ and $V = 2t$ as the derivative of $D = t^2$. (This is just a shorthand way of saying, "If the speedometer reading at time t is $2t$, then the odometer reading at time t will be $t^2 + $ some constant" and "If the odometer reading at time t is $t^2 + $ a constant, the speedometer reading will be $2t$ at time t.")

(**EXAMPLE 1**)

(a) How fast does the surface area of a sphere increase as the radius increases?

(b) How fast does the volume increase?

$V = \frac{4}{3}\pi r^3$

$SA = 4\pi r^2$

Solution

(a)   The surface area of a sphere is given by the formula $A = 4\pi r^2$. The derivative of $A = 4\pi r^2$ is $8\pi r$. (We have already established that the derivative of r^2 is $2r$, and if you multiply a graph by a constant you multiply its derivative by the same constant too.)

(b)   The volume of a sphere is given by the formula $V = \frac{4}{3}\pi r^3$, and its derivative turns out to be $4\pi r^2$.

If you examine the rates of growth of the volume and area as the radius increases, you find that

1.   The area is growing twice as fast when r becomes twice as large, three times as fast when r becomes three times as large, and so on.
2.   The volume is growing four times faster when r becomes twice as large, nine times faster when r becomes three times as large, and so on.

So although the volume and the area both get larger as the radius increases, they don't get larger at the same rate. As the radius gets larger, the ratio of volume to area gets larger as well.

The same principle applies to any three-dimensional figure (such as a cube, a pyramid, or a human being) that increases its dimensions proportionately without changing shape; its volume increases faster than its surface area.

(**EXAMPLE 2**)

A ball is thrown into the air from the ground, and its distance from the ground t seconds later is given by the formula $D = 64t - 16t^2$.
(a)   When will the ball reach its highest point?
(b)   How high will the ball go?
(c)   If you were in a building and reached out a window to catch the ball a split second before it reached its highest point, would you be likely to hurt your hand?
(d)   Find the height of the ball k seconds before and k seconds after it reaches the top. Show that these two heights are the same.
(e)   When will the ball hit the ground?

(f)   Show that it takes as long to go up as to come down.
(g)   Show that it hits the ground with the same speed that it had when it left the ground.

Solution

If we call the upward direction "plus" and the downward direction "minus," this problem reduces to the automobile trip problem we discussed earlier. The formula that relates the distance to time .is given, and the formula for the velocity can be calculated by finding its derivative.

We learned earlier that the derivative of $D = 64t$ is $V = 64$, and in the previous problem we mentioned that multiplying a graph by a constant has the effect of multiplying its derivative by the same constant, so the derivative of $D = -16t^2$ is $V = -32t$. There is a theorem which states that the derivative of a sum of two graphs is the sum of their derivatives; so if $D = 64t - 16t^2$, then $V = 64 - 32t$.

(a)   The ball will be highest when the velociy, $64 - 32t$, is zero, that is, when $t = 2$.
(b)   When $t = 2$, $D = 64$; this is as high as the ball will go.
(c)   No; the ball is (almost) instantaneously at rest.
(d)   At a time k seconds before the ball reaches the top, the distance from the ground will be $64(2 - k) - 16(2 - k)^2 = 16(4 - k^2)$. A similar calculation shows that this is the distance from the ground when $t = 2 + k$ as well.
(e)   The ball will hit the ground when $64t - 16t^2 = 0$, that is, when $t = 4$.
(f)   We conclude from answers (a) and (e) that it takes 2 seconds to reach the top and the same time to come down.
(g)   When $t = 0$ the velocity is 64, and when $t = 4$ the velocity is -64. This means that the ball rose from the ground with the same speed at which it hit the ground coming down (although it was traveling in opposite directions, so that the velocities were of different sign).

EXAMPLE 3

It costs a publishing firm $1000 + 2N$ dollars to print N books. If the book is priced at S dollars, the firm will sell $N = 2000 - 50S$ books. What should it charge for a book to make the most profit? What is this maximum profit?

Solution

If the selling price of a single book is S, then the profit, P, is given by the formula

P = (Number of books sold) (Selling price per book) − (Total cost)

$= (2{,}000 − 50S)\,(S) − [1{,}000 + 2\,(2{,}000 − 50S)]$

$= −50S^2 + 2100S − 5000$

Our job is to find the value for S that yields the maximum value of P.

If the profit, P, is to be a maximum, then the derivative of the profit, $−100S + 2100$, must be zero; and the derivative is easily seen to be zero when $S = \$21$. Substituting $S = 21$ into the expression for the profit, we find that the maximum profit is

$$P = −50\,(21)^2 + 2100\,(21) − 5000 = \$17{,}050$$

EXAMPLE 4

The operating efficiency of a battery is highest when it is neither too warm nor too cold. The voltage, V, at temperature T is given by the formula $V = 1.5 + .002T − T^2/50{,}000$. At what temperature do you obtain the greatest voltage? What is that maximum voltage?

Solution

If the voltage is a maximum, the derivative of the voltage is zero. The derivative of the voltage is $.002 − T/25{,}000$, so T must be 50. At that temperature the voltage, V, is 1.55.

EXAMPLE 5

The rate at which a quantity Q is increasing is proportional to the quantity itself. When $t = 0$ the quantity is 1, and when $t = 1$ the quantity is 2. What is the quantity when $t = 10$?

Solution

If we think in terms of our original automobile trip model, this would mean that the further the car traveled, the faster the speed. If it was going 40 miles per hour after 10 miles, it would be going 120 miles per hour after 30 miles.

It turns out (although we cannot prove it here) that the quantity is described by the equation $Q = 2^t$. When $t = 10$, $Q = 2^{10} = 1024$.

This same model serves for several different applications. For example, the rate of growth of a bacterial colony is proportional to the size of the colony itself (until growth is restricted by lack of food or by natural enemies). The spread of a disease or a rumor is roughly proportional to the number of people who are sick or who know the secret, at least initially. And the rate of growth of money compounding in a bank is proportional to the amount that's already there.

A similar problem arises when the rate of decay is proportional to the quantity that still remains. The rate at which a sugar cube dissolves is roughly proportional to the amount of sugar remaining. The cube gets smaller but at a slower rate. If the volume of the cube was 1 when $t = 0$ and $\frac{1}{2}$ when $t = 1$, then the volume would be $V = \left(\frac{1}{2}\right)^t$ at time t.

This kind of decay is common to radioactive materials. The time the material takes to decay from a fixed quantity to half that quantity never changes and is called the *half-life* of the substance.

EXAMPLE 6

A rectangular enclosure is to be built near a river with the river serving as one of its sides. (See Figure 18.) If there is 200 feet of fencing, how large an area can you enclose?

Figure 18

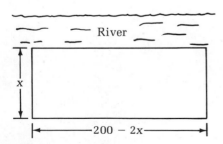

Solution

If x is the length of a side perpendicular to the river, then the side parallel to the river must have length $200 - 2x$. Thus the area is given by the formula $A = x(200 - 2x) = 200x - 2x^2$. The area is maximized when the derivative, $200 - 4x$, is zero, which occurs when $x = 50$. The maximum possible area is 5000.

1 We found earlier that the distance graph $D = t^2$ and the velocity graph $V = 2t$ correspond to one another. For each interval shown below calculate (a) the average velocity and (b) the instant during the interval at which the instantaneous speed is equal to this average velocity.
  (i) $t = 0$ to $t = 4$
  (ii) $t = 2$ to $t = 3$
  (iii) $t = 1$ to $t = 5$

2 There is a geometric interpretation of the Mean Value Theorem which says that for any two points A and B on a (smooth) curve there is a third point, C, with this property: The tangent line at C and the secant line connecting A and B are parallel, that is, they have the same slope. What is the relationship between this geometric theorem and the explanation of the Mean Value Theorem in terms of average and instantaneous speeds that we gave earlier?

3 Identify each of the statements below as a consequence of (i), (ii), (iii), or (iv):

  (i)   The Intermediate Value Theorem applied to the distance graph.
  (ii) The Intermediate Value Theorem applied to the velocity graph.
  (iii) The Mean Value Theorem.
  (iv) The fact that when a car reaches an extreme (northernmost or southernmost) point in the middle of a trip, the car is instantaneously at rest (that is, the speedometer reads zero).

  (a) An assassin plans to attach a bomb to his victim's speedometer so that it will explode when a velocity of 40 miles per hour is reached. It follows that the car can never reach an instantaneous velocity of 50 miles per hour and its average speed over any interval (before the explosion) must be less than 40 miles per hour.

  (b) The same assassin sets a land mine 40 miles north of where the car started moving. If the odometer has an initial reading of 0 miles, the odometer reading will never show 50 miles.

  (c) A car with an odometer reading of 0 starts out on a trip and then returns to its starting point. At no time was it more than 70 miles north of its starting point, and it always traveled in a north–south direction. When it was 70 miles north of its starting point, it hit a deer. Show that it couldn't have injured the deer very badly.

  (d) In a certain part of town there is a shuttle train that travels over a single set of tracks. After the train arrives at one end of the tracks, the train simply reverses direction without turning around. At 4 P.M. on a certain day, a robbery was committed adjacent to the tracks.

When it was noticed by the motorman, he reversed direction before reaching the scene of the crime and returned to his starting point (not wanting to be involved with the crime). The detective investigating the robbery told the motorman that he was holding him as a material witness, since Mr. Brown had seen him close to the crime and Mr. Smith had seen him going slowly. His response was, "I was close to the robbery and I was going slowly, but at two different times. When I was closest to the robbery, I was going so fast I couldn't see." "You're lying," said the detective, and indeed he was.

(e) An insurance company, observing that there had been an excessive number of accidents because of speeding, decided to supervise its policy holders more carefully. They installed an alarm next to the speedometer of each insured car; the alarm went off whenever the magnitude of the instantaneous velocity was 65 miles per hour (but not at greater or lesser magnitudes). An insurance inspector observed that it took one of the policy holders 4 hours to travel 260 miles. "That's pretty fast," admitted the driver, "but the alarm never went off." "We found the alarm to be in perfect order," said the inspector, and the driver responded, "I guess the magnitude of the velocity jumped up and down but never actually hit 65 miles per hour." The insurance agent wasn't convinced and canceled the policy. Why?

4   Oil is pumped in or out of a tank through a single pipe. If the quantity of oil in the tank at time t, $Q(t)$, and the velocity of oil traveling through the pipe at time t, $V(t)$, are graphed, $V(t)$ is the derivative of $Q(t)$ and (of course) $Q(t)$ is the integral of $V(t)$. Each of the statements below is a consequence of (i) the Intermediate Value Theorem (for either velocity or quantity), (ii) the Mean Value Theorem, (iii) the fact that when a graph has a maximum or minimum, its derivative is zero. State which.
(a)   At the instant when the tank was at its lowest point for a given day (at midday), there was no flow in or out of the tank.
(b)   When the tank reached its highest point during the day (at midday), the instantaneous flow through the pipe was zero.
(c)   The quantity of oil increased by 4000 gallons between 2:00 P.M. and 4:00 P.M. At some time during those four hours, oil was flowing into the tank at an instantaneous rate of 2000 gallons per hour.
(d)   At one time oil was being pumped into the tank, and one hour later it was being pumped out of the tank. At some point between these times the oil in the pipe was instantaneously at rest.
(e)   There were 4000 gallons in the tank at two different times. At some point between these times the oil in the pipe was instantaneously at rest.

5   If you put \$1 in a particularly generous bank, you get back \$$T^2$ after T years. (T need not be a whole number.) How much faster is your money growing after 15 years than it is after 5 years?

An uncut baked potato loses heat slowly. In contrast, a cut potato has a great deal of surface area relative to volume. Do you think it will lose heat fast or slowly? (See Exercise 6.)

6 When a potato is kept at a high temperature in an oven for an extended period of time, the heat that it contains is proportional to its volume. The rate at which it loses heat when it is taken from the oven is proportional to its surface area.
    (a) Would a large potato or a small potato have a faster drop in temperature when removed from the oven?
    (b) Would the potato cool faster if you cut it up into quarters? Why? (*Hint*: See Example 1 again.)

7 Assume that the heat lost by the body because of sweating is proportional to the surface area of the body. Using the same approach that you used in Exercise 6, guess whether a large animal or a small animal would be more uncomfortable in hot weather. (You may assume that an animal's comfort increases as it loses heat because of sweating.)

8 It can be shown that the area of a circle grows more rapidly than its perimeter as the size of the circle increases. Suppose two countries keep armies that are proportional in size to their areas. If both countries are in the form of a circle and each must defend its border (the perimeter of its circle), which country, the larger or the smaller, will find it easier to patrol its borders?

9 A ball is thrown into the air from a hill 100 feet above sea level with an initial velocity of 50 feet per second and hits the ground (at sea level) some time later. The height of the ball above sea level, H, in feet t seconds later is given by the formula $H = 100 + 50t - 16t^2$.
    (a) When will the ball reach its highest point?
    (b) How high will the ball go? How fast is the ball traveling at the top?
    (c) How fast is the ball going when it passes the top of the hill (that is, when $H = 100$ feet) on the way down?
    (d) When will the ball hit the ground?
    (e) How long does it take to hit the ground after passing the top of the hill on the way down?

10 In Example 3, show that the profit obtained when $S = \$21$ is \$17,050 and that the profit obtained when S is either \$20 or \$22 is \$17,000.

11 In Example 4, show that the voltage when T is either 40 or 60 is 1.548.

12 If a bank pays 5% interest annually, your money doubles about every 14 years. If you invest one penny in the bank for N years, you would accumulate about $2^{(N/14)}$ pennies. How much money would you have after 560 years?

13 Suppose that 10 pounds of radioactive material is decaying with a half-life of 1000 years. When will there be $1\frac{1}{4}$ pounds? Will there ever be none left (in theory)?

SOLUTIONS TO SELECTED EXERCISES

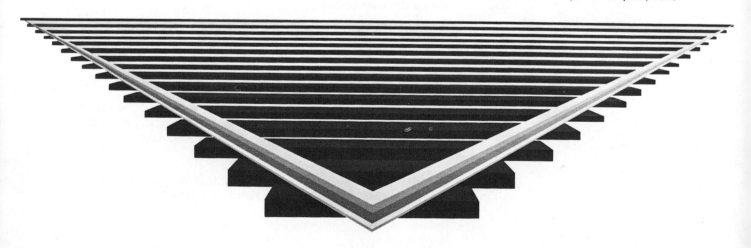

Patrick Hughes, ''Infinity,'' Screenprint, 1976

Solutions to odd-numbered Exercises, to selected even-numbered Exercises,
and to most Extras for Experts and Puzzles to Ponder follow.

WHY SPEAK MATHEMATICALLY?

INTRODUCTION

exercises

page 14

1 (a) The relevant factors would probably be (i), (ii), (iii), (iv), (vii), (ix),
and (x). Factor (v) might be important to ski manufacturers, and (vi)
might be important to manufacturers who supply the material for
dresses.
  (b) The relevant factors would probably be (ii), (iii), (iv), (v), (ix), (x), (xi),
(xii), and (xiii).
  (c) Factor (x) would seem to be of doubtful relevance; the remaining
factors might very well be included in the model.

2 (a) If job A requires that job B be completed first, draw an arrow from A
to B.
  (c) If B is A's best friend, draw an arrow from A to B.
  (e) If species A feeds on species B, draw an arrow from A to B.
  (g) Draw an arrow from A to B if A is the superior of B.

3 In each case the waiting time is reduced by an additional investment (in a
telephone switchboard, tollbooth, airplane, vendor, etc.)

extras for experts

page 16

1 (a) The jar will be half full the second before the end of the hour.
  (b) The bird will have traveled 180 miles.
  (c) Each tile covers one plain and one shaded square, so 31 tiles must
cover 31 plain and 31 shaded squares. But the two deleted squares
are of the same type, and thus the 31 tiles must cover a different
number of plain and shaded squares.
  (d) Clear from the hint.
  (e) If x is the amount of wine remaining in the red cup, then $1 - x$ must
be the amount of wine in the blue cup, since there was a pint of wine
at the start. Then x must be the amount of water in the blue cup,
since there was a pint of liquid in the blue cup initially and there
must be a pint of liquid after adding and subtracting a teaspoonful
of liquid.

The answers to (f), (g), and (h) are clear from the hints.

THE FOUNDATION

SET THEORY

 Sets

exercises
page 31

1 (a) A set consisting of those elements that are in either of the original sets.

(b) A set consisting of those elements that are in both of the original two sets.

(c) A set consisting of those elements that are not in the original set.

(d) ∅

(e) E

3 (a) A is always a subset of B, and it will be a proper subset if there is an element in $C \cap D'$.

(b) A will be a subset of B if $D \subseteq C$; it cannot be a proper subset of B.

(c) A is both a subset and a proper subset of B if and only if $C = \varnothing$.

(d) A is a subset of B but not a proper subset.

5 (a) (i)   (b) (iii)   (c) (i) and (iv)   (d) (iii)

(e) (iv)   (f) (i), (ii), and (iii)   (g) (iii)

(h) (i), (ii), and (iii)   (i) (i), (ii), and (iii)

(j) (v)   (k) (i), (ii), and (iii)

9 (a)   (i)   19, 20

(ii)   3, 6, 9, 12, 15, 18

(iii)   1, 2, 3, 4, 5, 6, 7, 8, 9

(b)   (i)   All the even integers in E.

(ii)   The four smallest and the four largest elements in E.

(iii)   The four smallest elements in E.

13 Use a Venn diagram, for example.

15 Neither rule is stricter. If you were in financial arrears and on academic probation, you couldn't play this year but you could last year with the Dean's permission. If you were in financial arrears but not on academic probation, or if you were on academic probation but not in financial arrears, you could play this year but last year you couldn't without the Dean's permission.

Infinite Sets

exercises

page 40

1 One way is to find a 1-to-1 correspondence between the set and a proper part of itself. If you can do so, the set is infinite.

3 (a) Finite
(b) Finite
(c) Finite (B may be the set of odd integers and C the set of even integers) or infinite (B and C may both be the set of even integers)
(d) Infinite
(e) Finite (if $B = E$) or infinite (B may be the even integers)
(f) Finite (B and C may both be the set of even integers) or infinite (B may be the set of odd integers and C may be the set of even integers).

5 Match the freshmen to a proper subset of the sophomores and the sophomores to the juniors. This will match every freshman to some junior, and there will be juniors left over (the mates of sophomores who are not matched with a freshman).

7 Let P be the set of positive integers and B be the set of integers (positive and negative). Match each positive integer i in B with $2i$ in P and each negative integer i in B with $-2i - 1$ in P.

extras for experts

page 41

2 Assign the sum of its avenue and street numbers to each corner. Policemen are assigned to low corners first. The first policeman goes to 1st St. and 1st Ave., the next two go to 1st St. and 2nd Ave. and to 1st Ave. and 2nd St., the next three go to the corners whose sum is four, and so on.

puzzles to ponder

page 42

1 This is one of the many paradoxical situations associated with the infinite which makes it repelling and fascinating at the same time. It *is* true that the number of patrons keeps increasing as closing time approaches and it is *also* true that at closing time no one is left.

2 There is nothing whatever wrong with the logic, and there really is no paradox. The logic only seems absurd because it is much easier to confirm that "all crows are black" than to confirm that "all objects that aren't crows aren't black," and this is so because of the size of the sets involved. If we had a different problem, if we were trying to verify that 200 people with measles in a city of 4 million were quarantined, for example, then it actually would be easier to confirm that all the people with measles were quarantined than to check that all people not quarantined didn't have measles.

4 This exercise demonstrates that long lines, or even infinite ones, have no more points than short ones, in the sense that there exists a 1-to-1 correspondence between the points on any two lines.

 # METHOD IN OUR MADNESS

PROBABILITY

 ## Sample Spaces and Sample Points

exercises page 52

1 Sample points might be (a) days of the week; (b) a type of restaurant in conjunction with a type of entertainment.

2 A sample point might be (a) a pair of languages or the language that was studied longest; (b) the list of winners of each game or the order in which the players finished.

3 (a) The event "two odd dice" is in both sample points.
  (b) The event "both numbers are equal" is in neither sample point.
  (c) If one number is greater than 4 and the other less than 3, the event would be in both (i) and (ii).

5 Heads or tails; each number from 1 to 6 on the face of the die might be a sample point.

7 You (i) received, (ii) did not receive your own letter. A sample space of 24 possible letter distributions.

8 There are 32 acceptance/rejection patterns for a single student, and these may be used as sample points. There are (32)(32) = 1024 patterns for the two students.

9 The winners might be
  ACE, ACF, ADE, ADF, BCE, BCF, BDE, BDF

11 Ten:
  ABC, ABD, ABE, ACD, ACE, ADE, BCD, BCE, BDE, CDE

Probability Measures of Sample Points and Events

exercises

page 58

1   (a)   (i) $\frac{13}{52}$ (ii) $\frac{20}{52}$ (iii) $\frac{4}{52}$ (iv) $\frac{26}{52}$

    (b)   (i) $\frac{1}{6}$ (ii) $\frac{3}{6}$ (iii) $\frac{2}{6}$ (iv)   0

    (c)   (i) $\frac{4}{6}$ (ii) $\frac{4}{6}$ (iii) $\frac{1}{6}$

    (d)   (i) $\frac{4}{10}$ (ii) $\frac{6}{10}$ (iii) $\frac{3}{10}$

    (e)   (i) $\frac{3}{10}$ (ii) $\frac{6}{10}$ (iii) $\frac{1}{10}$

2   (a)   (X, Y) is a sample point if X and Y are the cards selected from the two decks.
    (b)   (52)(52) = 2704
    (c) $\frac{1}{2704}$
    (d)   (i) $\frac{1}{169}$;   (ii) $\frac{1}{4}$

3   (a)   (X, Y) where X and Y are the numbers turned up on the dice.
    (b)   36   (c) $\frac{4}{36}$ (d) $\frac{6}{36}$ (e) $\frac{15}{36}$

5   (a) $\frac{1}{6}$ (b) $\frac{1}{6}$ (c) $\frac{2}{3}$
    (d)   Yes; (a), (b), (c) form a sample space.

7   (a) $\frac{1}{2}$, since as many students are selected as not.
    (b) $\frac{1}{6}$

9   (a) $\frac{2}{100}$ (b) $\frac{4}{100}$

11   (a) $\frac{1}{2}$ (b) $\frac{1}{6}$ (c) $\frac{1}{12}$ (d) $\frac{1}{2}$

13   (a) $\frac{1}{3}$ (b) $\frac{1}{4}$ (c) $\frac{1}{2}$ (d) $\frac{1}{4}$

15   $\frac{7}{12}$

17   (a) $\frac{1}{24}$ (b)   Yes; it would be $\frac{1}{20}$

19   (a) $\frac{1}{6}$ (b) $\frac{2}{3}$ (c) $\frac{1}{6}$

21   $\frac{1}{8}$

puzzles to ponder

page 62

There is no way of defining the probability of picking a particular integer which will avoid a contradiction. If the probability is defined to be zero, then the probability of picking any integer is also zero (it must be 1); and if the probability is positive, then the probability of picking any integer becomes infinite.

Probabilities of Compound Events

exercises
page 64

1 $\frac{3}{4}$

3 $\frac{11}{36}$

5 (a) 55%   (b) 10%   (c) 15%

7 $\frac{4}{9}$

9 You come out even on Friday.

11 25%

13 .3

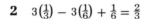

extras for experts
page 66

2 $3\left(\frac{1}{3}\right) - 3\left(\frac{1}{6}\right) + \frac{1}{6} = \frac{2}{3}$

 Permutations and Combinations

exercises
page 76

1 (a) 120   (b) 20

3 (a) 10   (b) 4

5 C_3^{100}; C_2^{100}

6 The same answers as in Exercise 5.

7 N

9 (a) $\frac{1}{15}$ (b) $\frac{8}{15}$ (c) $\frac{6}{15}$

11 (a) $\dfrac{C_5^{40}}{C_5^{50}}$ (b) $\dfrac{C_4^{40}C_1^{10}}{C_5^{50}}$ (c) $\dfrac{C_3^{40}C_2^{10}}{C_5^{50}}$

13 (a) $\dfrac{C_{18}^{98} + C_{20}^{98}}{C_{20}^{100}}$ (b) $\dfrac{C_{18}^{98}}{C_{20}^{100}}$

15 $\dfrac{C_4^{13}C_9^{39}}{C_{13}^{52}}$

17 (a)   455   (b)   2730

19 (a) $\frac{1}{336}$ (h) $\frac{1}{50}$

21 $\frac{4}{35}$

**extras for experts
page 78**

1 (b)   These are all possible ways of picking 3 students from a class of 9.
       (c)   These are all possible ways of picking K students from a class of N.

2 Any subset of a set of size N must have no elements, or 1 element, or . . . ,
or N elements.

3 Observe that $(C_i^N)^2 = C_i^N C_{N-i}^N$ and set $K = M = N$ in Question 1(c) above.

4 The expression is C_{K+1}^N, so it must be an integer.

5 (a)   (i)   1   (ii)   3   (iii)   3   (iv)   1
       (b)   8

6 (a) $2^2 = 4$ (b) $2^4 = 16$ (c) $2^5 = 32$.

 ## Conditional Probability and Independence

**exercises
page 85**

1 $\frac{1}{4}$

3 $\frac{1}{6}$

5 .210

7 (a) $\frac{3}{4}$ (b) $\frac{1}{4}$

9 (a) $\frac{1}{4}$ (b) $\frac{1}{2}$ (c) $\frac{1}{10}$

11 (a)   0   (b)   0   (c) $\frac{1}{6}$ (d) $\frac{2}{3}$ (e) $\frac{5}{6}$ (f)   0

13 (a) $\frac{1}{4}$ (b) $\frac{1}{13}$ (c) $\frac{1}{4}$ (d) $\frac{1}{13}$

15 (a) $\frac{1}{3}$ (b) $\frac{2}{9}$

17 (a) $\frac{1}{4}$ (b) $\frac{1}{3}$

19 B and C are the only two that are not independent.

21 (a) Yes   (b) No

23 $\frac{1}{3}$

25 $P(A \cap B \mid B) = P(A \cap B \cap B)/P(B) = P(A \mid B)$

27 (a) .36   (b) $1/\sqrt{2} \cong .707$

**extras for experts
page 88**

1 (a) If $P(A \mid B) = P(A)$,

then $P(A \cap B) = P(A)P(B)$

and $P(B) = P(A \cap B)/P(A) = P(B \mid A)$.

(b) $P(A \mid B') = \dfrac{P(A \cap B')}{P(B')} = \dfrac{P(A) - P(A \cap B)}{1 - P(B)} = P(A)$

by independence.

2 If the sum is 10, the probability that a 4 turned up is $\frac{2}{3}$.

3 $\frac{1}{3}$; A could not win the first three matches, or the game would be over.

**puzzles to ponder
page 89**

1 (a) $\frac{1}{197}$ (b) $\frac{1}{99}$

2 If $P(A \mid B) > P(A)$,

then $P(A \cap B) > P(A)P(B)$

and $P(A) - P(A \cap B') > P(A)P(B)$,

so $P(A) > P(A \cap B')/[1 - P(B)] = P(A \mid B')$

Bayes's Theorem

**exercises
page 96**

1 $\frac{2}{11}$

3 $\frac{2}{3}$; the white-and-white card shows white twice as often as the black-and-white card.

5 $\frac{3}{11}$

7 He is the "50%" salesman with probability .2 and each of the other salesmen with probability .4.

9 $\frac{5}{24}$

11 (a) $\frac{4}{13}$ (b) $\frac{1}{37}$

13 (a) $\frac{4}{7}$ (b) $\frac{76}{93}$

15 (a) $\frac{4}{23}$ (b) $\frac{16}{377}$

extras for experts
page 98

1 If A is pardoned, B and C each have an even chance of having their name given to A. If one of the others is pardoned, A will receive the name of the remaining person with certainty. When A hears B was not pardoned, C is twice as likely to have been pardoned as A.

2 $\frac{6}{101}$

The Binomial Distribution

exercises
page 102

1 The trials must be independent; they aren't if the mice are learning.

3 (a) $\frac{1}{2}$ (b) $\frac{3}{8}$ (c) $\frac{5}{16}$ (d) $\frac{95}{128}$

5 $\frac{496}{729}$

7 The same in each case.

9 (a) $\frac{1}{216}$ (b) $\frac{125}{216}$ (c) $\frac{2}{9}$ (d) $\frac{1}{8}$

10 (a) .729   (b) .243   (c) .027   (d) .001

11 (a) $\frac{27}{128}$ (b) $\frac{3}{64}$ (c) $\frac{81}{256}$

13 (a) $\frac{7}{27}$ (b) $\frac{125}{216}$

extras for experts
page 104

1 If you add up the probability of getting k heads for k from 0 to N, you get 1 (since these sample points make up a sample space). Therefore $1 = \left(\frac{1}{2}\right)^N (C_0^N + C_1^N + \cdots C_N^N)$; multiply both sides of the equation by 2^N.

3 (a) 2 or 3   (b) 1 or 2   (c) 1   (d) 1

THE ART OF INFERENCE

STATISTICS

1   Expected Value

1   (a)   7, 9, 8   (b)   9, 7, 8   (c)   7, 7, 7   (d)   7, 7, 7

3 $190, 0, 0

5 $500/49, $5, $10

7   1110, 800, 1000

9 $2

12   (a)   Expected value
      (b)   Mode
      (c)   Median
      (d)   (i)   cannot   (ii)   may   (iii)   must

13   Median

15   (a)   40th Street
      (b)   Any place between 40th Street and 45th Street
      (c)   The median

17 $3\frac{1}{2}$

19   (a) $\frac{2}{13}$ (b) $\frac{2}{13}$

21 $49

23   74°

(a)   80, 80, 76

(b)   The mode drops to 70.

Variance

exercises

page 124

1 Exercise 1:

    (a) $\frac{170}{11}$,   (b) $\frac{168}{9}$,   (c) $\frac{606}{9}$,   (d) $\frac{1130}{9}$

    Exercise 2:

    (a) 376,   (b) 158.64,   (c) 284,   (d) 49,   (e) 376

    Exercise 3: 1,193,900;

    Exercise 4: 10,109.91;

    Exercise 5: 33.6;

    Exercise 6: 1.02;

    Exercise 7: 161,900;

    Exercise 8: 103.4375;

    Exercise 9: 1

3 (a) 5 minutes late   (b) 129   (c) 9

5 (a) $\frac{3}{2}, \frac{1}{4}$ (b) $\frac{5}{2}, \frac{5}{4}$ (c) 4, 4

7 p, $p(1-p)$; $(1-p)$, $p(1-p)$

9 (a) 12,000

    (b) 13,600,000

    (c) The expected value would be doubled and the variance would be quadrupled.

    (d) The expected value would increase by $2000 and the variance would not change.

11 The truck should arrive 11 A.M., and he should expect to pay $59.

13 (a) 6, $\frac{20}{3}$ (b) 6, 8

extras for experts

page 125

Since the sum of the grades remains unchanged, the average must be unchanged as well.

    The old variance minus the new variance is

$$.05\{[(20)^2 - (20 - x)^2] + [(20)^2 - (x - 20)^2] + [(0)^2 - (-x)^2] + [(0)^2 - (x)^2]\}$$
$$= .05(80x - 4x^2), \text{ which is greater than 0 if } 0 < x < 20.$$

Testing Hypotheses

exercises

page 133

1 (a) .5, (b) .1; if you hire fair prospects: (a) .2, (b) .5

3 (a) You may conclude that the dice have 2 red faces when they have 5; probability: $1 - \left(\frac{5}{6}\right)^6$. You may conclude that the dice have 5 red faces when they have 2; probability: $\left(\frac{1}{3}\right)^6$.

   (b) You may conclude that the dice have 2 red faces when they have 5; probability: $\left(\frac{1}{6}\right)^6$. You may conclude that the dice have 5 red faces when they have 2; probability: $1 - \left(\frac{2}{3}\right)^6$.

5 $\frac{11}{1024}$; no, you must know how much better than 50% of the time she guesses correctly.

7 You mistakenly call the dice fair; probability: $\frac{1}{4}$. You mistakenly call the dice biased; probability: $\frac{11}{36}$.

9 $(.4)^6$

11 $(.1)^4 + 4(.1)^3(.9) + 6(.1)^2(.9)^2 = .0523$

13 (a) $N = 3$ (b) $(.4)^4 + 4(.4)^3(.6) = .1792$

puzzles to ponder

page 136

1 (a) $\left(\frac{1}{2}\right)^8 = \frac{1}{256}$; this record seems quite persuasive.
   (b) No; with 250 companies you would expect one to have a perfect record by chance.

2 If you test enough people, one is likely to do well by chance. Once you've "discovered" your talented subject, you must throw out the pilot experiment that uncovered him or her and start again.

Puts and Calls

exercises

page 143

3 (a) $750   (b) $375

5 (a) (i) 0   (ii) 0   (iii) $500
   (b) (i) $1000   (ii) $500   (iii) 0

7 $61\frac{2}{3}$

9 (a) (i) $1300   (ii) $350   (iii) $50
   (b) (i) 0   (ii) $1050   (iii) $2750

A Word of Caution

exercises
page 147

1 (a) Your sample may be systematically distorted if you overlook families that are not at home (they may be at work, for example).
    (b) Newspapers do not choose stories at random; they select those of unusual interest.

5 There may be fewer deaths for every mile traveled; this is the more important measure, since you spend more time traveling when you go slower.

Miscellaneous Problems

page 149

1

					Combined	Combined
					Class	
	1a	2a	1b	2b	1a and 1b	1b and 2b
Number of Boys	5	4	23	44	28	48
Number of Girls	20	21	2	6	22	27
Percentage of Boys	20	16	92	88	56	64

3 Suppose A takes 50 seconds 35% of the time and 60 seconds 65% of the time, B takes 55 seconds 50% of the time and 58 seconds 50% of the time, and C always takes 57 seconds. Then A, B, and C would win a three-way race 35%, $32\frac{1}{2}$%, and $32\frac{1}{2}$% of the time.

5 You're unlikely to lose, but when you do lose your losses are excessive. You must start with a finite amount of capital. If you start with $1021, for example, it will take 10 consecutive losses to wipe you out, but you only net $1 when you win and you are out $1021 when you lose.

7 Let Q be the probability that A survives. A can survive by hitting B on his first shot or by missing and having B miss him in turn, at which point A will have his original probability of survival; that is, $Q = \frac{1}{3} + (\frac{2}{3})(\frac{3}{7})Q$, so $Q = \frac{7}{15} = A$'s chance of survival.

9 It might be best not to fire at all. For example, if $a = .75$, $b = .8$, and $c = .85$, A should fire at C if he fires at all; if he hits, his chance of winning in the two-way duel with B in which B fires first is $\frac{3}{19}$. On the other hand, if C fires at B and hits him, A's chance of survival in his two-way duel with C is $\frac{60}{77}$, and it is even higher if B shoots C. Under these circumstances no one should fire at anyone.

GAMES ADULTS PLAY

GAME THEORY

 ## Some General Comments About Games

exercises
page 158

1 (a) Submarine and destroyer.
   (b) Survival and nonsurvival for the submarine.
   (c) The players' interests are opposed.

2 (a) Retailer and wholesaler.
   (b) The profit to the players as determined by the selling price.
   (c) The players have both common and opposed interests.

3 (a) Companies A, B, and C.
   (b) The payoffs are the potential profits.
   (c) The players have both common and opposed interests.

4 (a) Two competing gas stations.
   (b) The payoffs are the stations' profits.
   (c) There are both common and opposed interests.

5 (a) The prosecutor and the defendant.
   (b) The defendant's payoff is some jail term or freedom; the prosecutor's
      payoff is an improved or worsened conviction record.

6 (a) Husband and wife.
   (b) The payoff is the money won.
   (c) The players have identical interests.

exercises
page 163

 ## Two Person, Zero-Sum Games with Equilibrium Points

1 (a) (ii, B); -2 (c) (i, B); -1

3 (a) A dominates D; (i) dominates (ii) and (iii); A dominates B and C.
      (i, A) are in equilibrium; -1 is the value of the game.

5 (i) dominates (iii) and (iv); B dominates A and C;   any combination of (i), (ii) and B, D is in equilibrium, and the value of the game is 3.

7 (iv) dominates (ii); C dominates B and A;   (i) dominates (iii) and (iv). (i), C are in equilibrium, and the value of the game is 0.

extras for experts
page 166

1 Let $P(x, y)$ be the payoff when players I and II use strategies x and y, respectively. If (a, b) and (c, d) are two equilibrium strategy pairs, then $P(a, b) \leq P(a, d) \leq P(c, d)$ and $P(a, b) \geq P(c, b) \geq P(c, d)$ so $P(a, b) = P(c, d)$.

2 (a) If $a \geq 10$, then C dominates D for player II. If $a \geq 6$, then A dominates B for player I. If $a < 6$, then neither player has a dominating strategy.
   (b) If $a \geq 4$, then B dominates A. If $a \geq 6$, then D dominates C. If $a < 4$, then neither player has a dominating strategy.

3 (a) $3 \geq a$ and $b \geq 8$ (with at least one inequality proper)
   (b) $a \geq 3$ and $8 \geq b$ (with at least one inequality proper)
   (c) $(a \geq 3$ and $b \geq 8)$ or $(3 \geq a$ and $8 \geq b)$
   (d) $b \geq 3$ and $8 \geq a$ (with at least one inequality proper)
   (e) $3 \geq b$ and $a \geq 8$ (with at least one inequality proper)
   (f) $(3 \geq b$ and $8 \geq a)$ or $(b \geq 3$ and $a \geq 8)$
   (g) If a and b are either both less than 3 or both greater than 8.

4 (a) q and v are less than or equal to 5 and s and t are greater than or equal to 5.
   (b) No

5 (a)   (i) (B, ii);   payoff 5
         (ii) (B, i);   payoff x
         (iii) (A, i);   payoff 7.
   (b) If $5 < x < 7$, then (i, B) is an equilibrium strategy pair; and if $x < 5$, then (ii, B) is an equilibrium strategy pair.
   (c) If $5 < x < 7$, then (B, i) will be the equilibrium strategy pair with payoff x; and if $x > 7$, then (A, i) is the equilibrium strategy pair with payoff 7.

Two-Person, Zero-Sum Games without Equilibrium Points

exercises
page 175

1 $\left(\frac{1}{4}\right)(2000) + \left(\frac{3}{4}\right)(-1000) = \left(\frac{1}{4}\right)(-400) + \left(\frac{3}{4}\right)(-200) = -250 = \left(\frac{1}{16}\right)(2000) + \left(\frac{15}{16}\right)(-400) = \left(\frac{1}{16}\right)(-1000) + \left(\frac{15}{16}\right)(-200)$

3 $\left(\frac{1}{8}\right)(-100 + x) + \left(\frac{7}{8}\right)(5) = \left(\frac{1}{8}\right)(5 + x) + \left(\frac{7}{8}\right)(-10) = \frac{(x - 65)}{8}$

$= \left(\frac{15 + x}{120}\right)(-100 + x) + \left(\frac{105 - x}{120}\right)(5 + x)$

$= \left(\frac{15 + x}{120}\right)(5) + \left(\frac{105 - x}{120}\right)(-10)$

The game is fair if $x = 65$.

5 $\left(\frac{1}{5}\right)(0) + \left(\frac{4}{5}\right)(2) = \left(\frac{1}{5}\right)(8) + \left(\frac{4}{5}\right)(0) = \frac{8}{5} = \left(\frac{4}{5}\right)(0) + \left(\frac{1}{5}\right)(8) = \left(\frac{4}{5}\right)(2) + \left(\frac{1}{5}\right)(0)$

7 $\left(\frac{4}{5}\right)(5) + \left(\frac{1}{5}\right)(1) = \left(\frac{4}{5}\right)(4) + \left(\frac{1}{5}\right)(5) = \frac{21}{5} = \left(\frac{1}{5}\right)(5) + \left(\frac{4}{5}\right)(4) = \left(\frac{1}{5}\right)(1) + \left(\frac{4}{5}\right)(5)$

9 $\left(\frac{3}{4}\right)(2) + (0)(12) + \left(\frac{1}{4}\right)(6) = \left(\frac{3}{4}\right)(3) + (0)(-15) + \left(\frac{1}{4}\right)(3) =$
$\left(\frac{3}{4}\right)(6) + (0)(12) + \left(\frac{1}{4}\right)(-6) = 3 = \left(\frac{1}{2}\right)(2) + \left(\frac{1}{3}\right)(3) + \left(\frac{1}{6}\right)(6) =$
$\left(\frac{1}{2}\right)(12) + \left(\frac{1}{3}\right)(-15) + \left(\frac{1}{6}\right)(12) = \left(\frac{1}{2}\right)(6) + \left(\frac{1}{3}\right)(3) + \left(\frac{1}{6}\right)(-6)$

11 The mixed strategy (i) with probability $\frac{1}{3}$ and (ii) with probability $\frac{2}{3}$ dominates (iii). If II plays A and B each half the time, this will dominate C. The value of the game is $-\frac{60}{23}$.

Calculating Optimal Mixed Strategies

exercises

page 185

1 (a) A with probability $\frac{5}{6}$; C with probability $\frac{1}{2}$; value is $3\frac{1}{2}$.
   (c) A, B probability $\frac{1}{3}$, $\frac{1}{2}$; D, E probability $\frac{2}{60}$, $\frac{37}{60}$; value is 13.

2 (a) (i) probability $\frac{1}{4}$; A probability $\frac{1}{2}$; value is 7.
   (c) (i) probability $\frac{3}{4}$; A probability $\frac{1}{2}$; value is $2\frac{1}{2}$.

3 (a) $x = 0$ (c) $x = \frac{5}{4}$

5 (a) The offensive and defensive strategies are similar: A, B, C, and D are selected with probability .48, .24, .16, and .12. The probability of repelling the attack is .296.
   (b) New York with probability $\frac{1}{4}$ for Smith, probability $\frac{1}{3}$ for Jones; Smith gets 24 votes on the average.
   (c) Q and R should each be played $\frac{1}{3}$ of the time; the value is $\frac{1}{3}$.

7 Engel plays (i) and (ii) with probabilities $\frac{48}{209}$ and $\frac{76}{209}$, and Mendez plays A and B with probabilities $\frac{89}{209}$ and $\frac{54}{209}$; the value is $\frac{668}{209}$.

9 (a) S and T are played with probability $\frac{11}{20}$ and $\frac{2}{20}$; P and Q are played with probability $\frac{2}{7}$ and $\frac{3}{7}$; the value is 0.
   (b) S, T, and U are played with probability .1, .2, and .3; O, P, and Q are played with probability $\frac{59}{304}$, $\frac{135}{304}$, and $\frac{69}{304}$; the value is 1.

**extras for experts
page 189**

4　You should bid the small slam and your opponent the grand slam each with probability .6;　you should win the tournament 76% of the time.

 Two-Person, Non-Zero-Sum Games

**exercises
page 195**

1　(a)　The pair (55, 55) seems to be very stable. If someone shifts to 56, there doesn't seem to be any reason why the competition should follow (since its profit is greater when it undersells the other store), so the first store's price would likely drop to 55 again. (Or would it?)

　(b)　If the current prices were (56, 56), either company would gain initially by lowering its price, and its competitor would then be better off if it lowered its price as well. Thus (56, 56) seems to be very unstable.

　(c)　(55, 55) seems to be the most stable and yields the least joint profits. (56, 56), the payoff that seems to be least stable, maximizes joint profits and is more desirable to *both* companies than (55, 55). So the least desirable pair seems most stable.
　　　Returning to part (a), it appears that the stores may persist in obtaining a lower payoff.

3　While Tina's position is doubtful, Janet is certainly better off; she will get her first choice. Ordinarily one or the other of the travelers would have to make do with something less than their first choice, and in fact both might if they couldn't agree. Since Janet has no choice, Tina won't resent going along (as she otherwise might). The answer to question 5 seems to be yes.

TO PICK AND CHOOSE

AN APPLICATION TO VOTING

 The Voting Mechanism

**exercises
page 212**

1　With our assumptions Buckley would lose a two-way match with either Goodell or Ottinger.

3   In the initial two-way race, there is always one committee member who has the two competing cities as his first two choices. If he votes for his first preference, he will eventually get his last preference. By voting for his second preference instead, he will actually get it.

5   (a)   If everyone votes naively, the two means of transportation that take part in the first vote will lose.

    (b)   Alice and Barbara should vote for trains if buses and trains are compared first; Evelyn will lose out, but she's helpless. If buses and trolleys are compared first, Evelyn should vote buses, guaranteeing their eventual victory; in this case Charles and David lose out, but they have no recourse. Finally, if trains and trolleys compete on the first vote, David should back trolleys, and they will win; Alice and Barbara are then stuck with their last preference.

7   (a)   If the initial pairing is Lamont–Karp, the vote will be 31 to 21. If the initial pairing is Karp–Brown, the vote will be 30 to 22. If it is Brown–Lamont, the vote will be 31 to 21. The person winning the first vote must lose the second (with naive voting), since each candidate outvotes and is outvoted by one of the others.

    (b)   (i)   The members who hold a Brown–Lamont–Karp preference ordering will vote for Lamont on the first round, although they prefer Brown; and the Karp–Lamont–Brown people will vote for Brown, although they prefer Lamont.

        (ii)   With sophisticated voting Karp will win the first vote; the Lamont–Karp–Brown people will vote for Karp, and the Brown–Karp–Lamont people will vote for Lamont; the rest will vote sincerely. The vote for Lamont and Karp in the first race will be 22 to 30.

      (iii)   The first vote will be won by Brown 31 to 21. The Karp–Brown–Lamont people will vote for Brown, and the Lamont–Brown–Karp people will vote for Karp. Everyone else votes sincerely, and in all cases everyone votes sincerely on the second vote.

**puzzles to ponder
page 216**

The logic seems reasonable except that a case might be made on the basis of this logic for the chairman, Klein, choosing Acapulco (especially if there is a prevote discussion).

Some Applications

**exercises
page 219**

1   If everyone voted sincerely, Wilson would win in Example 2 and no bill would pass in Examples 3 and 4.

3 Wilson is helpless against a hostile coalition, although he could try to get his second rather than his third choice by dealing with Taft. In Examples 3 and 4 the Republicans are helpless and must accept their third preference.

Logrolling

exercises

page 222

1 Each bloc has its way on its third and fourth most important bills (which it didn't when everyone voted sincerely) but now loses on the two bills it considers most important as well as the two it considers least important.

3 A player who sees the trap cannot avoid it by refusing to trade; he must persuade others not to trade as well. If he refuses to trade (and everyone else trades as before), he will do even worse. The dilemma faced by voters here is similar to the prisoner's dilemma discussed in Chapter 5 (Game Theory).

Arrow's Impossibility Theorem (*Optional*)

exercises

page 231

1 (a) There must be at least three alternatives; here there are only two.
(b) (i) If the preferences of the inspectors are

	Inspector		
	I	*II*	*III*
First preference	A	C	C
Second preference	B	B	B
Third preference	C	A	A

then *A* would be preferred to *B* since *B* is never first and *A* is first once. If inspectors II and III drop *C* to third place without changing the relative positions of *A* or *B*, *B* will be preferred to *A*; this violates condition 3.
(ii) Here there is a dictator, which violates condition 5.
(iii) This violates condition 4, since the decision is prejudged.

(iv)   In the preference ordering

Inspector

	I	II	III
First preference	A	B	C
Second preference	C	A	B
Third preference	B	C	A

all restaurants are equivalent but if inspector II changes C's rating from last to first, B will be preferred to A; this contradicts condition 3.

2   If you support BS, you will be defeated by $\overline{BS}$.
If you support $B\overline{S}$, you will be defeated by $B\overline{S}$ or BS.
If you support $\overline{B}S$, you will be defeated by BS.
If you support $\overline{BS}$, you will be defeated by $B\overline{S}$ or $\overline{B}S$.

Power

exercises

page 240

1   In each case all six permutations are shown and the pivotal player is underscored.

(i) $AB\underline{C}$ $AC\underline{B}$ $BA\underline{C}$ $BC\underline{A}$ $CA\underline{B}$ $CB\underline{A}$

(ii) $AB\underline{C}$ $AC\underline{B}$ $BA\underline{C}$ $BC\underline{A}$ $CA\underline{B}$ $C\underline{B}A$

(iii) $AB\underline{C}$ $A\underline{C}B$ $BA\underline{C}$ $BC\underline{A}$ $C\underline{A}B$ $C\underline{B}A$

(iv) $A\underline{B}C$ $A\underline{C}B$ $B\underline{A}C$ $BC\underline{A}$ $C\underline{A}B$ $C\underline{B}A$

(v) $AB\underline{C}$ $A\underline{C}B$ $BA\underline{C}$ $B\underline{C}A$ $\underline{C}AB$ $\underline{C}BA$

3   (a)   The vote is determined by two out of three of the Liberals, Tories, and New Democrats, since the two smaller parties control a majority of the vote.

5   It is clear that the vote of the four largest parties determines the outcome so (a)   the Democratic-Socialists and Independents are dummies.

    (b)   The minimal winning coalitions are the largest party and the second, third, or fourth largest parties or the second, third, and fourth largest parties together. If you put the four largest parties in sequence, the largest will be pivotal if it is in second or third position, so it has half the power; the other three parties share the remaining half equally.

ON CONSTANT CHANGE

MARKOV CHAINS

 Introduction

exercises

page 250

1 A rat deciding which way to turn generally makes use of all of its past experiences, so a Markov chain would not be an appropriate model. If the rat had a peculiar form of amnesia so that it had precisely the same state of mind each time it arrived at the critical junction, a Markov chain might be used after all.

2 If you accept the doctrine of "the maturity of chances," you would not be able to use a Markov chain as a model, since according to that theory the coin "remembers" its past history, in effect. But the fact is that coins do not remember, and you can use Markov chains.

3 The group of economists who believe that the current price of a stock contains all the necessary information may use the Markov chain as a model; the other economists may not.

4 (a) The steps are the generations. There are two states: having an IQ less than 110 or greater than 110.
The transition probabilities are

	This generation	
	IQ > 110	IQ < 110
IQ > 110	.65	.4
IQ < 110	.35	.6

Next generation

(b) There is a new step each year; there are three states: accident, no accident with collision insurance, and no accident without collision insurance. The transition matrix is

This Year

	Accident	No accident with insurance	No accident without insurance
Accident	.1	.1	.1
No accident with insurance	.855	.765	.225
No accident without insurance	.045	.135	.675

It is possible to break the state "accident" into two states (accident with insurance and accident without insurance), but the two columns would be identical.

(c) There is a new step each year. There are four states: Cheating with or without an audit and not cheating with or without an audit. The transition matrix is

This Year

	Cheat with audit	Cheat with no audit	No cheating, no audit	No cheating with audit
Cheat with audit	.025	.05	.03125	.0125
Cheat with no audit	.175	.35	.21875	.0875
No cheating, no audit	.7	.525	.65625	.7875
No cheating and audit	.1	.075	.09375	.1125

5 (a)

This year

	Winner	Loser
Winner	$\frac{3}{4}$	$\frac{1}{4}$
Loser	$\frac{1}{4}$	$\frac{3}{4}$

(b)

| | Present capital |
	M
Capital after next bet — X − 1	$\frac{4}{5}$
X + 2	$\frac{1}{5}$

(c)

| | Price of stock today |
	X
Price of stock tomorrow — X + 2	.15
X + 1	.25
X	.25
X − 1	.2
X − 2	.15

7

		Action of last person		
		Turned down	Repaid loan	Did not repay loan
Action on next person	Turns down	.52	0	1
	Lends money and is repaid	.24	.5	0
	Lends money and is not repaid	.24	.5	0

9

		Today		
		Uses honor system and is cheated	Uses honor system and is not cheated	Doesn't use honor system
Tomorrow	Uses honor system and is cheated	.15	.4	.3
	Uses honor system and is not cheated	.15	.4	.3
	Doesn't use honor system	.7	.2	.4

11

                                          Last term

	Better than B average	Less than B average
Next term Better than B average	.6	.75
Less than B average	.4	.25

Transition Probabilities

exercises
page 258

1 $P_{ap}^{k+1} = P_{aa}^{k} P_{ap} + P_{ap}^{k} P_{pp}$

3 $P_{aa}^{3} = P_{aa}^{2} P_{aa} + P_{ap}^{2} P_{pa}.$
Now $P_{aa}^{2} = P_{ap} P_{pa} + P_{aa} P_{aa} = (.1)(.2) + (.9)(.9) = .83,$
so $P_{aa}^{3} = (.83)(.9) + (.17)(.2) = .781.$

5 $P_{pa}^{2} = P_{pa} P_{aa} + P_{pp} P_{pa} = (.2)(.9) + (.8)(.2) = .34$

      $P_{pp}^{2} = P_{pa} P_{ap} + P_{pp} P_{pp} = (.2)(.1) + (.8)(.8) = .66$

      $P_{pa}^{3} = P_{pa}^{2} P_{aa} + P_{pp}^{2} P_{pa} = (.34)(.9) + (.66)(.2) = .438$

      $P_{pp}^{3} = P_{pa}^{2} P_{ap} + P_{pp}^{2} P_{pp} = (.34)(.1) + (.66)(.8) = .562$

7   (b)   If we abbreviate by "l," "t," and "n" the states "parked legally," "ticketed," and "not ticketed," respectively, then
        $P_{lt}^{2} = P_{ll} P_{lt} + P_{lt} P_{tt} + P_{ln} P_{nt} = (.5)(.1) + (.1)(.05) + (.4)(.2) = .135$
     (c)   About **3000** tickets

9   (a)   If there were 1000 patients who were either serious or critically ill today, you would expect to have an average of 700 in these categories the following day and 490 the day after (since 30% of each category either die or are cured each day on the average).
     (b)   The "healthy" and "deceased" states can be entered but never left.

11   (a)   (i) A B (ii) A B

                     A $\frac{1}{4}$ $\frac{3}{16}$ A $\frac{5}{12}$ $\frac{7}{18}$

                     B $\frac{3}{4}$ $\frac{13}{16}$ B $\frac{7}{12}$ $\frac{11}{18}$

     (b)   (i) A B (ii) A B

                     A $\frac{3}{16}$ $\frac{13}{64}$ A $\frac{29}{72}$ $\frac{43}{108}$

                     B $\frac{13}{64}$ $\frac{51}{64}$ B $\frac{43}{72}$ $\frac{65}{108}$

Stable Probability Distributions

1 $\frac{8}{15}$

3 $\frac{37}{137}$

5 $\frac{10}{13}$ of all cars park illegally.

7 If we let H, S, C, and D represent the states healthy, serious, critical, and deceased, we have $H + .2S = H$; $.4S + .2C = S$; $.3S + .5C = C$ and $.1S + .3C + D = D$. The first and last equations yield $S = C = 0$. Any H and D such that $H + D = 1$ is stable.

9 (a)

This weekend

		Ski good weather	Ski fair weather	Ski poor weather	Don't ski
Next weekend	Ski good weather	.04	.03	.01	.02
	Ski fair weather	.2	.15	.05	.1
	Ski poor weather	.16	.12	.04	.08
	Don't ski	.6	.7	.9	.8

(b) $(2000)(\frac{20}{97}) = 412.37$

1 You can verify that the probability of having 0 at step $2N + 2$ is the sum of the probability of having 0 at step $2N + 1$ and half the probability of having 1000 at step $2N + 1$. The odd steps and similar calculations for the other states may be confirmed in a similar way.

2 The long-term number of workers on the morning, evening, and graveyard shifts will be $bc/(ab + ac + bc)$, $ac/(ab + ac + bc)$, and $ab/(ab + ac + bc)$, respectively. If $c = 1$, the numerator of the graveyard fraction, ab, can be no greater than the other numerators, which become a and b (the denominators are the same).

3 The chance that you will eventually win if you start with i dollars is the sum of (i) the chance that you win the first bet multiplied by the chance that you will eventually win if you start with $i + 1$ dollars and (ii) the chance that you will lose your first bet multiplied by the chance that you will eventually win if you start with $i - 1$ dollars.

(b) If you have \$0, you have already lost; and if you have \$$N$, you have already won.

(c) Since $\left(\frac{1}{2}\right)^{100}$ is very small (it is less than 10^{-30}), the formula effectively becomes $r_1 = 1 - \left(\frac{1}{2}\right)^i$.

(e) Bet large rather than small amounts against a house that has the odds in its favor.

4 In the long run the state will be S $b/(a + b)$ of the time and T $a/(a + b)$ of the time. If initially the probability of being in state S is $b/(a + b) + y$ and in state T is $a/(a + b) - y$, then the respective probabilities become at the next step $b/(a + b) + (1 - a - b)y$ and $a/(a + b) - (1 - a - b)y$. If both a and b are greater than 0 and less than 1, then $(1 - a - b)$ is between -1 and 1.

NUTS AND BOLTS

COMPUTERS

 The Anatomy of a Number

exercises
page 287

1 (a) 1010   (b) 1100100   (c) 1001101101
(d) 10000000001

3 10^N

5 (b) 2 switches with 3 positions can represent 9 different numbers.

extras for experts
page 288

1 (a) The binary number with N 1's represents $2^N - 1$; a 1 followed by N 0's is 2^N.

2 (a) $3^2 = 9$ (b) (i) $4^2 = 16$ (ii) P^2 (c) P^8

3   (a)   (i)   110   (ii)   111   (iii)   1111
                         100   11   1100
                       ─────   ─────   ───────
                       11000   111   111100
                                                    111   1111
                                                  ─────   ───────
                                                  10101   10110100

(b)   Multiplying by powers of 2; to multiply by 2^N, shift the binary point N places to the right.

4   (a)   20 binary places are roughly equivalent to 6 decimal places.
        (b)   10N binary places are required to write a number with 3N decimal places.

5   (a) $\frac{1}{2}$ (b) $\frac{5}{8}$ (c) $\frac{1}{3}$

The Representation of Numbers

exercises
page 294

1   (a) 462×10^9 (b) $52{,}000{,}098 \times 10^6$ (c) 4×10^4

3   (a) 345×10^{-6} (b) 1.765×10^{-19} (c) $2{,}100{,}002 \times 10^{-11}$

5

Memory	Fixed-Point	Floating-Point
-56123453	$-56{,}123{,}453$	$-.561234 \times 10^3$
$+23000048$	$+23{,}000{,}048$	$+.23 \times 10^{-2}$
$+01000000$	$+1{,}000{,}000$	$+.1 \times 10^{-51}$
-21000046	$-21{,}000{,}046$	$-.21 \times 10^{-4}$
$+12000056$	$+12{,}000{,}056$	$+.12 \times 10^6$
$+23122145$	$+23{,}122{,}145$	$+.231221 \times 10^{-5}$

7   (a)   20000049   (b)   20000038   (c)   32000050

9   (a), (b)   By subtracting 50 from the exponent the range of numbers you can represent becomes (roughly) 10^{-50} to 10^{50} instead of 1 to 10^{100}.

11   (a)   10; 10465652   (b)   1; 69684350   (c)   5; 46560051
         (d)   20000052

extras for experts
page 295

1   Add the J portions of both numbers and then subtract 50. Multiply the I portions (assuming that the decimal point is on the left) and then shift so that the leading figure is not zero (modifying the J portion to compensate).

2   Raise the J portion of the smaller number (shifting the I portion to the right correspondingly) so that both numbers have the same J value. Add the I portions. If there is no carry, the answer is the I sum and the common J value; if there is a carry, shift right once and increase J by 1.

3   The mesh would be 10 times as fine, but the range would be reduced by a factor of 10.

Errors

exercises
page 303

1   (a) .2, .02   (b) 0, 0   (c) .1, .025

3   (a) $\frac{1}{75}$ (b) .03

5   .24, .05

7   (a) .4 miles   (b) .0001
    (c) The answer to (a) would be $\frac{1}{5}$, but the answer to (b) would be unchanged.

9   (a) 3 miles; .03   (b) No

11 (a) Double the relative error in r.
    (b) Triple the relative error in r.

13 (a) The area inspector is twice as accurate as the length inspector; that is, his maximum relative error is half the other inspector's.
    (b) The volume inspector is three times as accurate.

15 .05; about 0; yes; the estimates represent the worst case.

17 You add relative errors when you take products. The relative errors will be halved when you take square roots.

extras for experts
page 305

1   (a), (b) The greatest errors occur when y has the largest relative error.
    (c) The relative error is greatest when the largest number in magnitude has the greatest relative error.

2   .005

3   In the first case we took the difference of two numbers that are almost equal (which magnifies the relative error enormously), and in the second case we took their sum.

6 (c)   An error of less than 20% in a induces an enormous change in all three of the variables x, y, and z.

7 (a), (b)   If you are multiplying, it makes no difference how the errors are distributed; but if you are adding, the most accurate answer will be obtained if the error is in the smaller number.

Iterative Methods (Loops)

**exercises
page 311**

1

Step	Pad F	Pad H	Pad B
1	4	101	10
2	5	101	10
3	5	101	10
1	5	101	15
2	6	101	15
3	6	101	15
1	6	101	21
2	7	101	21
3	7	101	21
1	7	101	28
3	8	101	28

2 (a)   Simply change the number on Pad H to 1001 or 1,000,001, respectively.

(b)   It would take the same time to write the program, but it would take the computer 10 times and 10,000 times as long to do the calculations, respectively.

3 There is more than one way of writing a program to accomplish (a) and (b). We offer one example of each.

(a)   Initially we put the number a into A(1),   0 into A(2),   0 into A(3), n into A(4), b into A(5),   and 1 into A(6).

A(10)   contains the command, "Add the number in A(1) to the number in A(2) and put the resulting sum into A(2), wiping out whatever was there before. Perform the command in A(11) next."

A(11)   contains the command, "If the numbers in A(3) and A(4) are the same, print the number in A(2) and stop. If the numbers are different, perform the command in A(12) next."

A(12)   contains the command, "Add the numbers in A(6) and A(3) and put the sum in A(3). Perform the command in A(13) next."

A(13)   contains the command, "Add the numbers in A(1) and A(5) and put the sum in A(1). Perform the command in A(10) next."

By keeping a record of the contents of locations A(1), A(2), and A(3) you can verify that the program stops when $a + (a + b) + (a + 2b) + \cdots + (a + nb)$ is obtained in A(2).

(b)   Initially put the number a into A(1), r into A(2), 0 into A(3), n into A(4), 0 into A(5), and 1 into A(6).

A(10)   contains the command, "Add the numbers in A(1) and A(3) and put the sum in A(3). Perform the command in A(11) next."

A(11)   contains the command, "If the number in A(5) is the same as the number in A(4), print the number in A(3) as the answer and stop. Otherwise, perform the command in A(12)."

A(12)   contains the command, "Multiply the numbers contained in A(1) and A(2) and put the product in A(1). Perform the command in A(13) next."

A(13)   contains the command, "Add the numbers in A(5) and A(6) and put the sum in A(5). Perform the command in A(10) next."

4   Command

location	A(1)	A(2)	A(3)	A(4)	A(5)	A(6)	A(7)
Initially	5	2	2.2360688	.001	4.4721377	2.2360688	.00202634
11					2.236067		
12					4.4721359		
13						2.2360679	
14							.0000009
15			2.2360679				
11					2.236068		
12					4.4721359		
13						2.2360679	
14						0	

5   Place the following numbers in each of the following locations:

Address	1	2	3	4	5	6	7	8
Number	$\frac{2}{3}$	3	10	1	2	0	3	0

Then perform the following commands:

A(11)   Multiply the numbers in A(1) and A(2) and put the product in A(8). Perform the command in A(12) next.

A(12)   Multiply the number in A(2) by itself and put the result in A(2). Perform the command in A(13) next.

A(13)   Multiply the number in A(7) by the number in A(2) and put the product in A(2). Perform the command in A(14) next.

A(14)   Divide the number in A(3) by the number in A(2) and put the quotient in A(2). Perform the command in A(15) next.

A(15)   Add the numbers in A(2) and A(8) and put sum in A(2). Perform the command in A(16) next.

A(16)   Add the numbers in A(4) and A(6) and put sum in A(6). Perform the command in A(17) next.

A(17)   If the number in A(5) is the same as the number in A(6), stop and print the number in A(2). Otherwise, perform the command in A(11).

Command location	A(1)	A(2)	A(3)	A(4)	A(5)	A(6)	A(7)	A(8)
Initially	$\frac{2}{3}$	3	10	1	2	0	3	0
11								2
12		9						
13		27						
14		.3703703						
15		2.3703703						
16						1		
17								
11								1.5802468
12		5.6186553						
13		16.855966						
14		.5932617						
15		2.1735085						
16						2		
17								

Stop. The answer after two iterations is 2.1735085. The answer to five decimal places is 2.15443, so the error after the first iteration was .2159403 and the error after two iterations was .0190785.

6   Start by putting A in A(1), 1.06 in A(2), 20 in A(3), 0 in A(4), and 1 in A(5). Then perform the command in A(10).

A(10)   Multiply the numbers in A(1) and A(2) and put the product in A(1). Perform the command in A(11) next.

A(11)   Add the numbers in A(4) and A(5) and put the sum in A(4). Perform the command A(12) next.

A(12)   Compare the numbers in A(3) and A(4). If they are equal, print the answer, which is to be found in A(1). If they are unequal, perform the command in A(10) next.

BUT CAN THEY THINK?

COMPUTER LEARNING AND GAMES

The Extensive Form of a Game

exercises
page 322

1 9! = 362,880

3 (a) Player I has a winning strategy by moving to *B* first, and then by moving to *J* if II moves to *E*. I will either win when II moves to *D* or at *J*. The winning strategy is clear from the matrix; when I plays LL, he wins whatever II does.

(b) Both players cannot have a winning strategy; if a player adopts a winning strategy, he wins *whatever* the other player does. There would be no consistent outcome if both played winning strategies.

(c) Player II can always lose if he plays LR.

(d) Two players cannot both have a losing strategy for the same reason that both cannot have a winning strategy (see (b)), but the same player can have both a winning and a losing strategy.

(e) If I plays LR, II can win with RL; if I plays RL, I will win with LL; and if I plays RR, II will win with LL. Whatever strategy II plays, I wins with LL.

Winning Strategies and Winning Positions

exercises
page 330

1 The winning positions (for the player who has the move) are *F*, *G*, *E*, *D*, *B*, and *A*. The only losing position is *C*. Player I's winning strategy is to move from *A* to *C* and then to move right if II moves left.

3 *G*, *A*, and *C* are winning positions, and *B* is a losing one. I wins by moving to *B*, and II must then move to a winning position for I.

5 Player I wins if (i) *V* is a win for I or (ii) both *U* and *T* are wins for I. Also, (iii) *U* and *S* are wins for I, and at least one of the positions *P*, *Q*, or *R* is a win for I, as well. In case (i) I moves to *V*. In case (ii) I moves to *B* and then to *T* when II moves to *C*. (If II moves to *U*, he loses.) In case (iii) I moves to *B*, then to *D*, and then to the winning position that follows *E*. If II detours to *U* or *S*, I will win.

7 (a) If the cumulative sum is from 89 to 93, you can add a number from 1 to 5 that will make the cumulative sum 94 and put your opponent in a losing position.

(b) By part (a), if you move when the sum is 88, your opponent can force you into a position in which the sum is 94.

(c) If the sum is x (for any of the values given) and it is your move, your opponent can put you in a position in which it is your move and the sum is x + 6. Therefore he can force you into each of these numbers in turn and finally, into 94 which we established earlier was a losing position.

(d) If it is your turn to move and the cumulative sum is 4 or 4 plus some multiple of 6 (4, 10, 16, . . . , 88, 94), you are in a losing position; otherwise, you are in a winning one. Thus, choose 4 if you move first.

(e) A winning strategy for the mover at 91 is: "Form 94. Your opponent must form a sum from 95 to 99. Add the number that will make this sum 100." (This number will be from 1 to 5 and therefore a legal move.)

(f) If you move first, the following is a winning strategy: "Add 4. At each turn, if your opponent picks x, add 6 − x. (Since your opponent's number is between 1 and 6, yours must also be.) Continue in this manner and you will win in exactly 15 moves."

Notice that each of the winning strategies in (e) and (f) spell out exactly what you will do in every situation that may arise and that the outcome is a win for the player using the winning strategy whatever his opponent does.

More Examples of Simple Games

exercises page 337

2 Player I wins by going to B first. If II goes to either F or G, I wins at the next move; if II goes to E, I wins by going right. (Or I adds 1; if II adds x, I adds 4 − x.)

3 The losing positions are N − K, N − 2K, N − 3K, ... If your opponent has a sum of N − iK and adds x, you add K − x.

4 (d) If I reduces (2, 2) to (2, 1), II should reduce this to (1, 1) and II must inevitably win. If I reduces (2, 2) to (2), II reduces this to zero and wins.

7 The player who moves need only delete that number to win.

extras for experts page 339

1 (a) From (1, 1) you can only move to (1), and your opponent must win.

(d) If you are at (N + 1, N + 1), and if you know (K, K) is a loss if K < N + 1, then your opponent can always put you into a losing position by forming (M, M) when you play (N + 1, M).

(e)   If $M \neq N$, then you can win from (M, N) by playing (M, M) if $M < N$ or (N, N) if $N < M$.

2   (a)   The player with the move wins if there are an odd number of 1's.

     (b), (c)   A position is a loss if there are both an even number of 1's and an even number of 2's, and is a win otherwise. If you move in a position with an even number of 1's and an even number of 2's and delete a 1, your opponent should do the same. If you delete a 2, your opponent should also do the same. If you take a 1 away from a 2, your opponent should take a 1 away from a different 2. (Since there are an even number of 2's there must be another 2.) This strategy will lead to a loss for you. If you must move in a position with an even number of 1's and an odd number of 2's by deleting a 2, your opponent will have to move in a losing position: one with an even number of both 1's and 2's. If you move in a position with an even number of 2's and an odd number of 1's, delete a 1. And if there are an odd number of 1's and 2's, take a 1 from a 2 and both the number of 1's and 2's will be even.

Evaluating Positions

exercises

page 347

1   (a)   Since $G(T) > C$, $G(S) > C$ and S is a winning position.

     (b)   Since $G(S) < C$, $G(T) < C$ and T is a losing position.

3   If S is a losing position, $G(S) < C$ and $G(T) > C$ for all positions T immediately following S (since they are all winning positions). So $M > C$. If S is a winning position, $G(S) > C$; there must be a losing position, T, immediately following S so that $G(T) < C$. Therefore $M < C$.

5   (a)   The EF might be an IEF with the cutoff value between 5 and 6.

     (b)   The EF might be an IEF, and the cutoff value would be between 4 and 5.

     (c)   The EF that generated these positional values could not possibly be an IEF. The position with value 7 would have to be a loss (looking at the positions that take on the values 9 and 12) and a win (looking at the immediately following positions).

extras for experts

page 350

1   (a)   If B were a losing position, then $G(B)$ would have to be less than $G(A)$, so $x < -2$. But then $G(E) < G(B)$ and E would be losing as well.

     (b)   If A is a winning position, D must be a losing position, since B is a winning position; hence $G(A) > G(D)$ and so $x > 4$. Hence $G(E) > G(B)$, so E is a winning position. Since B and E are winning positions, F must be a losing one.

2   A losing position has as its immediate successors only winning positions. A drawing position has as its successors at least one drawing position and no losing positions. A winning position is followed by at least one losing position.

3   Initially "X" has a win (by playing either 4 or 5), and the value of the position to "X" is 0. The subsequent positions that are losses for "O" are (i) and (iii), with values of -6 and -8. The draws are (ii) and (v), with values -4 and -3. The win is (iv), with value -1. From this much evidence the EF could be an IEF where a position is a win if its value is greater than C_2, a draw if it is between C_1 and C_2, and a loss if it is less than C_1, where $-6 < C_1 < -4$ and $-3 < C_2 < -1$.

4   Of the positions following (v), (va) is a draw with value -6 and the rest are wins with values 2, 1, and 5. Of the positions following (iv), (ivc) is a loss with value -10 and the rest are wins with values 0, 0, and 1.

    It appears that the EF is an excellent approximation to an IEF but is not perfect; position (i) is a loss and position (va) a draw, and both have the value -6. However, of the 14 positions considered, there were eight winning positions with values 0, 0, 0, 1, 1, 2, 5 and -1, three drawing positions with values -3, -4, and -6, and three losing positions with values -6, -8, and -10: a good correlation.

5   If this were an IEF, every position would be a win. By playing in the highest-numbered square remaining, a player could be sure that the average of the squares remaining would be lower on the following move than it was on his own, so the IEF value would decrease.

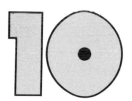

ANALYTIC GEOMETRY

     Geometry and Algebra: The Betrothed

exercises

page 363

1   (a) (i) $x = 2$ (b) (ii) $x = 2$ and (iii) $x = 3$
      (c) (i) $x = 0, y = -1$ (iii) $x = 2, y = 1$
      (d) All three are solutions.

3 $x = 0$, $y = 1$; $x = 2$, $y = 0$; $x = -4$, $y = 3$.   There are an infinite number of solutions.

5   The second equation is obtained by multiplying the first by 3, so they are, in effect, the same equation.

extras for experts
page 364

1   (a)   There are an infinite number of solutions; let x have whatever value you choose, and it is possible to find a y that will yield a solution.
(b)   One solution: $x = 5$, $y = -4$.
(c)   There is one solution; the solution to the first two equations (see (b)) satisfies the third equation as well.
(d)   There is no solution, since the one solution to the first two equations does not satisfy the third.
(e)   There is no solution to these two equations; if you add them, you obtain $0 = -1$.

2   There are an infinite number of solutions to these equations, but in all of them $x = 2$.

3   (a)   If b is negative   (b) $b = 0$ (c)   If b is positive

Analytic Geometry: A Union of Algebra and Geometry

exercises
page 369

1 A is at $x = 5$, $y = 2$; B is at $x = -5$, $y = -3$; C is at $x = -4$, $y = 2$; D is at $x = 4$, $y = -3$.

3   (a) $x = 0$ (b) $y = 0$ (c) $x = y$
(d) $x + y = 2$ (e) $y = 2$ (f) $x = 2$

5   (a)   oblique line   (b)   circle   (c)   vertical line
(d)   horizontal line   (e)   parabola

extras for experts
page 370

1   (a)   (i)   If the x-values are the same, one y-value is the negative of the other.
(ii)   If the y-values are the same, one x-value is the negative of the other.
(b), (c)   If (a, b) is on the graph, then $(-a, b)$ is also on the graph, so there is symmetry about the y-axis.

2   The graph is symmetric with respect to the line $x = 2$. (If $(2 + a, b)$ is on the graph $(2 - a, b)$ will be on the graph also.)

3 x = −1, y = 2; the solution to the equations are the coordinates of the point of intersection.

4 (a) They are symmetric with respect to the line y = x.
(b) They are symmetric with respect to the line y = 0.
(c) They are on a line through, and at equal distances from, and at opposite sides of, the origin.
(d) They are on the same side of a line through the origin, and one is twice as far from the origin as the other.
(e) One point is the projection of the other on the y-axis.

5 (a) A two-by-two square with sides parallel to the axes and the origin as center.
(b) All points above and to the left of the line y = x.
(c) Divide the plane into four parts by the lines y = x and y = −x; the regions above and below (but not to the right or left) of the origin.
(d) An infinite strip (of width about .707) at an angle of 45° to each axis.
(e) All points in the plane.
(f) No points in the plane.

Linear Equations

exercises

page 376

1 (a) is vertical, (e) is horizontal, and the rest are oblique.

3 (a) y = x + 1   (b) y = −3x − 1   (c) y = −3x
(d) y = −7x + 15

7 (a) $y = \frac{1}{2}x$ (b) $y = \frac{2}{3}x + \frac{1}{3}$ (c) $y = \frac{1}{2}x + 2$

9 You may not divide by zero.

11 k = 9

13 Zero; you would have to divide by zero.

extras for experts

page 378

1 (a) A line, a plane, or nothing
(b) A point, and all possibilities listed in (a).
(c) n ≥ 3

2 The lines are almost parallel, so a slight tilt changes their point of intersection radically.

3 (a) 10 seconds from now; 5 seconds ago.
(b) In 20 seconds they will both be 160 feet in the air.

4 (a) $y = 2x + a$ (b) $y = 2x + b$

5 If you subtract, you obtain $0 = a - b$, which can hold only if $a = b$.

The Distance Between Two Points

exercises
page 382

1 (a) 2   (b) 4   (c) 5   (d) $\sqrt{61}$

3 $a + b > \sqrt{a^2 + b^2}$ because if we square both sides we obtain $a^2 + 2ab + b^2 > a^2 + b^2$; this means your walk will be shorter if you go directly from one point to the other instead of walking parallel to the x axes.

5 If x is nonnegative, $|x|$ is defined to be x; if x is negative, $|x|$ is obtained by changing the sign of x, that is, $|x| = -x$.

7 The numerical values of x and $-x$ are the same, and the minus sign is discarded.

9 $|6 - (-4)| = |6 + 4| = 10$

extras for experts
page 383

2 The line joining (a, b) and (c, d) is $y = \dfrac{(d - b)x + bc - ad}{c - a}$; the perpendicular bisector of the line segment between (a, b) and (c, d) is $y = \dfrac{2(a - c)x + c^2 + d^2 - a^2 - b^2}{2(d - b)}$.

3 $(x - 2)^2 + (y + 1)^2 = 5^2$

4 (a) The part of $y = x$ and $y = -x$ above the x-axis.
   (b) The part of $y = x$ and $y = -x$ below the x-axis.
   (c) $y = x$ and $y = -x$
   (d) Same as (c)
   (e) the point $(0, 0)$
   (f) The line $x = -y$

5 $(x - 4)^2 + y^2 = 4(x^2 + y^2)$

Conic Sections and Other Graphs

exercises
page 393

1 (a) $x^2 + y^2 = 9$ (b) $x^2 + y^2 = 16$ (c) $x^2 + y^2 = 25$

3 One distance is $\frac{16}{5}$, and the other $\frac{34}{5}$.

5 One distance is $\frac{4}{3}$, and the other is $\frac{11}{3}$.

7 $x = y = 0$

9 $12x = 12 + y^2$

11 (a) $\sqrt{C/A}$; $\sqrt{C/B}$
(b) $C = 0$ (if neither A nor B is 0).

13 $Y = -C/B$ if $B \neq 0$ and $A = 0$.

**extras for experts
page 395**

1 Start with $(x + 4)^2 + y^2 = 9((x - 4)^2 + y^2)$

2 Start with $(x + 4)^2 = 4((x - 2)^2 + y^2)$

3 Start with $(x - 4)^2 + y^2 = 4(x + 2)^2$

5 It would be all points that are on either the graph in 4(a) or the graph in 4(b).

6 It would consist of all points that are on both of the graphs in 4(a) and 4(b).

11 MATHEMATICS OF CHANGE

CALCULUS

 ### The Description of a History: Distance and Velocity

**exercises
page 406**

1 The distances traveled in each of the first five hours are approximately 10, 40, 50, −50, and 60 miles.

3 Between the end of the second and fifth hours the car traveled 60 miles, so the average speed during this period is 60 miles ÷ 3 hours or 20 miles per hour.

5 (a) The average velocity during each of the first five hourly intervals is 10, 40, 50, −50, and 60 miles per hour, respectively. The average velocity over the entire 5-hour interval is 110 miles ÷ 5 hours = 22 miles per hour. The average speed over the 5-hour interval will be between the largest and smallest hourly average speeds.

   (b) If the average speed over five hours is equal to the largest hourly average speed, then all five hourly average speeds are the same.

7 (a) The greatest speed is 60 miles per hour; the least speed is 0.

   (b) The greatest speed is 38 miles per hour; the least speed is 25 miles per hour.

   (c) The greatest speed is 45 miles per hour; the least speed is −38 miles per hour.

   (d) The greatest distance you could have traveled in part (a) was 60 miles per hour × $\frac{1}{2}$ hour = 30 miles; in part (b), 19 miles; in part (c), $22\frac{1}{2}$ miles. The least distance you could have traveled in part (a) was 0; in part (b), $12\frac{1}{2}$ miles; in part (c), −19 miles (that is, 19 miles south). The average speed would always lie between the largest and smallest speeds.

The General Relationship Between the Histories

exercises

page 414

1 (b), (d), and (e) are true.

3 The velocity graph is always positive, so there is never any movement south;   (a) and (c) may be eliminated, and the answer is (b).

5 (a) and (iii), (b) and (i), (c) and (v), (d) and (iv), (e) and (ii)

7 (a) 5   (b) 5   (c) 10   (d) 10

Instantaneous Velocity

exercises

page 427

1 (a)   (i)   (140 − 40)/5 = 20 miles per hour
           (ii) (200 − 80)/6 = 20 miles per hour
           (iii) (120 − 60)/3 = 20 miles per hour

3 (a)   (i) $[(3\frac{1}{4})^2 − 3^2]/\frac{1}{4} = (25/16)/\frac{1}{4} = 6\frac{1}{4}$
           (ii) $[(3.1)^2 − 3^2]/.1 = 6.1$
           (iii) $[3.01^2 − 3^2]/.01 = 6.01$

(b) (i) $(3^2 - 1^2)/2 = 4$ (c) $[(3 + h)^2 - 3^2]/h = 6 + h$
     (ii) $(3^2 - 2^2)/1 = 5$ $[3^2 - (3 - k)^2]/k = 6 - k$
     (iii) $(3^2 - 2.9^2)/.1 = 5.9$ (d) $[t^2 - (t - k)^2]/k = 2t - k$

**extras for experts
page 428**

1   $\dfrac{(t + h)^2 + 4(t + h) - (t^2 + 4t)}{h} = \dfrac{2ht + h^2 + 4h}{h} = 2t + 4 + h,$ which is
close to $2t + 4$ when h becomes small.

2   $\dfrac{8(t + h)^2 - 8t^2}{h} = 16t + 8h,$ which becomes $16t$ when h becomes small.

3   $\dfrac{A(t + h)^2 + B(t + h) + C - (At^2 + Bt + C)}{h} = 2At + Ah + B,$ which
becomes $2At + B$ when h becomes small.

Some Techniques and Theorems

**exercises
page 435**

1   (i)   (a)   The average speed is 4.
          (b)   This is the instantaneous speed when $t = 2$.
     (ii) (a)   The average speed is 5.
          (b)   This is the instantaneous speed when $t = 2\frac{1}{2}$.
     (iii) (a)   The average speed is 0.
          (b)   This is the instantaneous speed when $t = 3$.

3   (a) (ii) and (iii)   (b) (i)   (c) (iv)
    (d) (iv) Since the train never passed the scene of the crime, it must
              have been instantaneously at rest when it changed direction,
              that is, when it was closest to the scene of the crime.
    (e) (iii)

5   Three times as fast.

7   Using the same reasoning as we did in Exercise 6, and assuming that the
    amount of heat a body contains is proportional to its volume, it would
    follow that large animals would be more uncomfortable than small ones.

9   (a) $\frac{25}{16}$ (b) $\frac{2225}{16}$;   the ball will be instantaneously at rest.

    (c) -50 (d) $\dfrac{25 + 5\sqrt{89}}{16}$ (e) $\dfrac{5\sqrt{89}}{16}$

13 $1\frac{1}{4}$ is one eighth of the original material. In 3000 years the material will
    have halved three times so that an eighth will be left. In theory, the
    material will never completely decay.

PICTURE CREDITS

INDEX

481